ASIAN POLITICAL, ECONOMIC AND SECURITY ISSUES

ASIA PACIFIC LIFE SCIENCES COMPENDIUM

Asian Political, Economic and Security Issues

Additional books in this series can be found on Nova's website under the Series tab.

Additional E-books in this series can be found on Nova's website under the E-book tab.

Public Health in the 21st Century

Additional books in this series can be found on Nova's website under the Series tab.

Additional E-books in this series can be found on Nova's website under the E-book tab.

ASIAN POLITICAL, ECONOMIC AND SECURITY ISSUES

ASIA PACIFIC LIFE SCIENCES COMPENDIUM

BANDIT CHUMWORATHAYI
EDITOR

Nova Biomedical
New York

For permission to use material from this book please contact us:
Telephone 631-231-7269; Fax 631-231-8175
Web Site: http://www.novapublishers.com

Additional color graphics may be available in the e-book version of this book.

Library of Congress Cataloging-in-Publication Data

ISBN 978-1-62257-338-7

Published by Nova Science Publishers, Inc. † New York

CONTENTS

PREFACE

This book compiles articles from the Asia Pacific Journal of Life Sciences. Topics discussed include clinical colposcopic practive guideline for colposcopists consulted by cervical-cancer-prevention via providers; knowledge increased after training via cryotherapy clinical-training-skill course of the Thai department of health; and benchmarking a novel linear regression model for evaluating the relationship between AHI and height in senior sleep apneics.

Chapter 1 - In the Genesis chapter of the Holy Bible, we can read that some people lived for more than 1,000 years. In allopathic medicine, there is no drug for prolonging life span. But there are several medicines for lengthening life period in naturopathy and unanimous treatments. In this work, the author has introduced the three hidden Indian alternative medicines; lion's tonic, shadowless trees, and sacred water.

Chapter 2 - The classical yoga was developed some 5,000 years ago. But the origin of Lemuran yogas dates back 15,000 years ago. The Lemuran yogas are still a hidden scientific mystery. In this work, the author introduced the Lemuran yogas.

Chapter 3 - Clinical practice guideline for cervical-cancer-prevention providers, a description of what they should do with clients, has been established and clearly taught in the visual inspection with acetic acid (VIA) and cryotherapy clinical-skills course. This describes when to refer clients to colposcopist for colposcopy. Although clinical colposcopic practice guideline, a description of what colposcopist should do with abnormal-Pap-smear clients, has been established in United Kingdom, it has never been established in Thailand. Moreover, if the authors consider upon the abnormal-VIA clients, this has never been established anywhere else in the world. Authors suggested six steps of colposcopy, modified from the UK's, that the colposcopists consulted by cervical-cancer-prevention VIA providers may be needed to do as follows: 1) Record reason for referral; 2) Record category of VIA abnormality; 3) Do VIA; 4) Do Colposcopy; 5) Record visibility of the squamocolumnar junction and/or lesion's findings; 6) Do a biopsy when any lesions were found, plus endocervical curettage when lesion extends into the cervical canal, otherwise, record reasons for not performing; 7) Appoint client for hearing-result visit and record the date; 8) Inform referring provider the final result.

Chapter 4 - The objective is to measure the improvement between pretest and posttest scores of participants who had been trained in VIA/Cryotherapy clinical-training-skill (CTS) course of the Thai Department of Health. Scores from knowledge's assessments, from the precourse- and midcourse-questionnaire, in a particular course, conducted in April-May 2007, were compared using paired t-test. Minimum passing levels of 85% (43/50) were used for the midcourse assessment.

The participants were 2 male physicians and 14 female registered nurses, aged 33.19±5.89 (range, 26-45) year-old. Knowledge's score was significantly increased from 34.31±0.69 to 47.25±0.76 ($p<0.0001$).

However, at the first midcourse assessments, there were 1/16 participants who did not pass. This study is the first reporting results of JHPIEGO VIA/Cryotherapy CTS course.

Although it was found to be effective in increasing the participant's scores achieved from knowledge's assessments, it was still not effective enough to increase all participants' scores to as high as 85%, which is the minimum passing level set by JHPIEGO. More recapitulations may be needed.

Chapter 5 - *Background and Aim:* This case-series study originally aimed to assess 124 elderly people's sleep disturbance. They aged 65 years and over. The setting was at the Sleep Medicine Laboratory in a Medical Center of Taiwan.

The 124 individuals with sleep disturbance were admitted for polysomnograph (PSG) from 2 January 2002 to 31 January 2003.The null hypothesis to be tested is "after controlling for sex, age, and BMI, there is no relationship between AHI and height, though some of the elderly were kypho-scoliotic" in the studied cohort.

Methods and Results: Among the 124 elderly individuals, there were 117 had height data available. Not only was the height of 76 from 117 subjects within one standard deviation of the mean height, it also revealed a bimodal-like appearance in distribution. The height and its distribution were examined and revealed that there was probably only one distribution in height, and probably not a bimodal distribution. However, there indeed was a mean and/or distribution difference with respect to height from the study data. A linear regression model with AHI as the dependent variable was established.

Conclusion: The proposed hypothesis has been successfully tested, and the null hypothesis stays. The alternative hypothesis is thus rejected, that is, there is no relationship between AHI and height in senior sleep apnea of the tested cohort in this study reported.

The conclusion is that any relationships with AHI that have been observed are based on an individual's weight alone - and not their height, which is the novelty to be presented.

Chapter 6 - The international response to the 2004 Boxing Day tsunami, as judged by the observed effects of the 17th July 2006 tsunami, deserves a mixed report card. The short-term response appears to have been reasonable, although efficient distribution of goods and co-ordination of effort were clearly problematic and ought to be closely scrutinised. Over the longer term, the installation of warning systems between the tsunamis was encouraging and the need to improve the speed with which warnings are communicated, and increase local peoples' knowledge on how to react to such warnings, have been clearly identified as issues. Representing the international community's greatest response failure, however, is the obvious unwillingness to seriously consider increased economic equality between developed and developing countries, as the most effective longer-term strategy to reduce harms associated with public health disasters. Despite the strategies that were implemented between the tsunamis, the 17th July 2006 tsunami still resulted in a loss of life in the order of 10 times that which occurred over 15 years ago as a result of the San Francisco and Los Angeles earthquakes. More needs to be done.

Chapter 7 - Objective: To study the prevalence and to identify the significant risk factors of pelvic and periaortic lymph node metastases in endometrial cancer patients at Lopburi Cancer Center, Muang, Lopburi, Thailand.

Method: A retrospective study of 184 patients who underwent primary surgical treatment of endometrial cancer at Lopburi Cancer Center between June 2002 and December 2008, was done. Analyses included demographics characteristics, stage, surgical procedure, histology, myometrial invasion, lymph-vascular space invasion, pelvic and periaortic metastases. Univariate and multivariate analyses were performed using logistic regression. P-values were obtained by chi-squared test to identify significant risk factors. The correlations of deep myometrial invasion, pelvic or periaortic lymph nodes metastases were obtained by odd ratio (OR) and its 95% confident interval (95%CI).

Result: Prevalence of pelvic lymph node and periaortic lymph node metastases were found in 20/175 (14.6%) and 12/130 (9.2%) of patients whose pelvic lymph or periaortic lymph nodes were dissected or sampled, respectively. In univariate analyses, endometrial histology, cervical stromal invasion, deep myometrial invasion, lymph-vascular space invasion, and adnexal metastases, were significantly correlated with pelvic lymph node metastases. Similarly, lymph-vascular space invasion and pelvic lymph node metastases were significantly correlated with periaortic lymph node metastases. In multivariate analyses, deep myometrial invasion and lymph-vascular space invasion were still significantly correlated with pelvic lymph node metastases, and only pelvic node metastases was still significantly correlated with periaortic node metastases.

Conclusion: The significant risk factors of pelvic lymph node metastases were endometrial histology, cervical stromal invasion, deep myometrial invasion, lymph-vascular space invasion, and adnexal metastases. Only lymph-vascular space invasion and pelvic lymph node metastases were significant risks for periaortic lymph node metastases.

Chapter 8 - Achilles tendon is a well known anatomical site prone to disabling injury with debilitating impact on daily functioning and quality of life. Chronic ruptures of Achilles tendons normally present four to six weeks after the original acute and often unrecognized rupture. Tendon ends are commonly retracted over time, with an atrophic appearance. In conservative management, a rehabilitation regime to ensure consistent elongation by the patients is fundamental. This case highlights the importance of incorporating the Calf, Achilles Tendon Stretcher (a simple therapeutic aid), and a collaboration care approach to impact an early patient self-management care and consequently, improve the efficacy of the rehabilitation regime.

Chapter 9 - Iron deficiency anemia (IDA) continues to be the major public health nutritional deficiencies that result in impaired growth and development, cognitive performance, scholastic achievement and increasing morbidity from infectious diseases. In the world, about 19% of the total population, and in India, 20% of the total population are adolescences, and faces a series of serious nutritional challenges, not only affecting their growth and development but also their live-hood as adult. Anemia is defined in the term of hemoglobin (Hb) level in blood, and serum ferritin (SF) is specific biochemical parameter that correlates well with total body iron stores. Overall, 97.2% of adolescent girls had some degree of anemia, in which 42.7% were mildly anemic, 50.7% were moderately anemic and 3.8% were severely anemic according to WHO criteria. Only about 3% of these girls had normal Hb concentration above 12.0 gm/dl. IDA in term of low SF level (<15ng/mL) found in 39.3% of adolescent girls who were anemic (Hb<12.0 gm/dl). The causative factors of anemia other than iron deficiency in the study population include bioavailability of iron from diet. Beside the dietary factor, worm infestations and communicable diseases, like malaria, play important role in reducing iron stores of the individual.

Chapter 10 - In scientific publications, authorship and disclosures are as essential elements as the content of the manuscripts. There is a need for novice researchers and new writers to be fully informed regarding what constitutes authorships. A clear demarcation on who should be included in the authorship list and who should be excluded, with a guideline on who to acknowledge, conflict of interests and disclosures should be disseminated to new writers. Lastly, ghost writers and simultaneous submissions should not be encouraged by the scientific community. Awareness of these essential component in writing can promote integrity and good publishing ethics amongst novice writers.

Chapter 11 – Background: The present study was designed to evaluate an ancient method of dressing using honey, ghee, and Curcuma longa, and to find the scientific background of using these materials in promoting tissue repair.

Method: In the present study honey, ghee and curcuma longa extracts were analyzed chemically. The honey compounds were separated and identified using Gas chromatography/Mass spectrometry (GC/MS). Honey phenolic and flavinoid compounds were analyzed by HPLC technique and the radical scavenging activity of honey sample extracts was measured using BrÜcker Elexsys 500 instrument operated at X-band Frequency. Fatty acids of ghee were assayed using a gas chromatograph equipped with a flame-ionization detector and the concentrations of vitamins C were measured spectrophotometrically. Analysis of Curcuma longa pure essential oils was carried out by gas chromatography (GC) and gas chromatography-mass spectroscopy (GC/MS) technique. The mixture of honey ghee and Curcuma longa dressings were mixed in clean vessels with help of an electric blender. The present study included thirty six patients in the surgical department at heliopolis hospital, Cairo, during the period between December, 2004, and December, 2007. The patients were divided in to two groups; the first group was treated with HGC-dressing, the second group was dressed with glycerin magnesia and/or povidine-iodine dressing.

Result: The chemical analysis of the mixture revealed that honey, ghee, and Curcuma longa extracts have many compounds which accelerate wound healing. The healing percentage is much better with H-G-C dressing ($p=0.001$).

Conclusion: This method of wound dressing has proved to be an effective method of encouraging healing and controlling infection. It was proved that honey, ghee, and Curcuma longa, contain eicosapentanoic, linoleic, α-linolenic, -linolenic, arachidonic acids, and vitamin C, which are important in controlling infection and promoting wound healing.

Chapter 12 - Objective: The present study aimed to investigate the neuroactivity effects of the methanolic extract of Phoenix dactylifera (MEPC) fruits and pits compared with fluoxetine. Method: MEPC of PC fruits and pits functional constituents was assayed. The levels of IL4, TNFα, IFNγ, lipid profile, and MDA in serum and 5HT, 5HIAA, IL1β, TNFα, IFNγ, were assayed in brain tissues of rats, the animal behavior, NK and LAK cell count were studied. Result: 5HT and 5HIAA levels in brain tissues were diminished in MEPC of fruit treated group than the other groups. The levels of the IL4, TNFα, IFNγ, in serum and IL1β, TNFα, in brain tissues were improved in MEPC fruit treated groups than the other groups. Moreover, the sucrose consumption was improved and SOD activities in liver tissues were significantly increased in group treated with PC fruit extract than other groups, and decreased MDA and NO levels lead to repair liver injury. Conclusion: MEPC of fruit and pits have effectiveness on inhibition of the selective serotonin reuptake, prevent the hyperactivity of the immune system, oxidative stress profile, lipid profile, and improved the sucrose consumption, which play a role in improving the etiology of depression.

Chapter 13 - The present work aimed to evaluate the therapeutic effect of oral co-administration of Nigella sativa (NS) oil, fish oil (FO), evening primrose oil (EPO) and vitamin E (Vit E) with chemotherapeutic agents in treatment of hepatocellular carcinoma (HCC) patients compared to control group and chemotherapy only group. 80 HCC male patients and 20 controls, the patients were divided into four groups G1 received only chemotherapy, G2, G3, G4 received an additional NS, EPO, and FO dose, respectively. VEGF, VEGF per platelet (plt) count, basic (b) FGF, TGF-β, IFN-γ-R1-mRNA, and prostaglandin E, were measured in studied groups, before and after treatment.

After receiving different types of treatment, serum VEGF and b-FGF levels were decreased. Mean of VEGF/plt count was also decreased significantly in all groups. There is a significant correlation between serum VEGF level and platelet count in different groups of study. Serum levels of TGF-β, which were high before treatment, showed a significant reduction reaching normal levels in patients receiving NS.

Moreover, basal levels of IFN-γ-R1-mRNA were lower in HCC patients compared to control, and its level was significantly increased in patients receiving NS and FO. The natural oil especially, Nigella sativa, can act as antiangiogenic agent and improve the antitumor effect and the performance status of chemotherapeutic agents in HCC.

Chapter 14 - Objective: The aim of the present study is to investigate the chemopreventive efficacy of nigella sativa, Bio Bran, and vitamin E, in the breast cancer patient.

Method: One hundred and sixty eight diagnosed breast cancer patients treated with six cycles of standard chemotherapy (CMF) followed by radiotherapy and Tamoxifen for at least one year. Subgroups of 42 patients each received in addition the above mentioned, treated with either Nigella sativa, Bio Bran, or vitamin E. 20 healthy subjects, served as a control group, were included in this study. The following parameters were investigated; IL-18, CEA, CA15-3, sFas, sFas-L, VEGF, IL-2, IL-6, IFN-γ. Result: The data so far obtained revealed that different supplements improve all parameters under investigated, especially in group treated with Nigella sativa.

Conclusion: Supplementation with Nigella sativa in the breast cancer patients is proved to be useful in reducing recurrence and improve disease free survival.

Chapter 15 - A study was conducted to determine the efficiency of lipid utilization by juvenile silver pomfret (Pampus argenteus). Duplicate groups of fish (34.3±1.2g) were fed three isonitrogenous (49% protein) diets containing 12, 16, and 20% lipid for 4 months. Significantly ($P<0.05$) higher weight gain and specific growth rates were observed in fish fed diet containing 16% lipid with a protein to energy (P:E) ratio of 22.6. A further increase in lipid level did not produce any improvement in the growth and feed utilization. There were no significant differences between the feed conversion ratio, protein efficiency ratio, apparent net protein utilization and energy retention in fish fed 16% and 20% lipid diets. Whole body lipid and fatty acid profiles were influenced by dietary lipid levels. Higher eicosapentaenoic acid (EPA) and docosahexaenoic acid (DHA) were accumulated in fish body fed 16 and 20% lipid diets. This study demonstrated that there is no beneficial effects in terms of growth and nutrient retention beyond 16% dietary lipid and a dietary lipid level of 16% in a 49% protein diet corresponding to a P:E ratio of 22.6 is optimum for better growth, feed utilization and whole body fatty acid profiles in silver pomfret.

Chapter 16 - Cypermethrin (CM) is a potent Type II pyrethroids pesticide extensively used in pest control. Studies have shown that cypermethrin induces genotoxicity in experimental animals. Antioxidants are known to protect the integrity of DNA from genotoxicants. In the present study, the authors have shown the potential protective effect of aqueous extract of fenugreek against cypermethrin induced DNA damage in femoral bone marrow lymphocytes, by *in vivo* micronucleus assay and alkaline comet assay. Male Wistar rats were treated with 1/10 LD_{50} (25 mg/kg body weight) of cypermethrin and 10% aqueous extract of fenugreek (GFaq) for 60 days. *In vivo* micronucleus assay showed that cypermethrin treatment caused a significant increase ($p<0.05$) in % MNPCE and a significant decrease ($p<0.05$) in % PCE and P/N ratio. Alkaline comet assay showed a significant increase ($p<0.05$) in tail length, % DNA in tail and Olive tail moment in cypermethrin treated rats, indicating the clastogenic effect of cypermethrin. Treatment with 10% GFaq showed significant reduction in DNA damage. The protective effect of GFaq could be attributed to the active flavonoids and polyphenols present in fenugreek.

Chapter 17 - A detailed microscopic study was carried out to see the changes in histological pattern of femoral artery in different age group ranging from new born to 70 years. 62 formalin-fixed specimens of either sex were studied. Thickness of tunica intima was minimal during infancy and it could not be measured, whereas in the age group 60 years and above, thickness of tunica intima ranged between 415 μs and 435 μs. While during infancy, mean thickness of tunica media was 115 μs [ranged 50-185 μs]. It reaches to the maximum mean thickness of 512 μs, in the age group 40-60 years. But in 60 years and above age group, mean thickness of tunica media decreased to 428 μs. All three layers of the arterial wall can be identified in infants. Tunica intima consists of internal elastic lamina and endothelium. Sub-endothelial layer is scarcely perceptible. Elastic fibers on tunica media are long and parallel, uniformly arranged and thicker than adult. In the earlier childhood, sub-endothelial connective tissue layer becomes more apparent. In adults, thickness of tunica intima increases while tunica media appears thin over regional intimal hypertrophy, but in the old age diffuse intimal hypertrophy is seen. Tunica intima become half, to two third of tunica media. Hyalination of collagen tissue of media occurs with loss of normal staining properties of elastic tissue. Elastic laminae become thinned and show fragmentation with occasional clumping. These changes are more prominent near inner media. Adventitia is thicker than media in infants and becomes thinner with age.

Chapter 18 - *Objective:* To evaluate safety and efficacy of local anesthesia for Lichtenstein hernioplasty in a general hospital. *Method:* After having permission from the Ethical Review Board of the hospital, any patients who had this procedure done by author, from July 2007 to February 2011, were prospectively assessed and recorded about their demographic data, intra-operative pain scores, complications, and additional anesthesia/ analgesia. *Result:* There were 170 male patients aged 18-97 year-old (mean 60.5) having this procedure done by the author during the mentioned period. Most of the patients had only initial pain in the first infiltration, with low intra-operative pain score (mean 2.3, range 1-4). No one needed additional anesthesia, neither regional nor general, or pre-operative/ intra-operative analgesia/ sedative, neither oral nor injected. Also, no one needed to be observed in the recovery room. Every patient was able to return to their residences immediately after surgery. No one needed re-admission from either surgical complications or post-operative pain. There was neither post-operative acute urinary retention nor wound infection found in all of the 170 patients. Only one complication, that author found, was minute wound hematoma, which could be treated by only supportive approach until it was completely healed. *Conclusion:* Local anesthesia for Lichtenstein hernioplasty can be done efficaciously and safely in a general hospital, although most patients in this study had one or more underlying medical diseases such as; coronary heart disease, diabetes mellitus, hypertension, which are commonly found in the elderly.

Chapter 19 - *Objective: Sida rhombifolia* subsp. *retusa* and *Embelia tsjeriam-cottam* were investigated for their potentials to exhibit cytotoxic, antibacterial, antitubercular and antimycotic activities *in vitro* for the first time. *Method:* Plants extracts were analyzed for phytochemical fingerprinting by High Performance Thin Layer Chromatography and High Pressure Liquid Chromatography, cytotoxic activity against Human Non-Small-Cell Lung Cancer (HOP 62) and Hepatocellular Carcinoma (Hep G2) cell lines by Sulforhodamine B assay, antimicrobial, antimycotic activity by agar well-diffusion method and antitubercular activity on Lowenstein-Jensen medium. *Result:* Phytochemical fingerprinting revealed Ecdysterone in *S. rhombifolia* subsp. *retusa,* Embelin in *E. tsjeriam-cottam* as chemotaxonomic markers and for routine quality control. Methanolic seed extract of *S. rhombifolia* and Embelin in fruits of *E. tsjeriam-cottam* exhibited marked cytotoxic activity against HOP 62, Hep G2 cells at $GI_{50} \leq 20$ µg/ml. They exhibited potent inhibitory effects against *Staphylococcus aureus, Diptheriae* sps, *Candida krusei, Aspergillus flavus* with inhibitory zones ranging from 18-26 mm and were also active to *Mycobacterium tuberculosis* strains. *Conclusion:* Present findings demonstrate potent cytotoxic, antibacterial, antitubercular and antimycotic activities in methanolic seed extract of *S. rhombifolia* subsp. *retusa* and Embelin in fruits of *E. tsjeriam-cottam.* This can be regarded as potential source of alternative natural products.

Chapter 20 - Strains which could exhibit enhanced production of protease and lipase were successfully produced from the wild strain *Bacillus pumilus* SG2 and enzyme production by the mutants was enhanced by 50% more than that of the wild strain. UV irradiation and EMS treatment was adopted to improve the wild strain and mutants which over-produced both protease and lipase were isolated. SGM4 was identified as the best mutant following UV irradiation. SGM4 was subjected to mutagenesis by EMS treatment and SGMM8 was identified as the potent producer of both protease and lipase. The enzymes produced by SGMM8 were subjected to ammonium sulphate fractionation and dialysis and the partially purified enzymes were characterized. Thus a stable mutant was which produced proteolytic and lipolytic enzymes was isolated and the enzymes were found to possess all the prerequisites of a detergent additive.

Chapter 21 - *Objectives:* 1) To explore the role of snoring affecting sleep disturbance 2) To examine the prevalence of abnormal body weight's influence on sleep disturbance 3) To examine the relationship between snoring and obstructive sleep apnea (OSA) (elevated AHI) in a community-dwelling older men and women as the studied subjects, with advancing age, age-related changes have been described for sleep-wakefulness and additional behavioral cycles. 4) To compare the trends in the relationship between elderly sleep disturbance and AHI 5) To signify the economic and cultural impacts on healthy weight and BMI maintenance with relationship to the prevalence of sleep disturbance 6) To review some of the related literature.

Methods: A retrograde single measures design was employed over nocturnal period. The study was conducted in a clinical sleep medicine laboratory of a Medical Center in Taiwan from January 2002 to January 2003. Totally 124 men and women were community-dwelling elders (mean age=71.85 years, SD=4.85 years) with complaint of sleep disturbance. Questionaires were asked. Polysomnography was conducted using a full 18-channel montage. Apnea was defined as a decrease in airflow of ≥ 90% for a minimum of 10 seconds. Hypopnea was defined as ≥ 30% decrease in airflow and desaturations required a ≥ 3% decrease in oxygen saturation for a minimum of 10 seconds. The apnea hypopnea index (AHI) was calculated as the sum of apneas and hypopneas divided by nocturnal hours of sleep.

Results: Sleeping apnea in this study was evaluated. AHI: Decile and the Quartile distributions were reported. A novel finding of AHI: a quantile-quantile plots (q-q plots) of AHI suggest deviations from normality, notwithstanding, none of the diagnostics of kurtosis and skewness on multiple linear regressions of AHI being significant or worrisome. For the correlation between AHI and BMI, after checking with Spearman's rank correlation coefficient (rs), it reveals 0.295. As to the relationship between AHI and Neck Circumference was noted to approximate with that between AHI and BMI, there was an insignificant effect on frequency of snoring by grouping factor of AHI. Spearman's rank correlation coefficient (rs) test reveals a mildly positive correlation between AHI and snoring (rs was 0.26). Characteristics of BMI in subjects with sleep disturbance were presented. Females had mean height, weight and cervical circumference compared with males. Note that the difference of mean BMI between non-snores and snorers is 2.02. AHI was noted as correlating with BMI, snoring, body height and weight, and cervical circumference respectively. Spearman's rank correlation test revealed that the relationship between snoring and BMI was highly significant. The effect of BMI grouping on the frequency of snoring as well as Asian BMI criteria was discussed.

Conclusion: The study of sleep disturbance in the elderly is a new area of scientific direction for fruitful areas of future research. Hopefully, this article provides some highlight in focusing on an individual patient-oriented approach to sleep disturbance. This is in contrast to a population-based approach in which the entire distribution of risk factors in total population in Taiwan will be expected in a very near future.

Chapter 22 - The heavy metal content from the non-polluted zone of the Panchganga River, Kolhapur (MS, India) water was found below detectable level. Samples from polluted zone showed variations in their heavy metal content in river water, sediment, and fish tissues. The bioaccumulation of heavy metals is found to be highest in liver, followed by gill, and muscle, of the selected fish species, from three selected stations of river Panchganga. From present study, it is concluded that the higher concentrations of heavy metals at selected stations is due to the heavy load of industrial waste in addition to domestic and agricultural waste. The industrial waste contains high amount of textile mill wastes, including dyes and untreated chemicals, which caused severe damage to the fisheries of Panchganga River.

Versions of these chapters were also published in *Asia Pacific Journal of Life Sciences,* Volume 4, Issues 1-3, edited by Bandit Chumworathayi, published by Nova Science Publishers, Inc. They were submitted for appropriate modifications in an effort to encourage wider dissemination of research.

In: Asia Pacific Life Sciences Compendium
Editor: Bandit Chumworathayi
ISBN: 978-1-62257-338-7

Chapter 1

LION'S TONIC, SHADOWLESS TREES, AND SACRED WATER: THE HIDDEN INDIAN ALTERNATIVE MEDICINES

Sennimalai Kalimuthu*
Kanjampatti PO, Pollachi Via, Tamil Nadu, India

ABSTRACT

In the Genesis chapter of the Holy Bible, we can read that some people lived for more than 1,000 years. In allopathic medicine, there is no drug for prolonging life span. But there are several medicines for lengthening life period in naturopathy and unanimous treatments. In this work, the author has introduced the three hidden Indian alternative medicines; lion's tonic, shadowless trees, and sacred water.

Keywords: lion's tonic; shadowless trees; sacred water; complimentary and alternative medicine (CAM)

ABOUT LION [1-4]

Scientific name:	*Panthera leo* (Was *Felis leo* until recently.).
Family:	Felidae.
Order:	Carnivora.
Class:	Mammalia.
Male:	Average Length: 9 ft. (2.7 meters).
Weight:	350-400 lbs. (157-180 Kg.).
Height:	36 in. (90 cm.).
Female:	Average Length: 8 ft. (2.4 meters).
Weight:	250-300 lbs. (112.5-136 Kg.).

* Email: newscience199@gmail.com

Height:	30 in. (75 cm.) (Lengths include the tail. Height measured at the shoulder.).
Number of teeth:	30.
Lifespan:	12 years in the wild, 20 years in captivity. Females typically live a few more years in the wild.
Maximum speed:	30 mph. over 50 yds. (48 kph. over 46 m.).
Biggest jump:	12 ft. (3.7 m.) vertical, 36 ft. (10.8 m.) horizontal.
Gestation period:	105-115 days.
Body temperature:	101 ^{0}F (38.33 ^{0}C).
Habitat: Grassy	plains and open woodlands.
Diet:	Carnivorous.
Predators:	Humans.

The female lion captures most of the mid-sized prey (wildebeast, zebra, etc.), but the males typically catch the really big prey (buffalo and giraffe). They can eat a lot at a single sitting: a male can eat 43 kg in a day; a female may eat over 25 kg. But their average intake is only about 8-9 kg per day. The roar of a lion spreads up to 8 kilometers.

LION'S TONIC

Lion's tonic is an alternative medicine. The origin of this drug belongs to the Lemuran continent. The Lemuran people were tall and strong. Normally, they lived for more than 100 years. An unknown Lemur, an expert, introduced lion's tonic. The lion is the king of the forest. The roaring of the lion spreads up to 8 kilometers. The wildness and physical powers of a lion can be watched on modern cable TVs like The Discovery Channel, Animal Planet and National Geographic. The consumer of this tonic was claimed to get equal physical powers of a lion. Ten kinds of herbal roots are the basic, optional raw materials for the preparation of lion's tonic.

The major basic raw material for this drug is the bones of an adult lion. Firstly, a pyramid shaped mini block should be prepared. Then the dead lion should be kept inside the block and it should be sealed. After 48 days, the dead lion should be taken outside of the pyramid and cleaned with pure water. Secondly, the herbal roots and the lion's bones should be mixed in the required ratio departed by the sages. Here, the application of number theory enters into the arena and dominates the preparation process. If the relevant number theory is missed, the tonic will be incomplete and not suitable for human consumption. Also, while making the pyramid shaped block, the application of sacred geometry should be followed. Taking into the relevant mathematical knowledge, the preparation of this biological tonic will become successful.

Here are the claimed benefits of consuming it: If one consumes lion's tonic for 41 days, he/she will live for 100 years; If one consumes lion's tonic for 82 days, he/she will live for 200 years; If one consumes lion's tonic for 123 days, he/she will live for 300 years; If one consumes lion's tonic for 164 days, he/she will live for 400 years; If one consumes lion's tonic for 205 days, he/she will live for 500 years, and so on.

The audience may raise a curious question. Modern allopathic medicine says that a human being can live up to a maximum of 110 to 120 years, provided that he/she consumes pure water and naturally cool and lives on a hill. Let us recall that in the Old Testament, people lived thousands of years without using any medicine.

In Jurisprudence, there is no judgment before argument. It applies here also. The author does not have any solid proof on hand, but this is a proposal for RandD. There is no treatment for chronic diabetic patients, but the author has an alternative herbal medicine which will definitely cure even chronic diabetic disease. So, in alternative medicine, there is no impossibility.

SHADOWLESS TREES

All the philosophical schools did not mention about shadow of matter. The south Indian sages were secular. They did not belong to any particular race or religion. Their preaching was very advanced, acceptable and applicable to each section of the people. They have mastered botanical, chemical, physiological, astronomical, and zoological science. They have never been told anything about gravity. These hermits explained in detail about shadows of matter.

It is known to everyone that in the presence of light, a matter has its own shadow. Although a matter has its own shadow, during the darkness it is invisible to the naked eye. The monks, like Kaalanginathar and Bogor, mentioned about shadowless trees. During their lifetime, these trees were traceable in Kolli Hills. These trees possess marvelous medicinal qualities. If one consumes the fruit of this tree, then one becomes immortal. The consumer always enjoys enlightenment.

If one can open his Third Eye through yogic exercises, then one's shadow will disappear in the presence of light. That is why the south Indian Tamil sages called this absence of shadow as the vanishing into the blue phenomena. The search for shadowless trees is still going on in Tamilnadu, Southern parts of India, but the alternative medicine scientists generally believe that such plants may be located in the Rocky Mountains, United States, and in the Amazon Forest, South America.

Focusing our efforts to the scientific arena, Schwartz and Green had to introduce shadow of matter for the consistence of their equations and theories in super-string theory. Before these co-workers, the scientific community used to laugh at the sage's scientific truth.

Finally, the success of joint papers of Schwartz and Green has made the research group silent. Nowadays, shadows of Schwartz and Green matter are a generally accepted physical phenomena.

Sacred Water

The South Indian seers gave guidelines on how to obtain "The sacred water." Sacred water is available only in the dense forests. Normally, it has been stored in rock caves several thousand years ago. The consumer of this water can enjoy and experience enlightenment and attain immortality. This will regenerate the human body from foot to head. At the opposite side of this sacred water, anti-sacred water has been hidden . Unlike the sacred water, this anti-sacred water cuts off human bodies upon touch. This can chop off even hard materials like iron and steel.

Discussion

The author has introduced the above three hidden Indian alternative medicines for favor of RandD. The first formula can be easily prepared, but it is quite difficult to locate the other two topics. However, let us recall what the great Emperor, Napoleon Bonaparte, said: "The word impossible must be taken away from the dictionary."

References

[1] http://www.asiatic-lion.org.
[2] http://www.redlist.org/search/details.php?species=15951.
[3] http://lynx.uio.no/catfolk/afrleo01.htm.
[4] http://news.nationalgeographic.com/news/2002/06/0612_020612_TVlion.html.

In: Asia Pacific Life Sciences Compendium
Editor: Bandit Chumworathayi
ISBN: 978-1-62257-338-7

Chapter 2

Hidden Yogas - A Remedy to Empty the Mind, to Sit Forgetting, and to Have Good Sex

Sennimalai Kalimuthu*
Kanjampatti PO, Pollachi Via, Tamil Nadu, India

Abstract

The classical yoga was developed some 5,000 years ago. But the origin of Lemuran yogas dates back 15,000 years ago. The Lemuran yogas are still a hidden scientific mystery. In this work, the author introduced the Lemuran yogas.

Keywords: yoga, Lemuran

On Classical Yoga

Yoga is an ancient Indian body of knowledge that dates back more than 5,000 years ago. The word "yoga" comes from the Sanskrit word "yuj" which means "to unite or integrate." Yoga then is about the union of a person's own consciousness and the universal consciousness. In yoga, the body is treated with care and respect because it is the primary instrument in man's work and growth. Yoga exercises improve circulation, stimulate the abdominal organs, and put pressure on the glandular system of the body, which can generally result to better health.

There are still a lot of misconceptions about yoga, for instance, yoga being a religion. Yoga is not a religion. It is more of a set of techniques for us to find spirituality. In fact, yoga is being practiced by a lot of people from different religions like Christians, Jewish, Buddhists, and Muslims. Another misconception is that yoga is an exercise, a way for us to keep fit. It is partly true, but if you think that yoga is just that then you are greatly mistaken.

* Email: newscience199@gmail.com

Yoga develops the body since a weak one is a hindrance to spiritual growth. It does not simply focus on the physical but on the mental and spiritual aspects as well. [1]

Yoga is a practical aid, not a religion. Yoga is an ancient art based on a harmonizing system of development for the body, mind, and spirit. The continued practice of yoga will lead you to a sense of peace and well-being, and also a feeling of being at one with their environment. This is a simple definition. The practice of yoga makes the body strong and flexible. It also improves the functioning of the respiratory, circulatory, digestive, and hormonal systems. Yoga brings about emotional stability and clarity of mind. In the practice of yoga, the ultimate aim is one of self-development and self-realization. Prevention and care is by far the best medicine there is. It all comes down to balance.

Yogis measure a person's age by the flexibility of the spine. After our mid-20s, the venous supply to the disks of the spine naturally atrophies. Through yogic exercises that bend and stretch the spine, fresh blood and nerve flow revitalize all the major glands and organs. Unlike other forms of exercise, which develop fitness by stressing the body, yoga builds strength and endurance while lowering the stress hormones that age us [2].

Patanjali is widely regarded as the founder of the formal Yoga philosophy. Patanjali's yoga is known as Raja Yoga, which is a system for control of the mind. He defined the word "yoga" in his second sutra, which is the definitional sutra for his entire work: "Yoga is the inhibition (*nirodhaḥ*) of the modifications (*vṛtti*) of the mind (*citta*)". As Osho tries to differentiate between these two paths, he quotes: Yoga is not a religion, but a way of living with health and peace of mind as its aims.

According to the Yoga *Sutras* of Patanjali, the ultimate aim of yoga is to reach "*Kaivalya*" (emancipation or ultimate freedom). This is the experience of one's innermost being or "soul" (the *Purusa*). Then one becomes free of chains of cause and effect (*Karma*) which tie us to continual reincarnation. In Kaivalya, one is said to exist in peace and tranquility, having attained absolute knowledge of the difference between the spiritual, which is timeless, unchanging and free of sorrows, and the material, which is not. Yoga is therefore a spiritual quest. However, along the path of yoga, the aspirant also gains health, happiness, tranquility and knowledge which are indicators of progress and an encouragement to continue their practice. Buddhism and other Eastern spiritual traditions use many techniques derived from Yoga [3].

Anyone regardless of age, sex, nationality, religion or social status can practice Yoga. It is not recommended that Children younger than 7-8 years practice yoga. Although younger children can playfully be introduced to yoga, they should never be forced as this can damage their spine and other joints because their bodies are still developing. There is no upper age limit and one is never too old to learn yoga.

Yoga does help in overcoming health problems. One needs to note that health is not just a disease-free state, but a state of physical, mental, emotional, and spiritual well-being.

Diseases may be dormant, interrupted or in a fully active stage and an individual is able to detect the disease depending upon his/her sensitivity. Most individuals realize that they have a problem only when the symptoms start showing.

In such cases, yoga asanas are taught in such a manner that the patient gets symptomatic relief. Later, the patient has to continue with their practice so as to get at the root of the disease. Practice of yoga also builds the character of tolerance in the practitioner, strengthens the nerves and quietens the mind and so, as Prashant Iyengar states, "Yoga helps cures what needs not be endured and endures what cannot be cured."

Lemuran Yogas

It is well known that yogas cure diseases, maintain good health and increase one's IQ.

Some yogas prolong the life span of human beings. Breathing techniques are a part and partial of yogas. One of the highest peaks of yogas and breathing techniques are that an expert can release the life and soul out of his/her body and then, after a particular period get back into the human body. The sages and siddhas attained salvation only through meditation, yogas and breathing techniques. Yogas and breathing techniques can reduce the mass of human beings to zero. Yogis of zero mass can attain supernatural and mystic powers. The siddhas teachings are focused on health, meditation and peace. Their mathematical biological yogas mainly rely on number theory and geometry.

Although these yogas are classical, they are not disseminated yet. Still it is a secret. The absence of this mathematics in yoga postures will not give the required result.

Here is the first example of Geometry in mathematical biological yogas: While standing, both the foot of the legs should be kept at 45^0 angle. This posture keeps the body cool, calm and relaxed. While raising the hands for performing particular yogas 45, 90, and 180 degrees must be applied. Secondly, while practicing these new yogas, number theory is applied. Let us go into the deep.

Everybody can practice these yogas. Kids, pregnant ladies, and aged people all can practice. The result is instant. One has to allot only ten minutes per day. One can easily overcome tension, worries, sorrows, all complexes and attain/experience TOTAL joy and pleasure.

Table 1. Power and glory of Lemuran yogas

S. No	Yogas	Benefits
1	Aga Oli	Hands, intestine, liver, heart, legs, pancreas, spleen and gall bladders are streamlined.
2	Naasi Naa	From top to bottom, good respiration is maintained. The general circulation becomes natural and normal.
3	Kooppagam	Foot, hands, shoulders are put in the right track. The inner heat of the body goes out.
4	Agaparimalam	Foot, buttocks, shoulders, spinal parts, thighs, face and eyes are regenerated.
5	Kaayamarutha l	The heart, neck and face are put in total action. The blood is purified.
6	Sama Nilai	The temperature of the body is normally maintained.
7	Saantham	The sexual organs and semen are purified.
8	Kunthagam	From bottom to top, the external parts become concrete.
9	Meivarthanam	Rectum, ureters, urethra, bladder, urine, and semen are purified.
10	Sugapanthana m	The coitus powers and duration of intercourse get prolonged.
11	Guru Vanakkam	The breathing is purely streamlined. The blood is totally purified. The life organs and forces are regenerated.
12	Mana Mounam	The purified body and soul experiences and enjoys TOTAL BLISS.

The important thing is that the duration of coitus marvelously increases. All the classical schools of thoughts preach: Sit and meditate and control your thoughts and mind. Really, after sitting and starting to meditate, the external thought not only disturbs the meditation but also ruins it. This is the plain and bitter truth. But in this type of yoga, the thoughts and disturbances are naturally controlled and the practitioner does NOTHING. Power and glory of these yogas are shown in Table 1.

Instructions of the Lemuran Yoga

Step 1: Place a pure cotton cloth on the floor. Stand on the cotton cloth. Keep the foot at 45^0. Slowly move the face towards the floor counting down 1, 2, 3, 4, 5. Before you pronounce 5, you must gently sit on the floor.
Step 2: Stretch the hands backward murmuring 1, 2, 3, 4, 5. While finishing 5, the hands should be kept on the floor.
Step 3: Move the legs forward repeating 1, 2, 3, 4, 5.
Step 4: Pronounce 1, 2, 3, 4, 5, and gently lay on the cloth looking at the sky.
Step 5: By spelling out 1, 2, 3, 4, 5, only turn the face 180 degree left side and move the left hand back side counting down 1, 2, 3, 4, 5.
Step 6: Repeat 1, 2, 3, 4, 5, 6, 7, 8, 9, 10, and turn the head to right horizontally and then keep the right hand backward counting down 1, 2, 3, 4, 5.
Step 7: Move the head from right to left side counting 1 to 10.
Step 8: Lift the left leg perpendicularly towards the sky and then fold the leg. Keep the left hand on the left knee.
Step 9: Repeat the above process keeping the head at the right side. Counting 1 to 10 is a must.
Step 10: Now lift the head and stomach up to 45 degree before finishing 1 to 5. Remain at this pose for 10 seconds.
Step 11: Stretch back the body on the floor by telling 1, 2, 3, 4, 5.
Step 12: Turn the head to left side horizontally and put down the left hand on the floor by repeating 1, 2, 3, 4, 5. Then place the left leg on the floor by saying 1, 2, 3, 4, 5.
Step 13: Perform the practice at the right side.
Step 14: Keep the face towards the sky for 20 seconds.
Step 15: Slowly awake and sit repeating 1, 2, 3, 4, 5.
Step 16: Lift the body and return to the initial pose by chanting 1, 2, 3, 4, 5.

Benefits of this Lemuran Yoga

The blood circulation becomes normal. The tissues are stream-lined. The breathing becomes quite natural. The body feels freshness. The mind stops wavering and the thoughts are concentrated. The performer experiences enlightenment.

Beyond this, the sexual organs are awakened and the vital fluid becomes thicker from thinner. The copulation period of the performer increases tremendously. After coitus, the body of the performer never gets tired. Practice of this mathematical biological yoga will make the white color of the semen into pale yellow color. And hence, the life span of the practitioner increases. The audience may think that author is boasting. The author politely proposes such people to practice this yoga. The performer will note the difference. No yoga literature on earth utilizes mathematics. The term mathematical biological yoga was first coined by the famous Thamizh (Tamil) sage Bohar. There are several yogas like this one.

ANOTHER LEMURAN YOGA

This yoga teaches us for good sexual intercourse. Sex is mental food for all living beings. The mating of birds and reptiles lasts only for a few seconds. Dogs enjoy coitus for several hours, say 1 to 7 hours. Several thousands of years back, there lived animals whose name was Yali. Yalies bodies were like that of an elephant but their heads were like a lion. Yalies enjoyed intercourse for several months. Like dinosaurs, the Yalies are extinct.

This mathematical biological yoga reveals the salient features of human copulation. If a man follows this yoga, he can satisfy several women simultaneously. It instructs how to perform push and pull of the male organ into the female organ. As we have already seen in the previous yoga, the angle of the male organ while pushing inside the female organ is very important. There are several types of pushing and pulling. During sexual enjoyment, the breathing should be as slow as possible, the more breathing the quicker ejaculation. While pushing and pulling, the secrets of inhaling and exhaling should be followed. At this stage, as we have already seen in the first yoga, the application of number theory should be adopted.

The over sexual thirst of the opposite can be easily quenched. In the court of the king, the sage Bohar performed this before the open audience. The sage challenged the audacious ladies and oppressed their lust. According to available allopathic scientific data, the maximum coitus duration is seven minutes and normally a normal man can not satisfy a normal woman. The R and D data reveals that while the male finishes his sex play, the lust of the opposite sex begins and increases. The fate is that nearly 95% ladies are dissatisfied sexually. This mathematical biological yoga is a remedy for these biological problems. The performer of this yoga can not only meet the requirements of his counterpart, but can also control and overcome his ejaculation. There are sources to prove this for the needy.

> Oh Lord leads me from the unreal to the real.
> Leads me from the darkness to light.
> Leads me from death to immortality.
> May there be peace, peace, and perfect peace.
> — Brihadaranyaka Upanishads (1.3.28) —

REFERENCES

[1] *http://www.abc-of-yoga.com/beginnersguide/whatisyoga.asp.*

[2] http://yoga.org.nz/benefits/physiological_benefits/yoga_aging.html.

[3] http://yoga.org.nz/benefits/physiological_benefits/yoga_aging.html.

In: Asia Pacific Life Sciences Compendium
Editor: Bandit Chumworathayi

ISBN: 978-1-62257-338-7

Chapter 3

CLINICAL COLPOSCOPIC PRACTICE GUIDELINE FOR COLPOSCOPISTS CONSULTED BY CERVICAL-CANCER-PREVENTION VIA PROVIDERS

Bandit Chumworathayi[1,*] and Nittaya Yoopakdee[2]

[1]Division of Gynecologic Oncology, Department of Obstetrics and Gynecology, Faculty of Medicine, Khon Kaen University, Thailand

[2]Department of Obstetrics and Gynecology, Health Promotion Center 1, Anusawaree, Bangkhen, Thailand

ABSTRACT

Clinical practice guideline for cervical-cancer-prevention providers, a description of what they should do with clients, has been established and clearly taught in the visual inspection with acetic acid (VIA) and cryotherapy clinical-skills course. This describes when to refer clients to colposcopist for colposcopy. Although clinical colposcopic practice guideline, a description of what colposcopist should do with abnormal-Pap-smear clients, has been established in United Kingdom, it has never been established in Thailand. Moreover, if we consider upon the abnormal-VIA clients, this has never been established anywhere else in the world. Authors suggested six steps of colposcopy, modified from the UK's, that the colposcopists consulted by cervical-cancer-prevention VIA providers may be needed to do as follows: 1) Record reason for referral; 2) Record category of VIA abnormality; 3) Do VIA; 4) Do Colposcopy; 5) Record visibility of the squamocolumnar junction and/or lesion's findings; 6) Do a biopsy when any lesions were found, plus endocervical curettage when lesion extends into the cervical canal, otherwise, record reasons for not performing; 7) Appoint client for hearing-result visit and record the date; 8) Inform referring provider the final result.

Keywords: VIA; cryotherapy; colposcopy; guideline

* Email: bchumworathayi@gmail.com

INTRODUCTION

An organized cervical-cancer-prevention (CCP) program, where screening and treatment of preinvasive lesions are combined as a single visit approach (SVA), has been implemented in Thailand since the year 2000. [1] Approximately 30 from 76 provinces are now using this approach. It has been widely known and as "VIA (visual inspection with acetic acid) and Cryotherapy CCP program".

The CCP providers are almost always registered nurses, but in other countries, these may be midwifes or other kinds of public health workers, according to their circumstances. Most of the time (90%), they can treat the positive-VIA clients by themselves, but sometimes (10%), they have to refer these positive or suspicious-cancer clients to a colposcopist for consultation or managements.

In Thailand, the consulted colposcopists are almost always gynecologists trained by the 3-day course of the Thai Department of Medical Service (TDoMS). The first reports of how the VIA nurses providers working in collaboration with these colposcopists, were published elsewhere [1,2,3].

Generally, the VIA nurses providers will refer their clients to a colposcopist when; 1) the lesion is suspicious for cancer, 2) the lesion extends into the cervical canal, 3) the lesion is larger than 2-mm zone around the cryotip. For the women who have Pap (Papanicolaou) smear instead of VIA because they are older than 45 year-old, younger than 30 year-old, or their squamocolumnar junction (SCJ) cannot be completely seen, the CCP nurses providers will refer them only when the results turn out to be abnormal.

In any sustainable program, periodic evaluations and feedbacks are generally needed as parts of quality assurance processes.

In 2004, The National Health Service Cervical Screening Programme (NHSCSP), in collaboration with the British Society for Colposcopy and Cervical Pathology, has published guidelines covering a variety of issues regarding quality assurance in cervical cancer prevention, including the standard requirements for colposcopy service [4].

Those are; 1) the following data should be recorded at colposcopy: (a) reason for referral (100%); (b) grade of cytologic abnormality (≥90%); and (c) visibility of the squamocolumnar junction (100%); 2) an excisional form of biopsy is recommended (≥95%) when: (a) colposcopic appearances indicate high-grade abnormality; (b) low-grade colposcopic change is associated with severe dyskaryosis or worse; and (c) a lesion extends into the canal; 3) reasons for not performing a biopsy must be recorded (100%); 4) all patients must have a biopsy or biopsies taken prior to local destructive treatment (100%); 5) biopsy should be carried out unless an excisional treatment is planned, when the cytology indicates persisting moderate dyskaryosis or worse, and when atypical transformation zone is present (100%); 6) of all biopsies taken (directed and excisional), more than 90% should be suitable for histologic interpretation; 7) if colposcopically directed biopsy is reported as inadequate for interpretation, it should be repeated if there is a residual colposcopic lesion (≥95%); and 8) for those with satisfactory colposcopic examination, the predictive value of a colposcopic diagnosis of a high grade lesion should be at least 65%.

In 2008, Manopunya et al. [5] audited routine colposcopy performance using the 8 standard requirements of the NHSCSP for abnormal cervical cytology between January and December 2008 at Chiang Mai University (CMU) Hospital, Thailand. They found that the standard requirements were not achieved in 2 practices: (1) the proportion of women who had recordings of visibility of the transformation zone (96.6%) did not achieve the NHSCSP requirement of 100%; and (2) the rate of excisional biopsy (87.8%) was lower than the 95% minimum required. In the future, colposcopists working in Thai CCP program may be audited in the same way of CMU hospital using the 8 standard requirements of the NHSCSP, but for abnormal cervical VIA instead of abnormal cervical cytology.

By considering these requirements, authors suggest eight steps of colposcopy for referred VIA clients, which must complete all the requirements mentioned above [4].

Suggested 8 Steps

1) Record reason for referral
2) Record category of VIA abnormality
3) Do VIA
4) Do colposcopy
5) Record visibility of the SCJ
6) Do a biopsy when any lesions were found
7) Appoint client for hearing-result visit
8) Inform referring provider the final result

Record Reason for Referral

This is required in 100%. [4] Reviewing and recording of this data will guide a consulted colposcopist what should be looking for, and where it is. This is also one of the most important data when reply the consulting CCP provider about her quality of management, if her referral is proper.

Record Category of VIA Abnormality

This is the substitution of grade of cytologic abnormality which is required in ≥90%. [4] Therefore, it is also required in ≥90% of all referred VIA clients, including the data of where the severest lesion is. If this information lacks, colposcopist must record its lack and inform the referring CCP provider, not only to help her improve her management quality but also to acquire that information from her.

Do VIA

Although a CCP provider must have done this before referring a client, a colposcopist should repeat it again. This may be done before or after colposcopy by washing the whole cervix with 3-5% acetic acid, waiting for 1 minute, and then see it with colposcopist's naked eye. Recording like a co-assessment, disagreement could acceptably happen in 20%. The referring CCP provider and her supervisors should also know this data. It will help improve both CCP provider's and colposcopist's quality of diagnosis.

Do Colposcopy

Some colposcopists may need to see the cervix under colposcope before soaking it with acetic acid. Therefore, this could be done before doing VIA without any deleterious effect. However, it may pose some bias to the forthcoming VIA test. In authors' opinion, this should be done after VIA to make the co-assessment fair and unbiased, with this reason, the result of colposcopy without acetic acid cannot be gained.

Record Visibility of the SCJ

This is also required in 100%. [4] Generally, the CCP provider would do Pap smear for clients without complete SCJs. Therefore, all referred VIA clients should have their SCJs completely seen. However, disagreement can also happen in this issue and the CCP provider should know this to help improve her diagnostic ability. In addition, visibility of the SCJ will guide the colposcopist when and where to do an endocervical curettage (ECC).

Do a Biopsy When Any Lesions Were Found

From the NHSCSP requirements, [4] an excisional form of biopsy is recommended (≥95%) when: (a) colposcopic appearances indicate high-grade abnormality; (b) low-grade colposcopic change is associated with severe dyskaryosis or worse; and (c) a lesion extends into the canal. In authors' opinion, (a) could be applied, (b) could not be applied, and (c) could be applied with either ECC or conization (either hot or cold). Reasons for not performing a biopsy must be recorded (100%), [4] especially when a lesion is seen.

Although a CCP nurse provider can do cryotherapy without performing any biopsies, colposcopist has no choice for denial. All patients must have a biopsy or biopsies taken prior to local destructive treatment (100%). [4] When any lesions persist on VIA or when atypical transformation zone is present, a biopsy must also be done (100%). [4] For the program's quality assurance, more than 90% of all biopsies taken (directed or excisional) should be suitable for histologic interpretation [4]. In addition, if a colposcopic directed biopsy is reported as inadequate for interpretation, it should be repeated if there is a residual colposcopic lesion (≥95%) and for those with satisfactory colposcopic examination, the predictive value of a colposcopic diagnosis of a high grade lesion should be at least 65%. [4] These mentioned indices will reflect service quality of the consulted colposcopist.

Appoint Client for Hearing-Result Visit

This is the must after colposcopy. Generally, when a biopsy is done, histological examination result will turn out within 3-4 weeks, not immediately after colposcopy. Therefore, any clients must know this fact, be appointed, and acknowledge the date of hearing-result visit. The colposcopist must also record this date for checking anyone who lost to follow-up, and contact her for coming back hearing result.

Inform Referring Provider the Final Result

The referring CCP provider also needs to know the result as well as her client. It is the responsibility of the consulted colposcopist to let her know her client's final diagnosis or management's result.

According to the program's quality assurance, she must know how she performs on VIA diagnosing, referring, and also 100% follow-up data from all clients she referred.

Conclusion

Providing cervical cancer screening should be considered and based on existing health care system. In Thailand, number of consulted colposcopists is still limited while most of them still do not know how to manage the referred VIA clients properly.

The suggested 8 steps of this CPG is only the advice, not the rule, for working with the referred VIA clients. Any colposcopists consulted by the VIA/cryotherapy CCP providers in anywhere of the world may follow this advice. Otherwise, using this as a guideline may be helpful when working in the VIA/cryotherapy CCP program.

References

[1] Royal Thai College of Obstetricians and Gynaecologists (RTCOG) and the JHPIEGO Corporation Cervical Cancer Prevention Group. Safety, acceptability, and feasibility of a single-visit approach to cervical-cancer prevention in rural Thailand: a demonstration project. *Lancet,* 2003; 361: 814-20.

[2] Chumworathayi, B., Srisupundit, S., Lumbiganon, P., et al. One-year follow-up of single-visit approach to cervical cancer prevention based on visual inspection with acetic acid wash and immediate cryotherapy in rural Thailand. *Int. J. Gynecol Cancer,* 2008; 18: 736-742.

[3] Chumworathayi B., Limpaphayom K. K., Srisupundit S., et al. VIA and Cryotherapy: Doing What's Best. *J. Med. Assoc. Thai.,* 2006; 89 (8): 1333-9.

[4] Luesley D, Lesson S. Colposcopy and Programme Management. Guidelines for the NHS Cervical Screening Programme. NHSCSP Publication, vol. *20.* Sheffield, UK: NHSCSP; 2004.

[5] Manopunya M., Suprasert P., Srisomboon J., et al. Colposcopy audit for improving quality of service in areas with a high incidence of cervical cancer. *Int. J. Gynecol. Obstet.,* 2010; 108: 4-6.

In: Asia Pacific Life Sciences Compendium
Editor: Bandit Chumworathayi

ISBN: 978-1-62257-338-7

Chapter 4

KNOWLEDGE INCREASED AFTER TRAINING IN VIA/CRYOTHERAPY CLINICAL-TRAINING-SKILL COURSE OF THE THAI DEPARTMENT OF HEALTH

***Bandit Chumworathayi*[1,*] *and Wachara Eamratsameekool*[2]**
[1]Division of Gynecologic Oncology, Department of Obstetrics and Gynecology, Faculty of Medicine, Khon Kaen University, Khon Kaen, Thailand
[2]Phanomphrai District Hospital, Phanomphrai District, Thailand

ABSTRACT

The objective is to measure the improvement between pretest and posttest scores of participants who had been trained in VIA/Cryotherapy clinical-training-skill (CTS) course of the Thai Department of Health. Scores from knowledge's assessments, from the precourse- and midcourse-questionnaire, in a particular course, conducted in April-May 2007, were compared using paired t-test. Minimum passing levels of 85% (43/50) were used for the midcourse assessment.

The participants were 2 male physicians and 14 female registered nurses, aged 33.19 ± 5.89 (range, 26-45) year-old. Knowledge's score was significantly increased from 34.31 ± 0.69 to 47.25 ± 0.76 ($p<0.0001$).

However, at the first midcourse assessments, there were 1/16 participants who did not pass. This study is the first reporting results of JHPIEGO VIA/Cryotherapy CTS course.

Although it was found to be effective in increasing the participant's scores achieved from knowledge's assessments, it was still not effective enough to increase all participants' scores to as high as 85%, which is the minimum passing level set by JHPIEGO. More recapitulations may be needed.

Keywords: medical education; VIA; cryotherapy; clinical training skill

[*] Email: bchumworathayi@gmail.com

Introduction

VIA (visual inspection with acetic acid) and cryotherapy has been implemented in combination as a single-visit-approach (SVA) cervical-cancer-prevention (CCP) program in many countries. This CCP program, developed by JHPIEGO, was established in Thailand since December 1999 by conducting a 5-days clinical-skill (CS) course for 4 clinical doctors-supervisors. This course was then followed by the 10-days CS course for 12 nurses-providers working for SAFE project in January, 2000. [1] After the SAFE project follow-up was completed in 2001, the model of 10-day CS course [2] has been successfully used for training nurses throughout Thailand.

At present, the Thai Department of Health (TDoH), who is taking responsibilities in conducting this course, has also been conducting the clinical-training-skill (CTS) course, [3] the training for trainers, as well. Over 1,000 nurses providers and over 200 nurses trainers, throughout Thailand, had been completely trained in these two kinds of competency-based course since 2001.

The TDoH has conducted these CTS courses for more than ten courses since 2001, however, there was no training result of this 10-day course, adopted from JHPIEGO, [3] reported in any medical education literatures. Basically, the 10-day course is designed for physicians, nurses, midwifes, or other appropriate health personnel in the country where it is implemented.

In Thailand, persons who are eligible to participate in this course are only physicians and registered nurses. Because there was no previous report of the effectiveness of this course, especially in Thailand, it still needs proving whether it is effective enough for its participants as they meant to be or not. In this study, the authors compare the individual scores of precourse-questionnaire assessment to those of midcourse-questionnaire assessment in a group of participants who attended this course during 23rd April - 4th May, 2007.

Method

This course used the reference manual, trainer's notebook, and participant's handbook, written by Sullivan et al. [3] Participants' performances were assessed according to three domains; knowledge, presentation skills, and clinical-teaching skills. Presentation skills' composed of lecturing by using audiovisual equipments in the real class, friend-participants acting as audiences.

Clinical-teaching skills composed of demonstrating, coaching, and evaluating of VIA/Cryotherapy related clinical skills, which all were performed on role plays and humanistic models, assessed by using checklists developed by JHPIEGO. Although no score was acquired from presentation skills' and clinical teaching skills' assessments, everyone must passed both checklists.

Knowledge assessments were performed first, before any lecture on the first day, by using true-false 50-item questionnaire and later on 10th day, by using multiple-choice questionnaire. Presentation skill assessments were performed on the 6th day. Clinical teaching skill-assessments were performed on the 7th day.

Table 1. The course schedule [3]

MODEL CLINICAL TRAINING SKILLS COURSE (10 Days, 20 Sessions)[1]				
DAY 1	**DAY 2**	**DAY 3**	**DAY 4**	**DAY 5**
A.M. (4 HOURS)	**A.M. (4 HOURS)**	**A.M. (4 HOURS)**	**A.M. (4 HOURS)**	**A.M. (4 HOURS)**
Opening: Welcome and introductions Overview of the course (Goals, Objectives, Schedule) Review course materials Identify participant expectations Precourse questionnaire Identify group and individual learning needs **Chapter 1: An Approach to Clinical Training** • Goal of clinical training • Mastery learning • Features of clinical training • Responsibilities • Selecting trainers	Agenda and opening activity **Chapter 3: Creating a Positive Learning Climate** • How people learn • Understanding group dynamics • Effective presentation skills • Introducing a presentation • Questioning techniques • Summarizing a presentation	Agenda and opening activity **Chapter 5: Delivering Interactive Presentations** • Illustrated lectures • Small group activities • Case studies • Role plays • Brainstorming sessions • Group discussions	Agenda and opening activity **Chapter 7: Developing Clinical Skills** • Skill transfer and assessment • The effective coach • Coaching • Using anatomic models • Clinical demonstrations **Demonstration**: Clinical role play showing effective demonstration and coaching skills.	Agenda and opening activity **Activity**: Participants will work in small groups to practice clinical skill demonstrations, coaching and skill assessment. **Activity**: Participants will plan an illustrated lecture using information from a clinical skill learning package. Participants will also prepare audiovisuals to support their presentations.
LUNCH	**LUNCH**	**LUNCH**	**LUNCH**	**LUNCH**
P.M. (3 HOURS)	**P.M. (3 HOURS)**	**P.M. (3 HOURS)**	**P.M. (3 HOURS)**	**P.M. (3 HOURS)**
Chapter 2: Planning for a Training Course • Planning considerations • Selecting participants • Course materials • Classroom arrangements • Clinic arrangements **Activity**: Participants will work in groups to plan a course. Review of the day's activities	**Chapter 4: Using Audiovisual Aids** • Writing boards • Flipcharts • Transparencies • Slides • Videotapes Review of the day's activities	**Chapter 6: Using Competency-Based Assessment Instruments** • Knowledge assessments • Skill assessments **Demonstration**: Introduction to anatomic models, and the participants practice care and maintenance of the models. Review of the day's activities	**Demonstration**: Clinical role play *(continued)* **Activity**: Participants will work in small groups to practice clinical skill demonstrations, coaching and skill assessment. Review of the day's activities	**Activity**: Participants will plan an illustrated lecture using information from a clinical skill learning package. Participants will also prepare audiovisuals to support their presentations. Review of the day's activities
Assignment: Chapters 1–4	Assignment: Chapters 5–6	Assignment: Chapter 7	Assignment: Plan presentation	Assignment: Plan presentation

Table 1. (Continued).

MODEL CLINICAL TRAINING SKILLS COURSE (10 Days, 20 Sessions)				
DAY 6	**DAY 7**	**DAY 8**	**DAY 9**	**DAY 10**
A.M. (4 HOURS) Agenda and opening activity **Activity**: Presentations (10–15 minutes) • Feedback from participants • Feedback from trainers • Videotape presentations for analysis if possible	**A.M. (4 HOURS)** Agenda and opening activity **Activity**: Presentations continued (if necessary) **Activity**: Individuals (or pairs) will plan for one or more of the following presentations: • Clinical demonstration using the learning guide • Coaching using the learning guide • Skill assessment using the checklist • Case studies • Role plays	**A.M. (4 HOURS)** Agenda and opening activity **Activity**: Presentations (10–20 minutes) • Feedback from participants • Feedback from trainers • Videotape presentations for analysis if possible	**A.M. (4 HOURS)** Agenda and opening activity **Chapter 8: Managing Clinical Practice** • Clients' rights • Opportunities for learning • Pre- and post-clinical practice meetings • Trainer as supervisor • Trainer as coach	**A.M. (4 HOURS)** Agenda and opening activity **Activity**: Midcourse questionnaire **Chapter 9: Conducting a Clinical Training Course** *(continued)* **Activity**: Individual goal setting, personal needs, interests, and training and cotraining plans of the participants. Development of action plans.
LUNCH	**LUNCH**	**LUNCH**	**LUNCH**	**LUNCH**
P.M. (3 HOURS) **Activity**: Presentations (10–15 minutes) • Feedback from participants • Feedback from trainers • Videotape presentations for analysis if possible Review of the day's activities	**P.M. (3 HOURS)** **Activity**: Individuals (or pairs) will plan for one or more of the following presentations: • Clinical demonstration using the learning guide • Coaching using the learning guide • Skill assessment using the checklist • Case studies • Role plays Review of the day's activities	**P.M. (3 HOURS)** **Activity**: Presentations (10–20 minutes) • Feedback from participants • Feedback from trainers • Videotape presentations for analysis if possible Review of the day's activities	**P.M. (3 HOURS)** **Chapter 9: Conducting a Clinical Training Course** • Preparing for training • Presenting a course overview • Conducting a training session • Giving knowledge and skill assessments • Qualification • Evaluating the course • Help and followup after training Review of the day's activities	**P.M. (2 HOURS)** **Activity**: Presentation of action plans **Course Summary** **Course Evaluation** **Closing Ceremony**
Assignment: Plan presentation	Assignment: Plan presentation	Assignment: Chapters 8–9	**Assignment**: Personal goals and plans	

RESULT

In this study, only scores from knowledge assessments were used for comparison, minimum passing level was 85%, because there were no scores from both presentation skills' and clinical-teaching skills assessments but every participant has to pass them all. All participants' scores, from the precourse- and midcourse-questionnaire, in this particular course were compared using paired t-test. P-value of less than 0.05 is considered significant. Data were analyzed using STATA version 10.0 software. Values are expressed as mean±SD or ±SE and 95%CI. The course schedule is shown in Table 1. The participants were 2 male physicians and 14 female registered nurses, aged 33.19±5.89 (range, 26-45) year-old. Knowledge's score was significantly increased from 34.31±0.69 to 47.25±0.76 ($p<0.0001$). However, at the first midcourse assessments, there were 1/16 participants who did not pass. Comparison of precourse-questionnaire assessment's and midcourse questionnaire assessment's scores are shown in Table 2. There were nine chapters in the reference manual, [3] therefore, questionnaire was divided according to these. These were; 1) An approach to clinical training, 2) Planning for a training course, 3) Creating a positive learning climate, 4) Using audiovisual aids, 5) Delivering interactive presentation, 6) Using competency-based assessment instrument, 7) Developing clinical skills, 8) Managing clinical practice, and 9) Conducting a clinical training course. Although scores from precourse assessment and midcourse assessment should not be compared because questionnaire types are different, it could be seen so clearly that there is no item with no correct in the midcourse assessment while there are three in the precourse assessment as shown in Table 3.

Table 2. Comparison of precourse-questionnaire assessment's and midcourse-questionnaire assessment's scores

Type of Assessments	Precourse-questionnaire assessment's scores Mean (95%CI)	Midcourse-questionnaire assessment's scores Mean (95%CI)	Paired t-test of scores' (N=16) P-value
Knowledge	34.31 (32.83-35.79)	47.25 (45.64-48.86)	<0.0001

Table 3. Details of scores on each items

Topic	Precourse assessment (T or F)		Midcourse assessment (4 choices)	
	Correct	%	Correct	%
An approach to clinical training	4	25.00	16	100.00
	7	43.75	14	87.50
	16	100.00	16	100.00
	11	68.75	16	100.00
	12	75.00	16	100.00
	13	81.25	14	87.50
	16	100.00	16	100.00
Planning for a training course	6	37.50	16	100.00
	16	100.00	16	100.00
	3	18.75	2	12.50
	16	100.00	12	75.00
	11	68.75	3	18.75

Table 3. (Continued)

Topic	Precourse assessment (T or F)		Midcourse assessment (4 choices)	
	Correct	%	Correct	%
Creating a positive learning climate	16	100.00	16	100.00
	0	0.00	14	87.50
	12	75.00	15	93.75
	14	87.50	13	81.25
	7	43.75	16	100.00
	16	100.00	11	68.75
Using audiovisual aids	4	25.00	13	81.25
	14	87.50	14	87.50
	14	87.50	6	37.50
	16	100.00	16	100.00
Delivering interactive presentation	3	18.75	12	75.00
	15	93.75	12	75.00
	16	100.00	15	93.75
	7	43.75	16	100.00
	5	31.25	13	81.25
	16	100.00	13	81.25
Using competency-based assessment instrument	15	93.75	13	81.25
	11	68.75	15	93.75
	15	93.75	14	87.50
	2	12.50	15	93.75
Developing clinical skills	0	0.00	14	87.50
	16	100.00	16	100.00
	16	100.00	12	75.00
	12	75.00	16	100.00
	16	100.00	15	93.75
Managing clinical practice	16	100.00	13	81.25
	13	81.25	15	93.75
	16	100.00	9	56.25
	4	25.00	16	100.00
	4	25.00	13	81.25
Conducting a clinical training course	0	0.00	6	37.50
	16	100.00	16	100.00
	8	50.00	14	87.50
	16	100.00	15	93.75
	6	37.50	14	87.50
	10	62.50	8	50.00
	15	93.75	2	12.50
	16	100.00	11	68.75

Discussion

This study is the first reporting results of JHPIEGO VIA/Cryotherapy CTS course. Although it was found to be effective in increasing the participant's scores achieved from knowledge's assessments, it was still not effective enough to increase every participant's scores to as high as 85%, which is the minimum passing level set by JHPIEGO, [3] as there was 1/16 participant unable to pass the first midcourse-questionnaire assessment. However, this participant passed the assessment at the end of the course by trainers' tutorial and the second assessment. This could be considered as the corrective measure.

To prevent any participant from such a corrective measure above, recapitulations may be needed each day of this training. Suggested point of improvement may be trying to use the recapitulation for at least 30 minutes each morning.

Conclusion

JHPIEGO VIA/Cryotherapy CTS course can definitely increase participant's knowledge scores, although not all participants can pass 85% level of midcourse assessment at once.

Acknowledgments

Authors would like to thank Dr Chalida Gespradit from the TDoH for her data support.

References

[1] Royal Thai College of Obstetricians and Gynaecologists (RTCOG)/JHPIEGO Corporation Cervical Cancer Prevention Group. Safety, acceptability, and feasibility of a single-visit approach to cervical cancer prevention in rural Thailand: A demonstration project. *Lancet,* 2003; 361: 814-20.

[2] Blumenthal P.D., McIntosh, N. Cervical Cancer Prevention Guidelines for Low-Resource Settings. Baltimore: *JHPIEGO,* 2005.

[3] Sullivan R., Blouse A., McIntosh N., et al. Clinical Training Skills for Reproductive Health Professionals. Baltimore: *JHPIEGO;* 1997.

In: Asia Pacific Life Sciences Compendium
Editor: Bandit Chumworathayi
ISBN: 978-1-62257-338-7

Chapter 5

BENCHMARKING A NOVEL LINEAR REGRESSION MODEL FOR EVALUATING THE RELATIONSHIP BETWEEN AHI AND HEIGHT IN SENIOR SLEEP APNICS

Bingh Tang
New York College of Traditiona, Chinese Medicine, Mineola, NY, US

ABSTRACT

Background and Aim

This case-series study originally aimed to assess 124 elderly people's sleep disturbance. They aged 65 years and over. The setting was at the Sleep Medicine Laboratory in a Medical Center of Taiwan.

The 124 individuals with sleep disturbance were admitted for polysomnograph (PSG) from 2 January 2002 to 31 January 2003.The null hypothesis to be tested is "after controlling for sex, age, and BMI, there is no relationship between AHI and height, though some of the elderly were kypho-scoliotic" in the studied cohort.

Methods and Results

Among the 124 elderly individuals, there were 117 had height data available. Not only was the height of 76 from 117 subjects within one standard deviation of the mean height, it also revealed a bimodal-like appearance in distribution. The height and its distribution were examined and revealed that there was probably only one distribution in height, and probably not a bimodal distribution. However, there indeed was a mean and/or distribution difference with respect to height from the study data. A linear regression model with AHI as the dependent variable was established.

Conclusion

The proposed hypothesis has been successfully tested, and the null hypothesis stays. The alternative hypothesis is thus rejected, that is, there is no relationship between AHI and height in senior sleep apnea of the tested cohort in this study reported.

The conclusion is that any relationships with AHI that have been observed are based on an individual's weight alone - and not their height, which is the novelty to be presented.

INTRODUCTION

The importance of sleep-disordered breathing (SDB) has been increasingly recognized over the last three decades. It is also associated with increased morbidity and mortality [1-3]. Surveys have estimated that more than 50% of community-living people aged 65 year and older experience sleep disturbances [4]. Based on the data archived, the author has another published report elsewhere mainly on total sleep time, except data of the height and its distribution.

The Reasons for Selecting this Age Group

The reasons for selecting this age group are as follow. First, this field is a rather understudied entity, especially in sleep research on such an age group mentioned in this article. Indeed, all patients are vulnerable, but it is so much as those who are elderly. Thus, the difficulty of sleep study is due to the problems faced in obtaining volunteers as well as the possible philosophical and theological under-tones that people in general associate with 'the sleep when they are at the end stage of their life'. Moreover, there is as well the misconception of sleep study on 'those who haven't many years left anyhow' in general. Next, as aforementioned, there are more than half of community-living people aged 65 year and over experience sleep disturbances. Third, sleep onset is often reported to be more difficult and nighttime awakenings more prevalent in the elderly. Certainly, the elderly sleep/wake pattern changes merit our attention. [6] On the contrary, someone may still comment 'why study old people, when it has been shown that Obstructive Sleep Apnea Syndrome loses its importance with respect to mortality in ages above 65 years [7].' [1]

Practically, the lack of information about possible confounders and treatment effects should be taken into consideration in the interpretation of these results. The last but not least, why study the relationship between Height and AHI? Height is an important component of anthropometric assessment. We all lose height when we age, even though when the height loss starts and how quickly it progresses vary quite somewhat among various people. General speaking, our height increases until our late forties and then decreases about two inches by age 80. Between these two age points, its midpoint is about at age of 65 years – the minimal age in this study [8].

It can thus be observed that studies like what is reported for the age group in this article have their own merit.

The other reason is that at an age of 65 years is an official age of retirement, which is the lowest limitation of age in the current study as well.

---

[1] Personal communication on 21 November 2007 with Professor Bengt Midgren, M D, Dept of Respiratory Medicine, University Hospital, Lund, Sweden.

The Hypothesis Proposed and Tested

The Hypothesis to be tested is that there is no relationship between AHI and height after controlling for sex, age, and BMI, though some of the elderly individuals were kypho-scoliotic.

CLINICAL SUBJECTS AND METHODS

All data used in accordance with the spirit and principles of Health Insurance Portability and Accountability Act regulations (HIPAA, 1996) [9]. All other data referred from other published studies with all their sources were respectively specified for the purpose of comparison. The Author reports no conflict of interest. Each individual gave a written informed consent. This author reports no conflict of interest.

Studied Subjects

During the period from 1 January 2002 to 31 January 2003 at the Sleep Medicine Center (Laboratory) of a Medical Center in Taiwan admitted no subjects with heart failure and chronic obstructive lung disease (the exclusion criteria 1 and 2 in the current study), as there was no physician at night in the Sleep Medicine Laboratory.

The Hospital Internal Research Board and Ethical Committee approved the study, which was in conformity with the Declaration of Helsinki. The author reports no conflict of interest. Under these circumstances, among the consecutive subjects who underwent nocturnal polysomnography (PSG) during that period, there were all together 1,087 cases of PSG performed.

The data belonged to individuals who were younger than 65 years were excluded (the exclusion criteria 3 in the current study). Hence the inclusion criteria were: 1] the chief complaint of sleep disturbance and 2] aged 65 years and over. As a result, out of total 1,087 cases of PSG taken during that period, there were 124 participants aged from 65 to 88.5 years and no one was demented. No pairs of individuals were related to each other. Among them, there were 117 individuals had body mass index (BMI) data available. Consents from the participants of this study include personal and medical data. In addition, at the time of their initial evaluation and enrollment, they all authorized clinical data in the research database to be used in this study.

Demographic Characteristics

The mean height is 153.07+- 8.87cms. The ratio of females to males was 1: 1.03. The ratio of females/males was 1: 1.7.

The mean age of male subjects was 71.6 ± 4.47, while that of females was 72.3 ± 5.47 years. All subjects had chief complaints of sleeping disturbance. (Table 1) Some of the elderly individuals were kypho-scoliotic.

Table 1. Demographic Characteristics of the Studied Cohort

Characteristics	*The studied Cohort*
Minimal age (years)	65 (sample size=124)
Mean of ages (years)	71.69±4.86
Mean of height (cms)	159.76±7.75
Mean of weight (Kgs)	65.5±11.96
Total number of subjects	124 78 (m): 46(f)
Ratio of males to females	1.7(m): 1(f)
Snorers/ total number of subjects	88.71%(110/124)
Diastolic blood pressure greater than 90 mmHg, and AHI greater than 5	32.7%(54/124)
Diastolic blood pressure great than 90 mmHg, and AHI less than 5	8.87%
Mean of BMI	25.68±4.54 (overweight: >25)
Mean for effect size	M2
Mean of cervical circumference (cms)	36.45±3.39
Mean of systolic blood pressures (mmHg)	140.08
Mean of diastolic blood pressures (mmHg)	79.87mmHg

Subjects' data under this study were collected during the period from 2002 to 2003. 2 A comparison was made with baseline characteristics of the elderly Taiwanese who underwent periodic health examination in Northern Taiwan, which was conducted from April to June 2001 – merely six months before the beginning of this study. In the study conducted merely six month earlier than this study, there were 2,060 individuals aged 73.0+_ 4.7 years. Their BMI was 24.3+_3.5. in comparison with the individuals in our study who had age 71.69 ± 4.86 years and BMI 25.68 ±4.54 (Noticeably, the criterion for overweight is BMI >25, with respect to the Far-Eastern Asian criterion for overweight.

The Nocturnal in-Laboratory PSG

In-laboratory nocturnal PSG recording was performed between 9 30 pm to 6 am. PSG (Alice 4 Sleep Diagnostic System, Respironics, Carlsbad, Calif., USA) comprised: (i) plethysmography for checking chest and abdominal wall motion; (ii) ECG; (iii) inspired and end-tidal carbon dioxide pressure (PETCO2);by mass spectrometry (model 1100 Medical Gas Analyzer, Perkin Elmer; Pomona, CA); (iv) combined oral nasal air flow, sampled with a

[2] It is true that the height measurements should have been obtained from the census data to have any validity in comparison. Unfortunately, for instance, the report of 'Elderly Nutrition and Health Survey in Taiwan (1999–2000): research design, methodology and content' was not published until 2005, two years and eleven months after the conclusion of this study. (It is noted that researchers generally much prefer to have narrower age categories and less year intervals from surveys but often have to be content with less useful information as reported in publications.) Noticeably, the WHO data do not include that of Taiwan.

three-pronged thermistor placed at the upper lip; (v) arterial oxygen saturation by pulse oximetry (model N 200, Nellcor, Hayward, Calif., USA); (vi) oximeter pulse wave form; (vii) electro-oculogram; (viii) EEG; (ix) chin electromyogram; (x) actigraphy; and (xi) microphone placed over the neck to monitor snoring. The transducers and lead wires permitted normal positional changes during sleep. Bedtime and awakening time were at each subject's discretion; the polysomnography was terminated after the final wakening in the morning.

The Classification of Obstructive Sleep Apnea Syndrome

The OSAHS was classified according to the criteria of severity of sleep apnea as follows. The degrees in severity of sleep apnea were defined, bases on the protocol of American Academy of Sleep Medicine Task Force (2007) as following:

1) zero (0) degree, apnea/hypopnea index (AHI) < 5,
2) first degree, 5 <= AHI < 15,
3) second degree, 15 <= AHI < 30, and
4) third degree, AHI > 30. [9] (Table 2)

The Sleep Staging

Sleep in the PSG findings were staged manually according to Rechtschaffen et al's criteria (1963) [10].

Statistical Methods

Data were analyzed with statistical software (SPSS 10.0). A two-tailed p value < 0.05 was recognized as significant. Spearman rank correlation tests were used between AHI and snoring, between AHI and height and between AHI and neck circumference. Linear regressions models were tested. Independent t tests were used for comparing between the studied and the comparison group merely for limited data. Kolmogorov-Smirnov test was performed for checking normality.

Results

1) The body height: Notice that in the current study, the ratio of female to male subjects was 46 to 78 (1 to 1.7). The average age of male subjects was 71.6±4.47 years, while that of females was 72.3±5.47 years. The mean height was 159.76 +_ 7.75 , and 95% confidential interval was from 158.3 to 161.2 cm. There were 64.96% of the observations (76/117) within + and - one standard deviations (SD) of the height distribution in the current study. Six subjects had the height of 163 cm, whereas five had that of 165 cm. The distribution does not appear normal. Instead, it has a long

tail to the lower heights. (Figure 1) Kolmogorov-Smirnov test revealed these height data unlikely were normally distributed, p = 0.01. They unlikely were log normally distributed, p = 0.001.

2) BMI and height: The Pearson's correlation coefficient between height and BMI was significantly different between males (- 0.227) and females (-0.0854).
3) Spearman rank correlation tests for the following variables are as the Table 4.
4) The relationship between apnea/hypopnea index (AHI) and height in the aged 65 years and over: results of analysis performed using a linear regression model
The model tests the hypothesis, "after controlling for sex, age, and BMI, there is no relationship between AHI and height." There was significantly more than zero % of variability in AHI that is explained by sex, age, and BMI three variables.
However, when height was added to see if it contributed explanatory power above and beyond what was already being explained in the model, it contributed nothing. The amount of variability in AHI explained by the model with age, sex, BMI and Height explains 15% of the variability in AHI, which is the exact same amount explained in AHI when only age, sex and BMI are in the regression. (Tables 3 and 4). Hence, this leads to the conclusion that any perceived relationship between height and AHI was accounted for by age, sex, BMI, or some combination of these. It is actually just sex and BMI, which are the only two covariates in the smaller model (without height) that are significant.

Table 2. Sleep variables with Apnea (changes of AHI)

Sleep variables with Apnea (changes of AHI)	Findings
Snoring	61.3%
Apnea (AHI >5)	81.3%(n=103)
Apnea (AHI 5~15)	25.8%(n=32)
Apnea (AHI 15~30)	22.6%(n=28)
Apnea (AHI>30)	34.7% (n=43)
both snoring and apnea simultaneously	50.8%(n=63)
PLMD (Periodic limb movement disorder)	41.1%(n=51)
both PLMD and apnea	32.3%(n=40)
Total sleeping time	322.5 ±91.3 minutes
Sleep efficiency	72.6%±17.9%
Sleep latency	23.9±39.6 minutes
The ratio of average time of stage 1 sleeps to total sleeping time	22.4%±13.5%
% of patients with disappearance of sleep stages 3 and 4	71%(n=88)
The ratio of Stage Rapid Eye Movement sleep time to total sleep time	10.1%±6%
3rd degree (30<AHI) severity of sleep apnea	34.68% of
2nd degree (15≤AHI<30) severity of sleep apnea	22.58%
1st degree (5≤AHI<15) severity of sleep apnea	25.81%
0 degree (AHI<5) severity of sleep apnea	16.94%

Table 3. Spearman's rank

Correlations

			BMI	SNORE	AHI
Spearman's rho	BMI	Correlation Coefficient	1.000	.511**	.295**
		Sig. (2-tailed)	.	.000	.001
		N	116	116	116
	SNORE	Correlation Coefficient	.511**	1.000	.260**
		Sig. (2-tailed)	.000	.	.004
		N	116	124	124
	AHI	Correlation Coefficient	.295**	.260**	1.000
		Sig. (2-tailed)	.001	.004	.
		N	116	124	124

**. Correlation is significant at the .01 level (2-tailed).

Table 4. Spearman rank correlation tests for the following variables

	BMI	Snoring	Neck Circumference	Height
AHI	r=0.295 p=0.001**	r=0.26 (0.09-0.42)	r=0.33 (0.16-0.48)	r= - 0.348 p=0.06 (See Tables 3 and 4)
Snoring	r=0.51 p=0.01*			

* denoting significant, while ** representing highly significant.

5) The hypothesis has been successfully tested and finally demonstrated that any relationships with AHI that have been observed are based on an individual's weight alone – but not their height. Such an important novel although negative finding is to be further explained in the forthcoming section.

Discussion

Among the 124 subjects, there were 117 whose height records were available.

If the heights were normally distributed, then 68.20% of all observations in the distribution would have fallen between the minus one standard deviation and plus one standard deviation of heights. Noticing the major result of this study, namely, the important and novel, although negative, findings are as follow.

Noticeably, height does not correlate with AHI. Other than the above, any relationships with AHI that have been observed are based on an individual's weight alone – but not their height. Such significantly important and novel, although negative, findings are to be explained as follow. In this age group, namely from 65 to 88.5 years, the reasons for height loss include the following:

- changes in posture
- changes in the growth of vertebrae
- a forward bending of the spine
- compression of the intervertebral discs
- increased curvature of the hips and knees
- decreased joint space in the trunk and extremities
- joint changes in the feet
- flattening of the foot arches

Noticeably, with more than 50% of the community-dwelling elderly having sleep disturbance, as aforementioned, with linear regression models, consider all 117 subjects together, any relationships with AHI that have been observed are based on an individual's weight alone – but not their height, in other words, height changes (basically losses but not increase) in the elderly was indeed not effective affecting AHI. The other key findings revealed that there was an observation on 76 from 117 individuals (64.96%), instead of 68.20% of the 117 individuals, within one standard deviation of the mean height (159.76 +_ 7.75 cm). In the study cohort, there were 66 subjects (66/117 = 56.41%) who were taller than the mean height of the group (159.76cm). The height of six out of those 66 subjects was 163 cm; and with five of these six individuals was 165 cm tall. On the other extreme, there were 9 individuals in the current study whose height was between 159.76 and 157.03 cm, and the height of another 42 subjects was even less than 157.03cms. (See Figure 1 of Height Distribution). The distribution of height presented in such an interesting and seemingly unusual fashion, which was worthwhile to be reported for following two main reasons:

(A) Is the Height Distribution Observed in the Current Study Bimodal? Why Such a Distribution of Height Exists?

No Western study is implying that apneics are taller than the comparison group, while on this issue of the height, certainly some more light could be shed by Western data, which shall be further discussed in one of the forthcoming sections. The seemingly bimodality derived from figure 1 may be a pure artefact. The bin width of the histogram is 1 cm leading to more than 30 bins. Rules of thumb for the bin size suggest about 10 in this case. What is viewed, might be merely and solely due to chance fluctuations. Moreover, the simplest factor that may cause bimodal distributions is the mixture between male and female patients. In the Figure 1, the histogram reveals the seemingly bimodal appeaqrance it would have disappeared with smoothing.

(B) In 1995, Taiwan Launched the National Health Insurance Plan, Which is a Compulsory Social Insurance Programme Embracing the Whole Population

All the insured are provided with right to equal access to healthcare services including various medical clinics. Due to a rather feeble gatekeeper referral system, patients have the complete freedom to select hospitals and and healthcare providers irrespective of the nature or severity of their illness; hence, almost all patients in our sleep medicine laboratory were referred from various medical clinics or self-referred. Taiwan provided and still provides the expense of PSG, with 1/10 of its co-payments to be paid by a patient per test. Therefore, our studied subjects were not particularly richer than the average, and then affordable for any full payment of the expensive PSG. Next, the population was well mixed there. Among them were the native Taiwanese, Han, the aboriginal and immigrants. There was no significant difference with regard to any ethnical or psychosocial influences on food habits, the nutritional adequacy, and the overall health status, which might affect body height alone.

It is noticed that, in general, the average height, as well as the craniofacial base observed in Asian individuals are respectively shorter than that in Caucasians [6 -13] certainly increase the risk of the narrow upper airway. Researchers [14 – 19] reported a worse severity of OSA in Chinese-Americans in California than in Caucasians. Ip et al [11-13] hypothesized some strong OSA risk factors, such as craniofacial features, which are more prevalent in Chinese than in Western populations that compromise the upper airway, must exist. Some, but not all, clinical observations of Asian patients with OSA support this hypothesis.

(C) Could Height Be an Independent and Sole Causal Factor for Sleep Apnea in Adulthood?

In light of the mean body weight of 65.5+_11.96 kgs, and the mean BMI of 25.68 Kgs/m2, as contrasted to that of a community population in Northern Taiwan, it suggests and probably implies that the height distribution of this study did not exclusively cause any independent and sole effect on the change of BMI. Another remote possibility is a causal factor for the height differences in the adulthood, and then sleep apnea must have taken place during the childhood of most individuals in order to alter their growth. Nevertheless, young patients in their childhood of sleep apnea are usually suffering from growth impairment, instead of growth acceleration. It does not appear plausible in the present context. It is noticed that characteristics of BMI as aforementioned are perfectly consistent with the literature [20-23]. BMI has been shown to have relevance to OSA (with the remembrance that upper limit of normal BMI in Far-Eastern Asian is 23.5 kg/m2 from current WHO data). [23-26]

A report from the Netherlands indicates that reduction of adult height (RAH) was calculated approximately in the adults' childhood for ALL survivors with and without cranial irradiation. [27] Such a reduction was in the preponderance of male and female childhood of ALL survivors who had received preventive cranial irradiation,. There was no such a history in any of the studied subjects.

THE LATENT CLINICAL FACTOR

After thorough evaluation with the process of elimination, another factor that is worthy of discussion in relation to the height distribution is the latent clinical factor, which is the rate people are shrinking due to ageing and hormonal effects [22, 24] on the loss of bone density. It is something clinically unobservable at the first glance on individuals. In the current context, not only was there predominantly an observation of the height loss, but also there was a lack of knowledge about their rate to shrink. On the parlance of the latent variable modeling, there might be a true nature, as whether a fast or slow shrinker - and this is what really causes subjects to fall into one or the other statistical distribution that one can possibly observe.

The other fact that apnea (elevated AHI) and snoring are associated with obesity and increased neck size has been widely reported in the literature, same as has the relationship of AHI to hypertension [20-22, 28- 36]

However, there are very few data available about how apnea and napping are related in aging samples such as the age group investigated in this study. Merely some data have fairly good AHI estimates (partly described in the Kripke et al. 2004 paper). There is indeed a need for systematically study on the actigraphic napping data in that sample, as opposing to the PSG data in this studied cohort.

The Physiological Rationale of this Study

The question is why should not height have an impact on abnormal breathing during sleep for those who suffer from kypho-scoliosis? This author has analyzed the data and though showed likely only one distribution of height, but the rationale for the study is absolutely indicated. There are specific anatomic presentations involving the neck (Marfan Syndrome), and the chest (Kypho-scoliose etc) that may impact on breathing. Among the 124 individuals studied, there was no one suffered from Marfan Syndrome; but there were some elderly individuals suffering from kypho-scoliosis. Such datum existed that showed that height had impact on respiration considering normal or abnormal breathing. BMI has been looked at in this study. Both weight and the calculation of BMI have been evaluated when considering breathing. It has been pointed out for many years now that BMI should be associated with measurement of neck circumference and hip-waist ratio, though the latter, hip-waist ratio, was not done in this study. Depending on weight distribution, there will be possibly an impact on breathing when supine and during sleep, but the weight distribution is a critical variable. Here there is a study, such as this one, of a variable, namely, height, which is part of the calculation of the BMI based on a specific formula, but selection of BMI compared to selection of weight alone was done to normalize the data and have weight data in some way that could be usable and that the body mass could be compared within individuals.

The fact that total height has nothing to do with breathing is an important finding though it is negative. Considering the presence of physiological rationale of kypho-scoliosis to perform this study, it is clear to see what the finding really brings to our understanding of normal or abnormal breathing at this age in life, regardless of the finding *per se* is negative or positive.

Limitations of this Study

BMI has been shown to have relevance to OSA (with the commemoration that upper limit of normal BMI in Far-Eastern Asian is 23.5 kg/m2 from current WHO data.)

It is noted that obesity in Asia, like in Singapore nationwide, is defined as BMI >23. Asian nutritionists are inclined to apply a limit of BMI > 23kg/m2 for obesity due to lower stature and body weight. But why should height have relevance in the problem is still unclear. Notwithstanding the fact that obstructive sleep apnea (OSA) may not be uncommon in Asian and Southern Pacific patients [16-19, 36, 37], compared with what has been reported for Western patients [23-35, 38, 39], there is hardly any information of large study. There is no prior publication of the height measurement alone related with AHI changes.

Conclusion

To strike a balance, based on the discussion predominantly relating to the parameter on height and its distribution, noticeably, there was probably only one type of height distribution in the studied cohort. It was probably not a bimodal distribution, as it would have disappeared with smoothing. However, there indeed was a mean and/or distribution difference from the population data.

In fact, there are more than 50% of the community-dwelling elderly having sleep disturbance. On the other hand, with linear regression models, with all 117 individuals being considered together, any relationships with AHI that have been studied were based on an individual's weight alone – but not their height, namely, height loss bears no influence toward AHI in this age group, which is the novelty to be presented, although the finding is negative, but it is important.

References

[1] Ancoli-Israel S. Sleep problems in older adults: putting myths to bed. *Geriatrics,* 52(1):20-30, 1997 Jan.

[2] Nortn , PG, Dunn, EV. Snoring as a risk factor for disease: an epidemiological survey. *British Medical Journal Clinical Research Ed.,* 291(6496):630-2, 1985 Sep 7.

[3] Young T, Palta M, Dempsey J, et al. The occurrence of sleep-disorded breathing among middle-aged adults. *N. Engl. J. Med.,* 1993; 32:1230-1235.

[4] Stradling JR, Crosby JH. Predictors and prevalence of obstuctive sleep apnea and snoring in 1001 middle aged men. *Thorax,* 1991; 46:85-90.

[5] Tang B. 2007: Living by a biological clock: age-related functional changes of sleep homeostasis in people aged 65-88.5 years. *Sleep and Biological Rhythms,* 2007;5: 179-194 (15 pages) Blackwell publishin.

[6] Schwab RJ, Gupta KB, Gefter WB, et al. Upper airway and soft tissue anatomy in normal subjects and patients with sleep-disordered breathing, significance of the lateral pharyngeal walls. Am J Respir Crit Care Med 1995 Lavie P, Herer P, Peled R et al. Mortality in sleep apnea patients: a multivariate analysis of risk factors. *Sleep,* 1995; 18: 149–57.

[7] Changes in height - http://www.healthinaging.org/ agingintheknow/ chapters_ch_trial. asp?ch=1#Changes%20in%20Height.

[8] ber, C.;Ancoli-Israel, S.;Chesson, A.; Quan, SF. for the American Academy of Sleep Medicine. Westchester, Ill: American Academy of Sleep Medicine; 2007. The AASM Manual for the Scoring of Sleep and Associated Events: Rules, Terminology and Technical Specifications.

[9] Rechtschaffen A, Wolpert EA, Dement WC, etr al. Nocturnal sleep of narcoleptics. *Electroencephalorgr Clin Neurophysiol,* 1963; 15:599-6.

[10] Olson LG, King MT, Hensley MJ, ct al, A community study of snoring and sleep-disordered breathing: prevalence. *Am. J. Respir. Crit. Care Med.,* 1995; 152:711-716.

[11] Ohayon M, Guillemenault C, Priest RG, et al. Snoring and breathing pauses during sleep: telephone interview survey of a United Kingdom population sample. *BMJ,* 1997; 314:860-863.

[12] Ip MSM, Bing Lam, Lawrence C.H. Tang, Ian J. Lauder, Toi Yan Ip, Wah Kit Lam. A community study of sleep-disordered breathing in middle-aged Chinese women in Hong Kong: prevalence and gender differences - clinical investigations. *Chest,* 2004; 125:127-134.

[13] Lam B, M S M Ip, E Tench and C F Ryan Thorax Craniofacial profile in Asian and white subjects with obstructive sleep apnoea. *Thorax,* 2005;60:504-510.

[14] Lam B, Lauder IJ, Tsang K, Chung K, Mok Y, Lam W. A community study of sleep-disordered breathing in middle-aged Chinese men in Hong Kong. *Chest,* 2001 ; 119: 62-9.

[15] Wing YK, Hui SH, Pak WM, Ho CK, Cheung A, Li AM, Fok TF. *A controlled study of sleep related disordered breathing in obese children Archives of Disease in Childhood,* 2003;88:1043-1047.

[16] Li, KK, Li, KK, Powell, NB, Kushida, C, Reily, RW, Adornato, B, Guilleminault, C. A comparison of Asian and Caucasian patients with obstructive sleep apnea syndrome. *Laryngoscope,* 199, 109:1937-1940.

[17] Lam B, Ip MS, Tench C F, Ryan. Craniofacial profile in Asian and white subjects with obstructive sleep apnoea. *Thorax,* 2005;60:504-510.

[18] Wong ML, Sandham A, Ang PK, Wong DC, Tan WC, Huggare J. Craniofacial morphology, head posture, and nasal respiratory resistance in obstructive sleep apnoea: an inter-ethnic comparison. *Eur. J. Orthod.,* 2005 Feb;27(1):91-7.

[19] Pawar R, Smith M, Wright JR, Strohl KP. Blood pressure regulation and sleep apnea. *Cardiovascular Reviews and Report,* 1996.

[20] Busetto L, Enzi G, Inelmen EM, Costa G, Valentina Negrin V, Giuseppe Sergi G, Vianello A. Obstructive Sleep Apnea Syndrome in Morbid Obesity. *Effects of Intragastric Balloon Chest,* 2005;128:618-623.

[21] Shahar E, Redline S, Young T et al. Hormone replacement therapy and sleep-disordered breathing. *Am. J. Respir. Crit. Care Med.,* 2003; 167:1186-1192.

[22] Groth, M. Sleep Apnea in the Elderly. *Clin. Geriatr. Med.,* 21 (2005), 701-712.

[23] Franco C, Brandberg J, LoBin L, Andersson B, Bengtsson B, Johannsson G. Growth Hormone Treatment Reduces Abdominal Visceral Fat in Postmenopausal Women with Abdominal Obesity: A 12-Month Placebo-Controlled Trial. *The Journal of Clinical Endocrinology and Metabolism,* 32 04;466-1474 20.

[24] Michael A. Grandner, BA and Daniel F. Kripke, MD. Self-reported Sleep Complaints With Long and Short Sleep: A Nationally Representative Sample.

[25] Yoon I-Y.; Kripke D.F.; Elliott J.A.; Youngstedt S.D.; Rex K.M.; Hauger R.L. Age-Related Changes of Circadian Rhythms and Sleep-Wake Cycles Journal of the American Geriatrics Society, *Volume,* 51, Number 8, August 2003, pp. 1085-1091.

[26] Bongers MEJ, Francken, A. B, Rouwe, C., amps, W. A. Postma, A. 2005: Reduction of adult height in childhood acute lymphoblastic leukemia survivors Golbin J, Somers V, and Caples S. Obstructive Sleep Apnea, Cardiovascular Disease, and Pulmonary Hypertension. The Proceedings of the American Thoracic Society, 5:200-206 (2008, Feb. 9) : 10.1513/pats.200708-143MG.

[27] Punjabi N. The Epidemiology of Adult Obstructive Sleep Apnea .

[28] The Proceedings of the American Thoracic Society, 5:136-143 2008 doi: 10.1513/pats.200709-155MG.

[29] Hoffstein MD, Chan CK, Slutsky AS. Sleep apnea and systemic hypertension: a causal association review. *Am. J. Med.,* 1991; 91: 190-96.

[30] Wilcox, I. Grunstein RR, Hedner JA, Doyle J, Collins FL, Fletcher PJ et al. Effects of nasal continuous positive airway pressure during sleep on 24-hour blood pressure in obstructive sleep apnea. *Sleep,* 1993; 16(6): 539-44.

[31] Hader Ca, Schroeder M., Hinzl Gh, Micklefield K, Rasche E. Sleep Disordered Breathing In The Elderly: Comparison Of Women And Men *J. Apple Physiol.,* 1998; 84: 1055-1062.

[32] Watanabe T, A Mikami, T Kumano-Go, N Suganuma, Y.

[33] The Relationship Between Esophageal Pressure and Apnea Hypopnea Index in Obstructive Sleep Apnea Sleep Research Online, 2000.

[34] Gozal D, Is It the Chicken or Is It the Egg? *American Journal of Respiratory and Critical Care Medicine Vol.,* 166. 2002, 1305-1306.

[35] The U S NIH definitions of Obesity: http://www.nhlbi.nih.gov/ health/public/ heart/obesity/lose_wt/risk.htm Tan WC, Koh TH. The prevalence and severity of OSA in Asian snorers: *Repir. Med.,* 2004 Jun; 98(6): 557-66.

[36] Duran J, Esnaola S, Rubio R, Iztueta A. Obstructive sleep apnea-hypopnea and related clinical features in a population-based sample of subjects aged 30 to 70 yr. Am J Respir Crit Care Med 2001; 163: 685-689.Ancoli-Israel S, Kripke DF, Klauber MR, Fell R, Stepnowsky C, Estline E, Khazeni N, Chinn A. Morbidity, mortality and sleep-disordered breathing in community dwelling elderly. *Chest,* Vol 109, 890-895, 1996.

[37] Lavie P, Lavie L, Herer P. All-cause mortality in males with sleep apnea syndrome: declining mortality rates with age. *Eur. Respir. J.,* 2005; 25:514-20.

In: Asia Pacific Life Sciences Compendium
Editor: Bandit Chumworathayi
ISBN: 978-1-62257-338-7

Chapter 6

ON TSUNAMIS, ECONOMIES AND ROMAN SENATORS

Anthony Shakeshaft*
School of Public Health and Community Medicine,
University of New South Wales, Australia

ABSTRACT

The international response to the 2004 Boxing Day tsunami, as judged by the observed effects of the 17th July 2006 tsunami, deserves a mixed report card. The short-term response appears to have been reasonable, although efficient distribution of goods and co-ordination of effort were clearly problematic and ought to be closely scrutinised. Over the longer term, the installation of warning systems between the tsunamis was encouraging and the need to improve the speed with which warnings are communicated, and increase local peoples' knowledge on how to react to such warnings, have been clearly identified as issues. Representing the international community's greatest response failure, however, is the obvious unwillingness to seriously consider increased economic equality between developed and developing countries, as the most effective longer-term strategy to reduce harms associated with public health disasters. Despite the strategies that were implemented between the tsunamis, the 17th July 2006 tsunami still resulted in a loss of life in the order of 10 times that which occurred over 15 years ago as a result of the San Francisco and Los Angeles earthquakes. More needs to be done.

Keywords: tsunami; economy; public health

INTRODUCTION

In the early hours of Boxing Day 2004, a gentle shore-break surged and receded, quietly soothing the coast-line of the Bay of Bengal. Certainly there was nothing to suggest the public health catastrophe for which the seemingly benign bay would shortly become an unwitting conduit. The extent of death, disease, psychological trauma and sheer physical

* Email: a.shakeshaft@unsw.edu.au; Phone: +610293850333. Fax: +610293850222.

carnage caused by that Christmas tsunami will most likely never be accurately quantified. But the measure that is simplest to define, and the most final, is also the most telling: an estimated 230,000 people were left dead or missing [1].

The subsequent outpouring of concern from around the world was both reassuring, from a humanitarian point of view, and impressive, from a logistical point of view. But the real effectiveness of the international community's response in helping re-build the lives of those devastated by that tsunami remained an open question, until a second major tsunami struck the Indonesian coast-line on 17th July 2006: an earthquake measuring 7.7 on the Richter Scale occurred 100 miles south of Java sending two to seven metre waves up to two kilometres inland. [2] Although of little reassurance to the families and communities of the estimated 700 people who lost their lives to this second tsunami, [2] it did provide a crucial sounding-board against which to examine the adequacy of the international response to the 2004 Boxing Day tsunami.

The short-term, and highly appropriate, international response to the occurrence of any disaster is for those relatively unaffected to help those affected. This usually manifests in the form of donations of goods (money, water, food, clothes and so on) and provision of skilled workers to boost local health-care efforts and re-build crucial infrastructure. Although generous, this assistance was not provided without problems in the aftermath of the 2004 Boxing Day tsunami: money and goods may not have arrived where they were most needed; [3] and disagreements about how relief efforts ought to be organised and funded arose [4].

To complement short-term efforts, it might reasonably be expected that initiatives were also pursued to ensure gains would be sustained over the longer-term. The idea that providing help over the longer-term is both necessary and difficult is hardly new. Concepts like the 'sponsor-a-child' programs attempt to tap into this very notion - finding ways to keep people engaged over a number of years - and have no doubt provided relief to thousands of children and their families. But there is a danger of becoming trapped into single-minded ways of providing aid that, by themselves, are less than optimal, particularly if they become reduced to self-perpetuating repeats of humanitarian relief. If attempts are only ever made to solve immediate problems, then well-intentioned aid efforts inevitably succumb to a cycle of trying to help with one problem, while awaiting the next. So, in response to the 2004 Boxing Day tsunami, what longer-term strategies might have been put in place?

What Should Be Done?

The most effective strategy, of course, would have been to simply prevent the occurrence of the second tsunami. This idea is essentially the equivalent of the health-care maxim that prevention is better than cure. In medicine, new medications and treatment procedures are continually developed in an effort to improve standards of health-care, but there is increasing recognition of the importance of prevention. [5] There is a subtle point here worth drawing out: nobody in the medical field seriously advocates prevention instead of treatment. Rather, the idea is to continue to develop the capacity to deal with injury and disease, while at the same time finding ways to reduce the frequency with which they occur.

Evaluating the possibility that the second tsunami could have been prevented requires some understanding of the causes of tsunamis. Short of a detailed geological lesson, it appears the explanation for the 2004 Boxing Day tsunami is that two sections of the Earth's crust were pressing against each other until they slipped. The effect was, according to British seismologist David Booth, "... just like moving an enormous paddle at the bottom of the sea ... a big column of water has moved ... This is an enormous disturbance." [6] To borrow a phrase from Kurt Vonnegut, suddenly billions of tons of water had to find somewhere else to go. And that they did. Given this geological basis for tsunamis, the only way to have prevented the second one from occurring would have been to have controlled the physical movements of the earth. The ability to control massive geological events that occur kilometres below the ocean floor, however, seems doubtful at best, certainly beyond the realm of reasonable possibility in the 18 months between the two tsunamis.

If it was impractical to have prevented the second tsunami from occurring, might it have been more effectively anticipated? This would at least have provided a mechanism for moving people out of the path of imminent disaster more efficiently, reducing the number of those killed. For earthquakes, however, it appears the location and time of their occurrence is essentially unpredictable. In any case, a daunting list of similarly unpredictable potential disasters is easily conceived, such as fires, stray comets and disease (including SARS, HIV and bird flu) [7].

If the second tsunami could not have been effectively prevented or predicted, the question becomes one of how the resultant harm might have been minimised once the earthquake that triggered the tsunami occurred. The most obvious strategy to achieve that outcome would be the construction of early-warning or monitoring systems to detect and communicate, at the earliest possible moment, the occurrence of such events. This idea is the basis of modern epidemiology (the study of epidemics), the origin of which is popularly credited to a British anaesthesiologist, John Snow.

In 1854 in Soho, London, an outbreak of cholera caused the deaths of hundreds of local residents. John Snow laboriously traced the source of the cholera to the water supply by mapping where the deaths occurred, convincing civic authorities to remove the handle from the centrally located water pump. He was not ultimately responsible for stopping the spread of cholera - by Snow's own admission everybody had moved out of the area and that stopped it - but he did show that once we understand how a catastrophe happens, action can be taken to minimise the effects of the problem, while simultaneously increasing the likelihood that future disasters might be averted, or their impact reduced. This idea raises a second subtle distinction: the aim is to minimise the scale of future disasters, not necessarily to stop them from occurring altogether. Like Snow's cholera victims, this approach is of little comfort to those devastated by the 2004 Boxing Day tsunami, but the principles of the approach could have been utilised to minimise the number of deaths caused by the second tsunami. The question is, were they?

EXISTED STRATEGIES

As it happens, tsunami warning systems do exist: Japan apparently has a "... highly advanced early warning system ..." in the Pacific Ocean, good enough that it was touted as the basis for a similar system that, it was argued after the occurrence of the 2004 Boxing Day tsunami, ought to be put in place in the Indian Ocean. [8] The estimated cost of this system is about AUD$40 million [8] - easily affordable, one would think, in global economic terms. This observation raises two specific issues. Firstly, prior to the 2004 Boxing Day tsunami, why did the Pacific Ocean have a tsunami early-warning system while the Indian Ocean did not? Secondly, did the Indian Ocean, specifically along the Indonesian coast-line, have such a system in place when the 17th July 2006 tsunami struck?

The existence of a tsunami early warning system in the Pacific Ocean, but not the Indian Ocean, prior to 2004 Boxing Day is troubling given it seems unlikely that the Indian Ocean would be solely immune to geological disturbance. Perhaps of greater concern is the admission from a helpful, although suspiciously non-identifiable, Japanese official after the occurrence of the 2004 Boxing Day tsunami that Japan and the US "... could provide tsunami warnings to countries around the Indian Ocean until their own system was in place..." [8] This seems to imply that countries not affected by this tsunami could have helped provide additional warning time, at least to Sri Lanka, India and the countries further to the west of them. So it appears existing technology to provide tsunami warnings in the Pacific Ocean prior to 2004 Boxing Day, could have been used to detect unusual activity in the Indian Ocean, but was not. Why? Perhaps we ought to give the benefit of the doubt and assume our helpful Japanese official is mistaken about the ability of Japan and the US to have provided tsunami warning systems in the Indian Ocean, or that he has been misquoted, or, indeed, that he does not actually exist.

If nothing else, it seems entirely reasonable that tsunami early warning systems would have been installed in the Indian Ocean after the 2004 Boxing Day tsunami. Reassuringly, this appears to have been the case. The UN, in co-operation with a number of Western countries, has worked to establish an early warning system off the coast of Indonesia and asserted in June 2006 that an interim system was operational. [9] Given the estimated 700 lives lost to the 17th July 2006 tsunami, however, it seems the effectiveness of this system was less than optimal. There is evidence that the physical monitoring system worked, prompting the chief of the Indonesian Meterorological and Geophysics Agency's technical department, Mr Fauzi, and the head of the International Tsunami Information Centre in Honolulu, Ms Laura Kong, to point out that the problem is now one of communication: how can warnings be most quickly passed on to those most vulnerable to an impending tsunami; and do potential victims know what to do in response to warnings? [9] Although the issue of communicating impending disasters is crucial, it is likely that even the most sophisticated and well-run systems will be less than optimal because they inevitably take time to implement: Mr Fauzi argues the estimated 20 minutes between the occurrence of the 17th July 2006 earthquake and the arrival of the waves on the Javanese coast would severely test the ability of any system to provide adequate warning to all those at risk of harm. [9] As such, despite the obvious desirability of the planned refinements, including installing a network of sirens to alert even the most remote villages, [10] early warning systems alone will always represent a less than optimal international response. What else might be done?

ADDITIONAL STRATEGIES

The UN has clearly, and reasonably, argued that harm from disasters would be further reduced if early-warning and monitoring systems were complemented by built environments and modes of transport designed to withstand pressures and natural assaults far in excess of what might reasonably be expected to occur, such as earthquake-proof seawalls and buildings. [11] It is relatively easy to demonstrate that, to some extent, such measures have been implemented in some countries. For example, New Zealand's Federal Parliament House in Wellington is state-of-the-art earthquake resistant, using rubber and lead piers to separate the concrete beams of the building from the foundations, to minimise structural damage should the foundations move. Japan has made some effort to construct breakwaters and earthquake-proof seawalls but has apparently failed, strangely enough, to test the integrity of these structures. [8] Further down the scale, numerous amateur pictures of the 2004 Boxing Day tsunami striking various coastlines bear stark testimony to the likelihood that in a number of other countries, such preventive measures have not been implemented at all.

If the technology exists to install effective early-warning and monitoring systems, as well as construct disaster resistant built environments, including adequate communication, transport and health-care infrastructure, and it is readily available, it is most likely to be utilised in those countries able to afford its additional cost. This observation reflects a practical, if morally questionable, economic imperative: money can not be spent on such reinforcements if there is only enough to provide basic necessities. If this is indeed the case, then it might be expected that fewer deaths will have resulted from disasters that have impacted on developed countries, more capable of meeting such expenses. Even a cursory glance at a relatively recent history of earthquakes suggests this is the case.

On the 17th October 1989, an earthquake measuring 7.1, on the Richter Scale, struck San Francisco, followed by a 5.2 aftershock 37 minutes later, killing 62 people. On the 17th January 1994, an earthquake measuring 6.6 hit Los Angeles in the US, taking 60 lives. Exactly one year later, 17th January 1995, an earthquake measuring 7.2 struck Kobe in Japan, killing 6,432 people, while on the 26th September 1997, an earthquake measuring 5.6 struck central Italy, killing 10 people.

In contrast, an earthquake measuring 6.1 on the Richter Scale struck Rostaq, Afghanistan on 4th February 1998, killing more than 4,000 people. On the 17th August 1999, an earthquake measuring 6.7 in Izmit, Turkey killed more than 17,000 people, while on 26th December 2003, an earthquake measuring 6.6 in Bam, Iran took approximately 26,271 lives. Finally, of course, the 26th December 2004 earthquake, which measured 9.0 and caused the tsunami, resulted in an estimated 230,000 deaths.

What do these figures indicate (other than that we ought to be very nervous on the days of the 17th and the 26th)? It is reasonably clear that the relationship between the magnitude of the earthquake and the number of lives lost would be mediated by the location of the earthquake, the location of the city relative to the fault line and the density of the populations affected: the closer the earthquake's epicentre to populated regions, the closer the fault line to the city and the greater the population density, the greater the number of lives will be lost. These factors would explain the discrepancy in the number of deaths in Japan, compared to the US and Italy: Kobe may well have had a larger population or higher population density and the city itself was directly ruptured by the fault-line. However, these factors alone do not adequately account for the data. Los Angeles and San Francisco, in particular, would have had reasonably high populations and population densities at the time of their earthquakes and yet the number of lives lost in those cities was relatively small. In contrast, the earthquakes in Izmit and Bam claimed relatively high numbers of lives, despite those cities having relatively small populations of approximately 225,000.

So if the magnitude of the earthquake, the location of the epicentre and fault lines, and the size and density of the population do not adequately explain the relationship between the occurrence of an earthquake and the resultant number of lives lost, what does? Aggregating these earthquake data presents us with an emerging pattern. Earthquakes that have caused damage to developed countries have averaged a magnitude of 6.6 (excluding the San Francisco aftershock) and claimed 6,564 lives (or if the obviously more extreme case of Japan is removed, an average shock of magnitude 6.4, claiming 132 lives).

Earthquakes that have caused damage to developing countries have averaged a magnitude of 7.1 and claimed 277,271 lives (assuming a final death toll from the 2004 Boxing Day tsunami of 230,000). This suggests developing countries, relative to developed countries, have experienced shocks of approximately 0.5 greater magnitude (7.1 versus 6.6) and a death toll in the order of 40 times greater (277,271 versus 6,564). The on-average greater intensity of the earthquakes to have struck developing countries seems to be of a magnitude that is disproportionately low, relative to the much higher death rates.

It is true that the tsunami was a different kind of disaster, compared to land-based earthquakes, such that the damage caused and the number of lives lost as a result of the tsunami may be atypical. However, the pattern of a disproportionately greater number of deaths in developing, relative to developed, countries holds when the recent tsunami is excluded from the comparison. In that scenario, developing countries have experienced earthquakes of a comparable average magnitude (6.5 versus 6.6), but have experienced a death toll in the order of seven times greater (47,271 versus 6,564).

Given the apparent strength of this pattern, it should hardly be surprising that it appears equally applicable to a range of public health disasters: in Sub-Saharan Africa, the estimated HIV/AIDS prevalence rate of 7.4% of the adult population compares to 0.2% in Oceania and 0.6% in North America; [12] the majority of SARS cases occurred in Asia, most notably the agriculturally based regions of Southern China; [13] while the potential bird flu epidemic [7] seems most likely to originate in those countries that have experienced the highest related death rates to date: Vietnam; Thailand; and Cambodia [14].

Proposed Strategies

These observations provide a compelling argument that the economic fragility of developing countries somehow increases their vulnerability to public health disasters disproportionately, and raises the possibility of economic development as an effective lever for minimising the impact of such disasters. There appear to be two possibilities: either reduce the economic inequalities between countries, or improve the public health infrastructure within developing countries. As the World Health Organisation has pointed out, however, these possibilities raise a chicken-and-egg dilemma: will economic growth in developing countries more or less automatically improve their public health status, or is it preferable to focus on improving dysfunctional public health systems, even within poorly managed economies? [15].

The principal problem with the former argument is that the additional economic burden imposed by the presence of eradicable diseases (which comprise the majority of their excess burden of disease) and damage to built environments, increases the difficulty of achieving economic development. The principal problem with the second argument is that devoting economic resources to improving public health infrastructure, as a primary means of more efficiently implementing proven interventions for eradicable diseases or improvements to built environments, reduces the resources available for acquiring proven interventions and appropriate building materials from the international market in the first place. The optimal solution, of course, is to try do both: encourage developing countries to create domestic conditions conducive to international financial assistance, the adequate provision of which facilitates the development of effective domestic public health infrastructure [15].

An examination of current financial assistance from economically developed countries to developing countries, identifies a number opportunities to reduce the extent of the existing economic inequality between them: fairer trade deals, if not compassionate deals that actually favour poorer countries; ongoing reduction or cancellation of Third-World debt; and provision of a more equitable level of aid, with a lower proportion of tied aid. Australia, for example, has committed 0.26 per cent of its Gross National Income (GNI) to overseas aid for 2004/2005 (or 26 cents for every $100 Australia earns), well below the UN target for developed countries of 0.7 per cent of GNI. [16]

That only five countries (Sweden, Luxembourg, The Netherlands, Denmark and Norway) meet this UN target suggests either the unlikely possibility that the target is unreasonable, or the more likely possibility that there is an opportunity for developed countries to increase their level of assistance.

Identifying opportunities for developed countries to further assist developing countries is not to argue that they are currently doing nothing. On the weekend of 5th-6th February 2005, for example, the finance ministers of G7 countries met in London and agreed, in principle, that existing Third-World debt should be cancelled [17].

In addition, in 2000, UN Member States identified and endorsed eight Millennium Development Goals for reducing global poverty. [15] Although clearly defining specific goals is a proven public health strategy to increase the likelihood they will be achieved, goal setting does not, of itself, represent actual economic development: whatever the precise nature of their complex interactions, developed countries exemplify the relationship between economic development and reduced poverty, civil unrest and displaced citizens. Perhaps a lesson from antiquity is reason enough to hope repeated calls for the exigent economic development of the Third-World will be acted upon more effectively: in 150BC a Roman Senator, Cato the Elder, was an ardent critic of Carthage, warning repeatedly of its potential threat to Rome by concluding every speech he made in the Senate, irrespective of the topic, with the demand that Carthage must be destroyed. It may have taken an influential senator four years, but in 146BC his wish was spectacularly fulfilled.

Crucially, the argument for greater assistance from developed countries does not necessarily mean that they will be relatively economically worse off then they are currently. It does not mean countries should not compete for advantage, nor continue to foster, as their primary priority, policies that they consider to be in their national interest: just as within a domestic economy some individuals will be wealthier than others, some countries will be more developed and wealthy than others. But it does mean that attempts need to be made to identify and advocate policies that elevate the poorest nations above a basic level of sustainability.

One way to demonstrate this is in terms of a Bell curve, applied to differing levels of economic development.

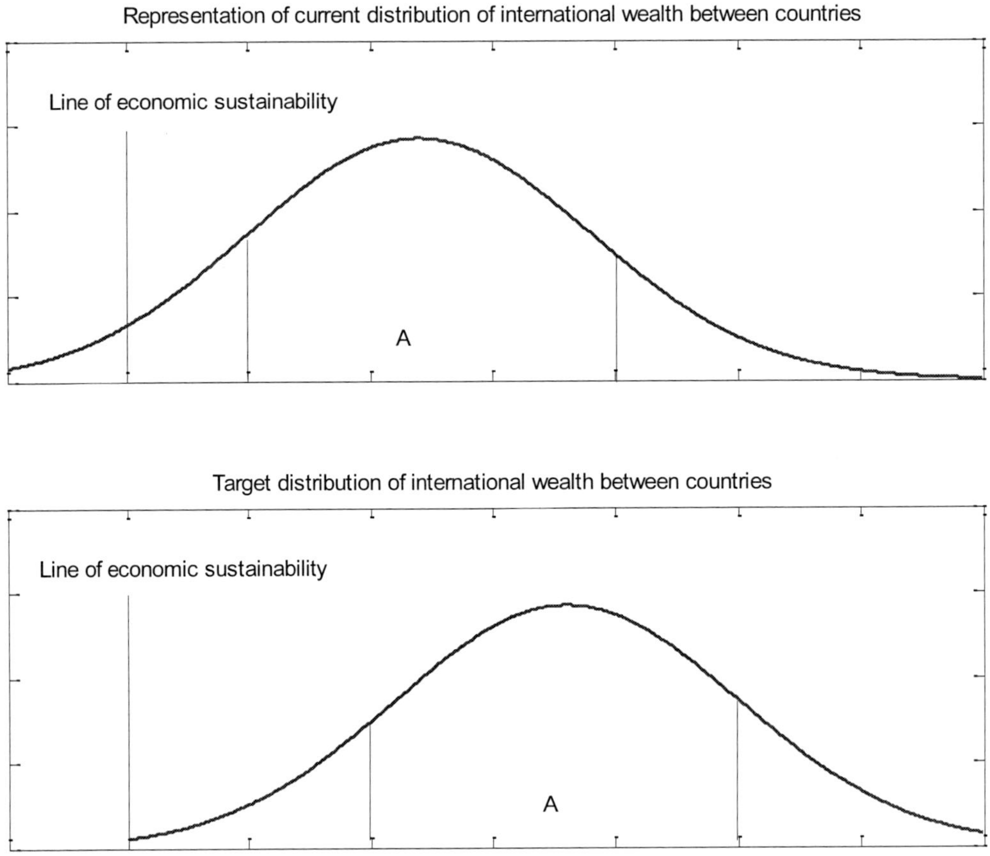

Figure 1. Current and target distributions of wealth between countries.

The Bell curves in Figure 1 delineate a basic relationship between the levels of economic wealth of different countries, where most countries fall into the middle, or average, section (section A), some countries are below average wealth (they sit on the curve to the left of section A) and some countries are above average wealth (to the right of section A).

The first graph shows that a number of countries of below average economic wealth are also below (that is, to the left) of the arbitrary economic sustainability line, while the second shows a shift of the whole curve to the right, such that no countries of below average economic wealth are also below the economic sustainability line. The important point here is that the shape of the curve is preserved: some countries are still more economically wealthy than others, but no countries are left below the line of economic sustainability. Given limited global resources, this shift may need to be balanced by a proportionately lesser shift, further to the right, for countries of above average wealth; a small price to pay, by any humanitarian measure. In essence, the target distribution curve in Figure 1 delineates a more equitable distribution of global wealth between countries. The notion of equity, both in relation to public health and more broadly, is difficult to define for at least two reasons. Firstly, defining exactly what might constitute an equitable distribution of resources necessitates agreement on a range of social and cultural values, which is difficult in multi-cultural communities like Australia's, and disproportionately more difficult between different countries. Secondly, there is more than one way of conceptualising equity: it can be defined as the equal distribution of resources to equal entities (termed horizontal equity), or as the unequal but equitable distribution of resources to unequal entities (termed vertical equity). [18] Despite the difficulty of defining these concepts, there is clear potential for applying greater vertical equity to the distribution of global wealth between nations, as an effective means of increasing the economic sustainability of the World's poorest countries and, as a consequence, decreasing their vulnerability to large-scale public health disasters.

In addition to a more equitable distribution of global wealth between countries, a second possibility for reducing the vulnerability of developing countries to public health disasters is for them to work to improve their domestic economies and political processes, thereby increasing the likelihood they will obtain more effective international economic assistance. This point was implied by the Australian Prime Minister, John Howard, at the World Economic Forum held in Davos, Switzerland, between 26th-30th January 2005, when he expressed the view that Third-World countries must reduce corruption within their governments before they could expect full assistance from the developed world. Such comments may be defensible, provided they do not place undue emphasis on the responsibilities of developing, relative to developed, countries: it would not be reasonable for assistance from developed countries to be exclusively dependent on the elimination of corruption in the domestic economies of developing countries. Corruption within public offices is most likely to be both a cause and a symptom of economic hardship and, indeed, the example of developed countries suggests corruption will be reduced, or at least become less visible, by economic development and exaggerated by continual economic hardship. Such comments are clearly indefensible if they are used to justify the maintenance of an obviously unfair allocation of global wealth.

CONCLUSION

In summary, the international response to the 2004 Boxing Day tsunami, as judged by the observed effects of the 17th July 2006 tsunami, deserves a mixed report card. The short-term response appears to have been reasonable, although efficient distribution of goods and co-ordination of effort were clearly problematic and ought to be closely scrutinised. Over the longer term, the installation of warning systems between the tsunamis was encouraging and the need to improve the speed with which warnings are communicated, and increase local peoples' knowledge on how to react to such warnings, have been clearly identified as issues. Representing the international community's greatest response failure, however, is the obvious unwillingness to seriously consider increased economic equality between developed and developing countries, as the most effective longer-term strategy to reduce harms associated with public health disasters. Despite the strategies that were implemented between the tsunamis, the 17th July 2006 tsunami still resulted in a loss of life in the order of 10 times that which occurred over 15 years ago as a result of the San Francisco and Los Angeles earthquakes. More needs to be done. It might be argued, finally, that advocating this kind of structural reorganisation in the international community is over-kill in response to events that could be categorised as sheer bad luck; unfortunate results from a combination of freak natural events and developing countries lying directly in the path of the ensuing tsunamis. If an example closer to home is needed, however, a reminder that Australia is not immune to the processes which led to the massive public health disasters created by the Asian tsunamis, has already been provided. At 8.12 on the evening of 2nd March 2005, Darwin, Australia's northern most capital city, was shaken by an earthquake measuring 6.8 on the Richter Scale, the epicentre of which was in the Banda Sea 800 kilometres north of Darwin. According to Dr Mark Leonard from Geoscience Australia, the Banda Sea gets an earthquake of a magnitude of about five, every two months or so and is "... one of the most seismically active areas in the world..." [19]. If a tsunami was to originate in the Banda Sea, East Timor is one country likely to be significantly affected. Australia has recently established a trade deal with East Timor, regarding gas and oil fields, with multi-million dollar benefits for Australia (the precise amount varies wildly, depending on the source of the estimate). The point of detailing this series of events is that damage resulting from a tsunami originating in the Banda Sea would not be a completely random result of a combination of accidental events, about which developed countries can do little: the Banda Sea is unstable; Australia is doing less than it could to help develop the economy of East Timor, which appears to be the best defence against potential public health disasters; East Timor is duly disproportionately and unnecessarily vulnerable to the instability of the Banda Sea. The instability of the Banda Sea can not be helped; the exact location or time of the next major earthquake or tsunami can not be predicted with any great accuracy; but with enough political will, strategies can be implemented that increase the economic development of East Timor, in turn leaving it less vulnerable without seriously compromising Australia's economy or the extent to which Australia is exposed to a tsunami in the Banda Sea. To optimistically invoke the indomitable spirit of Cato the Elder, the inadequate economies of Third-World countries must be developed.

Acknowledgments

Mr Stuart Gilmour (Statistician, University of NSW) produced the graphic. He, Associate Professor James C. Frankish (Assistant Director, Institute for Health Promotion Research, University of British Columbia, Vancouver, Canada) and Associate Professor Christopher Doran (Centre for Burden of Disease and Cost-Effectiveness, University of Queensland) provided valuable improvements to the manuscript.

References

[1] UN Office of the Special Envoy for Tsunami Recovery. The Human Toll. Available at http://www.tsunamispecialenvoy.org/country/humantoll.asp.

[2] World Health Organisation. Situation report # 9 Java Tsunami of 17 July 2006, 26 July 2006.

[3] Brummitt, C. Meagre food supply stunting tsunami kids. *The Sydney Morning Herald*, 30 January 2005.

[4] Moore M. Competition swallowing up tsunami aid money in Aceh. *The Sydney Morning Herald*, 10 March 2005.

[5] Buchan H, Sewell JR, Sweet M, Editors. Adopting best evidence in practice. *The Medical Journal of Australia,* 2004; 180: S41-S72.
[6] Associated Press. The geology behind the tsunami. *The Sydney Morning Herald*, 27 December 2004.
[7] World Health Organisation. Assessment of risk to human health associated with outbreaks of highly pathogenic H5N1 avian influenza in poultry. Communicable Disease Surveillance and Response Report, 14 May 2004. Available at http://www.who.int/csr/disease/avian_influenza/ assessment2004_05_14/en/.
[8] Associated Press. Tsunami warning systems examined. *The Sydney Morning Herald*, 19 January 2005.
[9] Walsh B. Without warning. *TIME Asia Magazine*, 23 July 2006.
[10] Smith-Spark L. Indonesia tsunami system 'not ready'. BBC News (International version), 19 July 2006.
[11] United Nations. Hyogo framework for action 2005-2015: building the resilience of nations and communities to disasters. World Conference on Disaster Reduction, 18-22 January 2005. Available at http://www.unisdr.org/news/OUTCOME-FINAL-as-separate-non-official-document.pdf.
[12] UNAIDS. Global summary of the HIV and AIDS epidemic in 2004, December 2004. Available at http://www.unaids.org/NetTools/Misc/DocInfo.aspx?LANG=enandhref=http%3a%2f%2fgva-doc-owl%2fWEBcontent%2fDocuments%2fpub%2fTopics%2fEpidemiology%2fSlides02%2f12-04%2fEpiCoreDec04Slide002_en%26%2346%3bppt.
[13] World Health Organisation. Summary of probable SARS cases with onset of illness from 1 November 2002 to 31 July 2003. *Communicable Disease Surveillance and Response Report*, 31 December 2003. Available at http://www.who.int/csr/sars/country /table2004_04_21/en/.
[14] World Health Organisation. Cumulative Number of Confirmed Human Cases of Avian Influenza A/(H5N1) since 28 January 2004. Communicable Disease Surveillance and Response Report, 2 February 2005. Available at http://www.who.int/csr/disease/avian_ influenza/country/cases_table_2005_02_02/en.
[15] World Health Organisation. Macroeconomics and Health: Investing in Health for Economic Development. Geneva: WHO; 2001.
[16] Australian Council for International Development (ACFID). Aid budget 2004-05 overview and analysis. ACFID, 2004. Available at http://www.acfid.asn. au/campaigns/aid/aid_budget_analysis_04_05.pdf.
[17] Crabb A, *et al.* Rich nations to write off $100bn debt. *The Sydney Morning Herald*, 7 February 2005.
[18] Mooney G. Here's a recipe for a more equitable health care system in Australia. *On Line Opinion*, 21 July 2003.
[19] Australian Associated Press. Darwin rocked by earthquake. *The Sydney Morning Herald*, 3 March 2005.

In: Asia Pacific Life Sciences Compendium
Editor: Bandit Chumworathayi

ISBN: 978-1-62257-338-7

Chapter 7

PREVALENCE AND RISK FACTORS OF PELVIC AND PERIAORTIC LYMPH NODE METASTASES IN ENDOMETRIAL CANCER PATIENTS AT LOPBURI CANCER CENTER

Suwaree Siripaopradit,* Thira Assavatharn and Somchai Chanwiset
Department of Gynecologic Oncology, Lopburi Cancer Center, Muang, Lopburi, Thailand

ABSTRACT

Objective: To study the prevalence and to identify the significant risk factors of pelvic and periaortic lymph node metastases in endometrial cancer patients at Lopburi Cancer Center, Muang, Lopburi, Thailand.

Method: A retrospective study of 184 patients who underwent primary surgical treatment of endometrial cancer at Lopburi Cancer Center between June 2002 and December 2008, was done. Analyses included demographics characteristics, stage, surgical procedure, histology, myometrial invasion, lymph-vascular space invasion, pelvic and periaortic metastases. Univariate and multivariate analyses were performed using logistic regression. P-values were obtained by chi-squared test to identify significant risk factors. The correlations of deep myometrial invasion, pelvic or periaortic lymph nodes metastases were obtained by odd ratio (OR) and its 95% confident interval (95%CI).

Result: Prevalence of pelvic lymph node and periaortic lymph node metastases were found in 20/175 (14.6%) and 12/130 (9.2%) of patients whose pelvic lymph or periaortic lymph nodes were dissected or sampled, respectively. In univariate analyses, endometrial histology, cervical stromal invasion, deep myometrial invasion, lymph-vascular space invasion, and adnexal metastases, were significantly correlated with pelvic lymph node metastases. Similarly, lymph-vascular space invasion and pelvic lymph node metastases were significantly correlated with periaortic lymph node metastases. In multivariate

* Email: siri_waree@hotmail.com

analyses, deep myometrial invasion and lymph-vascular space invasion were still significantly correlated with pelvic lymph node metastases, and only pelvic node metastases was still significantly correlated with periaortic node metastases.

Conclusion: The significant risk factors of pelvic lymph node metastases were endometrial histology, cervical stromal invasion, deep myometrial invasion, lymph-vascular space invasion, and adnexal metastases. Only lymph-vascular space invasion and pelvic lymph node metastases were significant risks for periaortic lymph node metastases.

Keywords: endometrial cancer; nodal metastasis; prevalence; risk factors

INTRODUCTION

Endometrial cancer is a common gynecologic malignancy. A patient with endometrial cancer typically presents with symptomatic bleeding, 75-80% of cases are diagnosis in stage I disease. When confined to the uterus, the prognosis for endometrial cancer is excellent. In 1988, FIGO adopted the surgical and pathologic staging system to be used for staging of endometrial cancer. Surgical staging was specified to include hysterectomy with bilateral salpingo-ophorectomy, peritoneal washings, biopsies of any suspicious intraperitoneal or retroperitoneal lesions and pelvic-periaortic lymphadenectomy. Pelvic and/or periaortic lymph node metastases affect the risks of recurrence and survival. Many studies found the benefits for removing reservoir of recurrence. [1,2,3] A number of factors have been shown as prognostic variables and found to be correlated with risk of pelvic and periaortic lymph node metastases such as; histologic type and grading, myomerial invasion, lymph-vascular space invasion. [1,2,3,4,5,6]. Almost all of endometrial cancer patients in Lopburi Cancer Center had been primarily treated with surgical staging procedure according to the recommendation of FIGO. However, we do not have any data about the prevalence and risk factors of pelvic and periaortic lymph node metastases that definitely affect patient survival. Therefore, the objectives of this study were to estimate the prevalence and to identify the significant risk factors of pelvic and periaortic lymph node metastases in endometrial cancer patients at Lopburi Cancer Center, Muang, Lopburi, Thailand.

METHOD

We performed a retrospective review of all endometrial cancer patients who underwent primary surgical treatment at Lopburi Cancer Center between June 2002 and December 2008. Cases which had complete surgery from other hospitals were excluded from this study. There were 184 patients who were surgically staged with peritoneal washings, total abdominal hysterectomy or modified radical hysterectomy, bilateral salpingo-oophorectomy and selective bilateral pelvic and/or periaortic lymphadenectomy. Pelvic and periaortic lymphadenectomy were done in 175 and 130 cases, respectively. Data regarding medical history and pathology were gathered from our outpatient medical records. Stratified analyses were performed to examine possible effect modification by known clinical risk factors. Univariate and multivariate analyses were performed using logistic regression analyses. P-values were obtained by chi-squared test to identify the significant risk factors. Their magnitudes of correlations associated with deep myometrial invasion, pelvic or periaortic lymph nodes metastases were obtained by odd ratio (OR) and its 95% confident interval (95%CI) calculation.

RESULT

Between June 2002 and December 2008, 184 endometrial cancer patients underwent surgical staging at Lopburi Cancer Center. Demographic characteristics of these patients are shown in Table 1. The mean$\pm$SD of their age was 56.01$\pm$0.68 year-old. 44 patients (23.91%) were nulliparous. High BMI (> 25 kg/cm^2) were found in 95 patients (51.63%). Almost all of them (180 patients, 97.83%) presented with abnormal vaginal bleeding. In 16 patients (8.7%), the first surgery was done in other hospitals.

By following FIGO 2009 for uterine cancer staging, there were 10 staged IIIA (5.4%), 1 staged IIIB (0.54%), 10 staged IIIC1, and 10 staged IIIC2. (Table 2) Most of them (126, 68.5%) were in stage I. From Table 3, total abdominal hysterectomy or modified radical hysterectomy together with bilateral salpingo-oophorectomy were done in all 184 cases (100%). However, unilateral or bilateral pelvic lymph node dissection or sampling were also done in only 175 patients (95.2%) while periaortic lymph node sampling or dissection were done too in only 130 patients (70.7%).

Table 1. Demographic characteristics of 184 patients. Values are either means or numbers and their standard deviations (SD) or percentages (%)

Characteristics	Means or Numbers	SD or %
Age in years (mean)	56.01	0.68
Gravida (mean)	2.29	0.14
Parity (mean)	2.07	0.13
Nullipara	44	23.91
Menopause	120	65.22
Late menopause > 52	36	19.57
Obesity BMI > 25	95	51.63
Diabetes mellitus	37	20.11
Hypertension	58	31.52
Smoking	2	1.09
Exogenous estrogen	2	1.09
Breast or colon carcinoma	2	1.09
Chief complaints		
1. Bleeding	180	97.83
2. Mass	2	1.09
3. Pain	1	0.54
4. Abnormal Pap smear	1	0.54
First surgery in other hospitals	16	8.70
Positive endocervical curettage	41	22.29

Table 2. Distribution of 2009 FIGO stages in 184 patients

Stages	Numbers	%
IA	89	48.4
IB	37	20.1
II	15	8.2
IIIA	10	5.4
IIIB	1	0.5
IIIC1	10	5.4
IIIC2	10	5.4
IVA	6	3.2
IVB	6	3.2

Table 3. Distribution of risk factors for nodal metastases

Risk factors	N	%
Surgical procedure		
TAH/mRHND + BSO	9	4.9
TAH/mRHND + BSO + unilat/BPND	45	24.5
TAH/mRHND + BSO + unilat/BPND + PAS	130	70.7
Histology from hysterectomy		
No residual tumor	11	6
Endometrioid adenocarcinoma	146	79.3
Clear cell adenocarcinoma	6	3.2
Serous carcinoma	14	7.6
Mixed carcinoma	5	2.7
Others	2	1.1
Histologic grading		
Grade 1	78	42.4
Grade 2	35	19.0
Grade 3	38	20.6
Cervical invasion		
No or glandular	147	80
Cervical stromal	37	20
Myometrial invasion		
< ½ myometrium	110	59.8
> ½ myometrium	74	40.2
LVSI		
Negative	163	88.5
Positive	21	11.4
Adnexal metastases		
Negative	164	89.1
Positive	20	10.9
Pelvic lymph node metastases		
Negative	149	85.1
Positive	26	14.6
Periaortic lymph node metastases		
Negative	118	90.7
Positive	12	9.2

Table 4. Univariate analysis of pelvic and periaortic lymph nodes metastases by risk factors

Risk factors	Pelvic LN				Periaortic LN			
	N	Neg	Pos	OR,CI,P	N	Neg	Pos	OR,CI,P
Histological grading				2.35,				0.37,
1 and 2	111	99	12	0.9-5.8,	83	76	7	0.05-2.9,
3	30	24	6	0.063	26	25	1	0.347
Endometrial histology				3.07,				2.22,
Endometrioid	146	126	20	1.2-7.9,	110	102	8	0.55-8.9,
Non-endometriod	29	22	7	0.02*	20	16	4	0.262
Cervical stromal invas.				3.56,				2.34,
No	140	126	14	1.5-8.6,	104	99	5	0.65-8.4,
Yes	35	23	12	<0.01*	26	19	7	0.195
Myometrial invasion				4.21,				1.89,
< 1/2 myometrium	104	97	7	1.7-10.3,	71	66	5	0.56-6.5,
> 1/2 myometrium	71	52	19	<0.01*	59	52	7	0.305
LVSI				8.35,				5.24,
Negative	155	139	16	3.1-22.7,	111	104	7	1.4-19.8
Positive	20	10	10	<0.01*	19	14	5	0.014*
Adnexal metastases				4.96,				1.80,
Negative	157	139	18	1.8-13.6,	115	105	10	0.36-8.9,
Positive	18	10	8	<0.01*	15	13	2	0.472
Pelvic LN metastases								22.96,
Negative					104	102	2	5.6-94.4,
Positive					26	16	10	<0.01*

LN=Lymph node, N=Number, Neg=Negative for metastases, Pos=Positive for metastases, OR=Odd ratio, CI=95% Confident Interval, P=p-value.

Table 5. Multivariate analysis of pelvic and periaortic lymph nodes metastases by risks

Risk factors	Pelvic LN				Periaortic LN			
	N	Neg	Pos	OR,CI,P	N	Neg	Pos	OR,CI,P
Histological grading				1.07,				0.11,
1 and 2	111	99	12	0.3-3.2,	83	76	7	0.0-1.3,
3	30	24	6	0.904	26	25	1	0.077
Endometrial histology				2.09,				1.56,
Endometrioid	146	126	20	0.7-6.6,	110	102	8	0.3-9.3,
Non-endometriod	29	22	7	0.207	20	16	4	0.628
Cervical stromal invas.				2.57,				1.17,
No	140	126	14	0.9-7.5,	104	99	5	0.2-7.2,
Yes	35	23	12	0.084	26	19	7	0.867
Myometrial invasion				3.29,				0.49,
< 1/2 myometrium	104	97	7	1.2-9.1,	71	66	5	0.1-2.8,
> 1/2 myometrium	71	52	19	0.021*	59	52	7	0.421
LVSI				8.00,				3.68,
Negative	155	139	16	2.5-25.8,	111	104	7	0.6-22.7,
Positive	20	10	10	<0.01*	19	14	5	0.160
Adnexal metastases				2.21,				0.42,
Negative	157	139	18	0.6-7.6,	115	105	10	0.0-4.3,
Positive	18	10	8	0.21	15	13	2	0.461
Pelvic LN metastases								35.81,
Negative					104	102	2	5.5-234,
Positive					26	16	10	<0.01*

LN=Lymph node, N=Number, Neg=Negative for metastases, Pos=Positive for metastases, OR=Odd ratio, CI=95% Confident Interval, P=p-value.

From post-operative surgical specimen histological examinations, there were endometrioid adenocarcinoma in 146 patients (79.3%), serous carcinoma in 14 patients (7.6%), clear cell adenocarcinoma in 6 patients (3.2%). 38 patients (20.6%) had histologic grade 3 and 66 patients (35.9%) had deep myometrial invasion. Stromal cervical metastases and lymph-vascular space invasion (LVSI) were found in 37 (20.1%) and 21 (11.4%) patients, respectively.

About extrauterine disease, the powerful prognostic factors in endometrial cancer, adnexal metastases were found in 20 patients (10.9%) while pelvic lymph node and periaortic lymph node metastases were found in 20 (14.6%) and 12 (9.2%) of patients whose pelvic lymph or periaortic lymph nodes were dissected or sampled, respectively.

In addition, we found that grade 3 carcinoma had statistically significant correlation with deep myometrial invasion when compared to grade 1 or grade 2 [OR(95%CI)=3.39(1.61-7.14), $p<0.01$], even when adjusted by cervical stromal invasion and endometrial histology [OR(95%CI)=3.31(1.56-7.03), $p<0.01$].

Univariate analysis (Table 4), in patients with pelvic or periaortic lymph node metastases (175 and 130 patients, respectively), showed that endometrial histology, cervical stromal invasion, deep myometrial invasion, lymph-vascular space invasion, and adnexal metastases, were significantly correlated with pelvic lymph node metastases. Similarly, lymph-vascular space invasion and pelvic lymph node metastases were significantly correlated with periaortic lymph node metastases. Multivariate analysis (Table 5) showed that 1) only deep myometrial invasion and lymph-vascular space invasion were still significantly correlated with pelvic lymph node metastases, adjusted for all other factors, and 2) only pelvic node metastases was still significantly correlated with periaortic node metastases, adjusted for all other factors.

DISCUSSION

In this retrospective analysis of risk factors for nodal metastases, we present our experiences of patients with primary surgically staged endometrial cancer at Lopburi Cancer Center. The value of complete surgical staging for endometrial cancer is not only diagnostic but also therapeutic presumably by removing potential reservoirs for future recurrence. [1,2,3] So, endometrial cancer patients at Lopburi Cancer Center have been surgically staged according to the recommendation of the International Federation of Gynecology and Obstetrics (FIGO).

In this study, pelvic and periaortic lymph node sampling/dissection were done in 95.2 and 70.7%, respectively. We found that the prevalence of pelvic lymph node metastases was 14.6%. In study of Chi et al [3] and Luanratanakorn et al [4], these were less than our report (9 and 11.6%, respectively). In addition, periaortic lymph node metastases were found in 9.2% (12 patients of this study). Surprisingly, in 2 of 12 patients (16.7%) whose periaortic lymph node metastases were found, their pelvic lymph nodes were negative for malignancy. This proportion was higher than reported by Abu-Rustum NR, et al [5] that isolated periaortic lymph node metastases were found in only 1%.

We found that histologic grade 3 was not significantly correlated with pelvic and periaortic node metastases, but there were still high associations (OR=2.35). However, significant correlation with deep myometrial invasion when compared to grade 1 or grade 2, even when adjusted by cervical stromal invasion and endometrial histology was found.

The associations between histologic grading and myometrial invasion, pelvic or periaortic node metastases, adnexal metastases, positive peritoneal washings cytology, local recurrence, and hematogenous spread, were commonly found in other studies [6,7]. This means that our study had some similar results, but our small sample size might lead to some statistically non-significant. The risk factors of pelvic lymph node metastases that were found to have significant correlation in this study were; endometrial histology, cervical stromal invasion, deep myometrial invasion, lymph-vascular space invasion, and adnexal metastases. The risk factors of periaortic lymph node metastases that were found to have significant correlation in this study were; lymph-vascular space invasion, and pelvic lymph node metastases. However, the 2 patients, who had isolated periaortic lymph node metastases, did not have any of the 2 risk factors. Similarly, others reports have also shown statistically significant correlations between high grade tumor, deep myometrial invasion, intraperitoneal metastases [7,8], lymph-vascular space invasion, and cervical invasion [6], with nodal metastases. Because almost all of risk factors for nodal metastases were pathological which could not be evaluated before operation, and frozen section could not be done in our hospital. So, pelvic and periaortic sampling/lymphadenectomy should be performed in every primary surgical staging for endometrial cancer. [5,9,10,11]. Furthermore, Todo Y et al [11] found that systematic lymphadenectomy (pelvic and periaortic lymphadenectomy) had significant survival effect for patients with intermediate and high risk for recurrence. In conclusion, we have shown that incidence of pelvic lymph node metastases was 14.6% and incidence of periaortic lymph node metastases was 9.2% in our hospital. The significant risk factors of pelvic lymph node metastases were endometrial histology, cervical stromal invasion, deep myometrial invasion, lymph-vascular space invasion, and adnexal metastases. Only lymph-vascular space invasion and pelvic lymph node metastases were significant risks for periaortic lymph node metastases. In the near future, the incidence and prevalence of recurrence, and also, survival, will be studied for including in the database of our hospital.

Acknowledgment

We would like to thank Dr. Pornsom Hutacharern and Dr. Wanchai Pongpipat for their help in methodology suggestion, and Assoc. Prof. Dr. Bandit Chumworathayi for his help in data analyses and statistic suggestions, and finally, all staffs in Lopburi Cancer Center.

References

[1] Creasman WT, DeGeest K, DiSaia PJ, Zaino RJ. Significance of true surgical pathologic staging: A Gynecologic Oncology Group Study. *Am. J. Obstet Gynecol.* 1999; 181: 31-4.

[2] Boronow RC, Morrow CP, Creasman WT, DiSaia PJ, Silverberg SG, Miller A, et al. Surgical staging in endometrial cancer: Clinical-pathologic finding of a prospective study. *Obstet. Gynecol.* 1984; 63: 825-38.

[3] Chi DS, Barakat RR, Palayekar MJ, Levine DA, Sonoda Y, Alektiar K, et al. The incidence of pelvic lymph node metastasis by FIGO staging for patients with adequately surgically staged endometrial adenocarcinoma of endometrioid pathology. *Int. J. Gynecol. Cancer* 2008; 18: 269-73.

[4] Luanratanakorn S, Yuenyao P, Chumworathayi B, Suwunnapang S. Prevalence of pelvic lymph node metastases in endometrial cancer patients at Srinagarind hospital. *Srinagarind Med. J.* 2007; 22: 346-53.

[5] Abu-Rustum NR, Gomez JD, Alektiar KM, Soslow RA, Hensley ML, Leitao MM Jr, et al. The incidence of isolated paraaortic nodal metastasis in surgically staged endometrial cancer patients with negative pelvic lymph nodes. *Gynecol. Oncol.* 2009; 115: 236-8.

[6] Tang X, Tanemura K, Ye W, Ohmi K, Tsunematsu R, Yamada T, et al. Clinicopathological factors predicting retroperitoneal lymph node metastases and survival in endometrial cancer. *Jpn J. Clin. Oncol.* 1998; 28: 673-8.

[7] Sakuragi N, Hareyama H, Todo Y, Yamada H, Yamamoto R, Fujino T. Prognostic significance of serous and clear cell adenocarcinoma in surgically staged endometrial carcinoma. *Acta Obstet. Gynecol. Scand* 2000; 79: 311-6.

[8] Creasman WT, Morrow CP, Bundy BN, Homesley HO, Graham JB, Helter PB. Surgical pathologic spread patterns of endometrial cancer: A Gynecologic Oncology Group study. *Cancer* 1987; 60: 2035-41.

[9] Piver MS, Lele SB, Barlow JJ, Blumenson L. Paraaortic lymph node evaluation in stage I endometrial carcinoma. *Obstet Gynecol* 1982; 59: 97-100.

[10] Ayhan A, Yarali H, Urman B, Günalp S, Yüce K, Ayhan A, et al. Lymph node metastasis in early endometrium cancer. *Aust N Z J. Obstet. Gynaecol.* 1989; 29: 332-5.

[11] Todo Y, Kato H, Kaneuchi M, Watari H, Takeda M, Sakuragi N. Survival effect of para-aortic lymphadenectomy in endometrial cancer (SEPAL study): A retrospective cohort analysis. *Lancet* 2010; 375: 1165-72.

In: Asia Pacific Life Sciences Compendium
Editor: Bandit Chumworathayi
ISBN: 978-1-62257-338-7

Chapter 8

A CASE OF CHRONIC ACHILLES TENDINOPATHY: INCORPORATING A PATIENT SELF MANAGEMENT APPROACH

***S. Y. Loh*[1,*] *and T. Z. Low*[2]**
[1]Department of Rehabilitation, Faculty of Medicine, University of Malaya, Malaysia
[2]Foot and Ankle Clinic, Kuala Lumpur Sports Medicine Centre, Bukit Damansara, Kuala Lumpur

ABSTRACT

Achilles tendon is a well known anatomical site prone to disabling injury with debilitating impact on daily functioning and quality of life. Chronic ruptures of Achilles tendons normally present four to six weeks after the original acute and often unrecognized rupture. Tendon ends are commonly retracted over time, with an atrophic appearance. In conservative management, a rehabilitation regime to ensure consistent elongation by the patients is fundamental. This case highlights the importance of incorporating the Calf, Achilles Tendon Stretcher (a simple therapeutic aid), and a collaboration care approach to impact an early patient self-management care and consequently, improve the efficacy of the rehabilitation regime.

Keywords: achilles tendinopathy; calf achilles tendon stretcher; home device; patient self-management

INTRODUCTION

Achilles tendon, at approximately two to six centimeters proximal to its insertion in the calcaneus, has long been known as an anatomical site which is prone to disabling injury. It is the site for common tendinous lesion occurring in 18 per 100,000 persons [1], and occurs in

*E-mail: syloh@um.edu.my

active individuals regardless of age. It is estimated that Achilles *tendinitis* accounts for 6-17 percent of all walking, sport-related injuries [2] and other accidental injuries like cuts via sharp household tools, penetrating injuries and road traffic accidents. The term tendinitis is used if degenerative tendinopathic changes like unorganized matrix, more fibroblasts and inflammatory cells, as well as increased vascularity are seen [3].

The most common clinical diagnosis of Achilles overuse injuries is *tendinopathy*, where its basic etiology is multi-factorial [4]. Archilles *Tendinosis* refers to a chronic non-inflammatory degenerative condition associated with overuse, and is characterized by a diffuse thickening of the tendon [5].

All these pathologies cause pain and disability, with the associated disruption of the normal gliding motion and loading characteristics of the Achilles tendon. The Cochrane review for Achilles tendinitis (i.e. when tendon, associated paratendon or bursa are inflamed without degenerative changes), between the year 1966-2000, found nine trials (n=697 patients) providing some evidence to support the role for oral non-steroidal anti-inflammatory drugs (NSAIDs), with weak evidence for heel pads, heparin injection, and peritendinous steroid injections. Evidence showed that early functional therapy protocols showed promising outcomes [6,7]. Achilles tendinopathy have been reported as a condition which may be resistant to both conservative and surgical intervention, along with persistent symptoms [8,9].

In chronic cases of tendinopathy, an ankle-foot orthosis can be beneficial for patients who undergo conservative management [10]. The retraction of the proximal stump leads to shortening of the proximal gastrocnemius-soleus. This limits the contractile force that the muscle can develop and causes biomechanical inefficiency which leads to weakness of ankle plantar flexion, manifested as a flat-footed, nonpropulsive gait on the affected side. Therefore, we advocate the importance of a self-management approach involving a gentle continuous stretching within a functional rehabilitation regime incorporated into patient's daily routines. Effective management of such cases calls for collaborative partnership with the 'expert patient' actively involved in the therapy protocol.

CASE REPORT

The patient is a 50 year old Chinese man presented with a painful right heel over time. He is an avid golf player and plays 18 holes once a week. His BMI was about 25 kg/m^2. In early 2009, he experienced pain over his right Achilles tendon when he walked into a hole (uneven ground) in the golf course. He reported feeling a sharp pain around the heel. He tried to continue the golf game but the injury was painful, and he limped to finish the last part of the game. Two weeks before that incident, he reported switching to two pairs of newly bought shoes---a walking shoe and the golf shoe which he wore during the accident. He stated that both shoes were rather tights and he did experience some degree of soreness over his ankle region after a week of usage. After the accident at the golf range, the pain gradually increased *especially with uphill walks.* He also complained of stiffness especially in the morning and after prolonged rest. He continues with his new shoes and reported he started to use a shoe horn as the shoes were tight.

When pain became unbearable and consistent, he consulted an orthopedic surgeon. His clinical examination showed tenderness on palpation, swelling and thickening over the

Achilles tendon, and mild erythema along the lower one third of the right Achilles tendon. There was a limited range of ankle movement---he could plantar flex his foot without resistance but was not able to perform a toe rise. There was no crepitus when the tendon was pressed or when the affected ankle was moved. Pain was experienced when his tendon was pinched with the index and thumb, and pain was strongest at two centimeter from insertion of tendon. Thomson test (when patient was prone and the calf muscle was squeezed, plantar flexion was noted) was, however, not positive [11]. Radiographs were taken at that time. Patient reported he was informed he may have a calcaneal spur observed from the radiographs. Patient was diagnosed with Achilles tendinopathy, prescribed NSAID and *was referred* for conservative therapy. He was prescribed relative rest and to stop playing golf for the time being. A night posterior splint with the foot at 90° was prescribed for the first 3 weeks, and a heel lift for his walking shoes was also provided to him by the occupational therapist. He attended outpatient physiotherapy once a week for over six months.

However, pain persisted and especially after prolonged sitting or lying down, and he had to take frequent break during his walks. After almost six months of unrelieved symptoms with poor functional outcomes, the occupational therapist designed the Calf Achilles Tendon Stretcher (CATS) and introduced it into his daily activities. His protective night posterior splint was also adjusted (from 90 degree angle) to 20 degree plantar flexion to allow more plantar flexion. The aim of the CATS was to ensure that a constant stretching and regular exercise can be incorporated into patient's daily routine for effective rehabilitation. The regime of continuous stretch was thought necessary since, the tendon ends normally would have retracted over time and thus needs to be elongated consistently. The patient was also provided with an understanding of the multifactorial causes and contributors of Achilles tendinopathy and assisted towards the goal setting for rehabilitation of his ankle problems. As the differential diagnoses was retrocalcaneal bursitis (possibly due to exertion from both pair of tight shoes, patient was persuaded to get a new 'one-size' larger walking shoe. He also started his 50 single-leg toes rises every morning before he gets out of bed, to build-up his Achilles tendon.

The CATS was designed with two features-i) a static version to enable the stretching regime to be incorporated into his daily routine-ii) a dynamic version to allow ankle joint exercises. The CATS-dynamic allowed a graded eccentric and concentric phases of one-legged (partial to full weight bearing) ankle dorsal and plantar flexion exercises at home.

The CATS-static provided rest to the affected foot at work. He continues using the heel lift (inserted into his new, one-size bigger shoes). The CATS-static was then introduced to replace the standard stretches (performed by leaning against a wall with the heel on the floor) in a continuous approach by asking patient to rest his right foot during his activities of daily living like dining and reading, watching television and whilst working on the desk.

The simple device designed by an occupational therapist with the patient, was aim at getting patients to self manage the gradual stretching of the Achilles tendon at work (with CATS static version) and at home (with CATS dynamic version). Patient performed three sets of 10 repetitions of eccentric exercise on the ankle boards. After three weeks of continuous use of CATS, he reported that he was 90 percent pain free with no joint limitations. The ultimate range of movement in his ankle was 55-70°. Two months after treatment, he reported marked improvement in his functional walking along with increased speed upon further reduction in pain.

He has thus resumed his golf activities in full swing. With time, he gradually resumed his golf activities over the next few months. The patient reported qualitative progress in symptom management (pain, limited range, and swelling) and in his functional activities and was very satisfied with the use of CATS in his rehabilitation regime. The long term results in the form of performing previous activities were excellent. The patient still utilizes the CATS static at work occasionally and concurred that the preventive stretches can help reduce the probability of another injury at the golf range.

Discussion

In this case study, the patient had to stop his golf games over several months, reported difficulty tailing his colleagues during lunch breaks, and the disruptions in many other activities had reduced his quality of living. Achilles tendinopathy have been showed to interfere markedly with patients daily activities.

Research evidence showed 56 percent of competitive track and field athletes with Achilles tendonitis discontinued all sporting activities for a minimum of four weeks to promote healing [12]. Untreated cases can turned into tendinosis and keep patients away from work and recreational activities. Complicating the poor recovery results is the fact that often the differential diagnoses are possibly presence as well. In this case study, retro-calcaneal bursitis is a strong preceding condition to the tendon injury. Retro-calcaneal bursitis and associated Haglund deformity which typically are classified as more distal to mid-substance injury are often caused by stiff shoe [3].

The effect is due to the inflammation from the compression (force from stiff shoe) of the Achilles tendon against the postero-superior calcaneal tuberosity.

It is also not uncommon to see insertional Achilles tendinopathy within and around the bony attachment onto the calcaneus, and with degenerative and calcified changes [3]. This is also possible in this patient as such pathology is commonly associated with aging processes. Furthermore, since recovery of injury to lower limb usually takes a long time, the typical and highly possible retraction of the tendon ends, warrants immediate, conscientious rehabilitation regimes.

This case study documented that recovery can be facilitated using a patient self-management approach [13]. Our over-stretched health care services are struggling to cope with the demands of acute care. Thus, patients with longer term health conditions should be intervened with this care approach that encourages active participatory role from patients.

In this case study, the poor outcome from rehabilitation over the lengthy nine months duration have aggravated towards a shortened and thickened Achilles tendon.

A partnership care approach incorporating the simple device (the CATS) has resulted in maximum benefits for the patient. The goal of CATS was to introduce and reinforce a routine of graded, consistent and continuous stretching and self-initiated exercises. It allows lengthening of the Achilles tendon fibers gradually but consistently, as well as, strengthening the gastronemius and soleus muscles of the affected calf.

Conservative treatment exercise, with a bias towards eccentric contractions has been shown to be promising [8,14], and can be reinforced by incorporating the CATS into patients' daily routine. Stretching the tissues throughout the foot and calf helps to reduce strain and promote healing around the area. The range of motion obtained was similar to other reports in the literature [15].

The equinus deformity (pronated ankle) improves with the gradual reduction of pain and the adapted shoe with a heel lift. Studies by Cook and colleagues [8] demonstrated that although tendon load is similar during the concentric and eccentric phase of ankle movement, the tendon vibrates at higher frequencies during the eccentric phase than during the concentric phases---which explain the mechanisms behind the effectiveness of eccentric exercises used in the treatment of Achilles tendinopathies.

The incorporation of CATS into the patients' activities of daily living increases the effectiveness of the rehabilitation phase. However, as the causes of Achilles tendinopathy can be multifactorial, the use of CATS with other modalities as well as shoe adaptation may be necessary and will vary from case to case.

This case study demonstrates the need for a comprehensive round the clock rehabilitation regime using a patient self-management approach to manage a long standing problem of Achilles tendinopathy. The CATS- static provides regular stretching and the CATS dynamics provides the ankle exercises.

The movement biomechanics between the eccentric and concentric phases of one-legged ankle dorsal and plantar flexion exercises, can be graded from partial to full weight bearing. The CATS provided a useful therapeutic rehabilitation aid for the management of Achilles tendinopathy/ tendinosis.

It is also a useful aid for prevention of Achilles tendon injury, especially for ladies who are constantly in high heels leading to a higher risk for Achilles tendon injuries. Future comparative study using RCT is needed on the efficacy of CATS with Achilles tendinopathy to guide clinicians toward an evidence-based approach to patient self-management collaborative care.

REFERENCES

[1] Leppilahti J, Puranen J, Orava S: Incidence of Achilles tendon rupture. *Acta Orthop. Scand.* 1996; 67: 277-9.

[2] Cretnik A, Frank A. *Incidence and outcome of rupture of Achilles tendon.* Wien Klin Wochenschrift 2004; 116 (Suppl 2): 33-8.

[3] Duerden JD, Keeling J. Disorders of the Achilles tendon. *Current Orthopaedic Practice* 2008; 19: 253-9.

[4] Järvinen TA, Kannus P, Maffulli N, et al. Achilles tendon disorders: etiology and epidemiology. *Foot Ankle Clin.* 2005; 10: 255-66.

[5] Mazzone MF, McCue T. Common conditions of the Achilles tendon. *Am. Fam. Physician* 2002; 65: 1805-10.

[6] Lansdaal JR, Goslings JC, Reichart M, et al. The results of 163 Achilles tendon ruptures treated by a minimally invasive surgical technique and functional after treatment. *Injury* 2007; 38: 839-44.

[7] Suchak AA, Spooner C, Reid DC, et al. Postoperative rehabilitation protocols for Achilles tendon ruptures: a meta-analysis. *Clin. Orthop. Relat. Res.* 2006; 445: 216-21.

[8] Cook JL, Khan KM, Purdam C. Achilles tendinopathy. *Man. Ther* 2002; 7: 121-30.

[9] Speck M, Klaue K. Early full weight bearing and functional treatment after surgical repair of acute Achilles tendon rupture. *Am. J. Sports Med.* 1998; 26: 789-93.

[10] Myerson MS. Achilles tendon ruptures. Instr Course Lect. 1999;48:219-30.

[11] Thompson TC. A test for rupture of the Tendon Achilles. *Acta Orthop. Scand.* 1962; 32: 461-5.

[12] Welsh RP, Clodman J. Clinical survey of Achilles tendonitis in athletes. *Can. Med. Assoc. J.* 1980; 122: 193-5.

[13] Barlow J, Wright C, Sheasby J, et al. Self-management approaches for people with chronic conditions: a review. *Patient Educ. Couns* 2002; 48: 177-87.

[14] Petersen W, Welp R, Rosenbaum D. Chronic Achilles tendinopathy: a prospective randomized study comparing the therapeutic effect of eccentric training, the airheel brace, and a combination of both. *Am. J. Sports Med.* 2007; 35: 1659-67.

[15] Porter DA, Mannarino FP, Snead D, et al. Primary repair without augmentation for early neglected Achilles tendon ruptures in the recreational athlete. *Foot Ankle Int.* 1997; 18: 557-64.

In: Asia Pacific Life Sciences Compendium
Editor: Bandit Chumworathayi

ISBN: 978-1-62257-338-7

Chapter 9

ASSESSMENT OF IRON DEFICIENCY ANEMIA AMONG ADOLESCENT GIRLS IN ALLAHABAD, UP, INDIA

Ajay Kumar,[1,*] Sweta Misra,[2] Renu Pandey,[1] Shalini Maheswari[2] and Balesh Kumar Tiwari[1]

[1]Centre for Biotechnology, University of Allahabad, Allahabad, UP, India
[2]Department of Obstetrics and Gynecology, MLN Medical College, Allahabad, UP, India

ABSTRACT

Iron deficiency anemia (IDA) continues to be the major public health nutritional deficiencies that result in impaired growth and development, cognitive performance, scholastic achievement and increasing morbidity from infectious diseases. In the world, about 19% of the total population, and in India, 20% of the total population are adolescences, and faces a series of serious nutritional challenges, not only affecting their growth and development but also their live-hood as adult. Anemia is defined in the term of hemoglobin (Hb) level in blood, and serum ferritin (SF) is specific biochemical parameter that correlates well with total body iron stores. Overall, 97.2% of adolescent girls had some degree of anemia, in which 42.7% were mildly anemic, 50.7% were moderately anemic and 3.8% were severely anemic according to WHO criteria. Only about 3% of these girls had normal Hb concentration above 12.0 gm/dl. IDA in term of low SF level (<15ng/mL) found in 39.3% of adolescent girls who were anemic (Hb<12.0 gm/dl). The causative factors of anemia other than iron deficiency in the study population include bioavailability of iron from diet. Beside the dietary factor, worm infestations and communicable diseases, like malaria, play important role in reducing iron stores of the individual.

Keywords: serum ferritin (SF); hemoglobin (Hb); iron deficiency anemia (IDA)

[*]E-mail: ajaykumarbiotech@rediffmail.com

INTRODUCTION

In the world, adolescent population is about 19% of the total population, and in India, this is about 20%. These people face a series of serious nutritional challenges not only affecting their growth and development but also their live-hood as adult. Adolescence is a particularly unique period in life (between 10-20 year-old), because it is a time to intense physical, psychological and cognitive development. Increased nutritional needs at this juncture relate to fact that adolescents gain up of 50% of their adult weight, more than 25% of their adult height adult skeletal mass during this period [1].

Teenage pregnancies are common in India, which produces a state of physiological stress of extra nutrient demand that consequently leading to maternal morbidity and mortality [2]. In India, anemia affects an estimated 50% of the population and is widely prevalent in infants, pre-school children, adolescent girls, pregnant women, and lactating mothers [3]. During adolescence, daily requirements of nutrients are greater, due to rapid expansion of tissue and red mass as well as menstrual blood loss [4].

Iron deficiency anemia (IDA) continues to be the major public health nutritional deficiencies that result in impaired growth and development, cognitive performance, scholastic achievement and increasing morbidity from infectious diseases [5,6]. Anemia is defined in the term of hemoglobin (Hb) level in blood and serum ferritin (SF) is specific biochemical parameter that correlates well with total body iron stores [7].

The government of India recognized the seriousness of the problem of anemia and initiated the National Nutritional Anemia Control program in 1970 to prevent and control nutritional anemia in pregnant women, lactating mothers and children. Under the program the adult beneficiaries receive the tablets containing 100 mg of iron as ferrous sulfate and 500 mg of folic acid (Iron plus Folic Acid: IFA) for a period of 100 days/years, while the children are provided IFA tablets containing 20 mg of iron and 100 mg of folic acid for 100 days/year. However, no any significant impact has been seen over period by the intervention program in reducing anemia [8]. Studies carried out in Uttar Pradesh (UP) have revealed that anemia is the most common nutritional problem in adolescent girls, thus this study had been conducted in Allahabad, UP, to evaluate the magnitude of this problem.

METHOD

Sample Size

A total of 213 adolescent girls who were enrolled during January 2009 to January 2010 at Department of Obstetrics and Gynecology, Moti Lal Nehru (MLN) Medical College, Allahabad (UP).

In this study, the age of the adolescent girls was assessed up to nearest completed year from the date of birth. In the absence of date of birth records, the age was ascertained from the mothers with the help of local events and a cohort of girls whose date of birth or exact age was known.

Clinical Examination

All adolescent girls were examined for presence of pallor (paleness of inner side of the eye lids, inner side of the mouth and nails) and koilonychias (papery thin bend upwards with concave or spoon shape nails of fingers and toes). In addition, history of morbidity during previous 15 days was elicited.

Blood Sampling

For Hb estimation by cyanmethemoglobin method, middle finger of left hand was cleaned with spirit and cotton and was allowed to dry for 30 seconds. Using a disposable sterile lancet, skin was punctured at right angles of finger tip. First drop of blood sample was wiped with tissue paper. Using a clean dry Hb glass pipette, blood was drawn slowly up to a little above the 20μl mark on pipette.

Blood was transferred slowly and completely on the coded Whatman No.1 filter paper in the form of circular spot with tip touching the filter paper. The filter paper with blood spot was folded diagonally, placed in a bread box and allowed to dry in shade. Theses samples were protected from sun light, flies and dust, dry sample were packed in polythene cover and transferred to the laboratory.

For estimation of serum ferritin, venous blood (1.0 ml) was collected on a sub-sample of 20% adolescent girls by selecting every 5th of subject systematically using a random start. On the day of sampling, serum was separated by centrifugation at room temperature for 10 minute at 2000 RPM and was stored frozen at -20 ^{0}C until analyzed.

Estimation of Hemoglobin

The blood was diluted in drabkin diluting fluid, in which Hb reacted with ferricyanide to form methemoglobin; it was further converted to cyanmethemoglobin, by action of cyanide. The solution obtained is examined in spectrophotometer.

Its absorbance is proportional to the amount of Hb in blood. The girls who had Hb $\geq$ 12.0 gm/dL were normal, 10.0-12.0 gm/dL were mild, 7.0-10.0 gm/dL were moderate and those having < 7.0 gm/dL were classified as having severe grade of anemia.

Estimation of Serum Ferritin

Serum ferritin was estimated by using ERBA LISA Hepatitis B kit (Transasia Bio-Medicals, Ltd). The ferritin concentration of unknown was determined graphically by plotting the mean absorbance of the ferritin standards against the ferritin concentration on the logarithmic graph paper. A standard curve was then drawn and used to determine the ferritin concentration.

Analysis of Data

Statical analysis was performed using SPSS 8.0 Window version. Appropriate statistical techniques (mean, median, and SD) were calculated. The adolescent girls were classified into different grades of anemia using standard WHO cut-off levels of hemoglobin [9]. Relative extant of iron stores on the basis of serum ferritin levels will be assessed using cut-off value of less than 15.0 ng/ml as an indicative of depleted iron stores [9].

RESULT

Age Distribution

The distribution of the adolescent girls studied according to their age is shown below (Table 1). The proportion of adolescent girls in the community ranged from 7.5% to 15.9% in between age 11 year-old to 19 year-old.

Hemoglobin Level

The mean Hb level was 9.85 ± 1.4 gm/dL among adolescent girls. Mean Hb level calculated for each age, between 11 to 19 year-old, ranged from 9.61-10.16 gm/dL. The mean Hb levels among adolescent girls according to age shown in Table 2 and Figure 1. Among these adolescent girls, Hb level was highest in the 11 year-old group and was lowest in the 15 year-old group. There was no constantly decrease or increase of Hb level among age groups.

Table 1. Number of adolescent girls according to their age

Age (year)	11	12	13	14	15	16	17	18	19	Total
Number	20	34	37	23	19	16	22	21	21	213
Percentage	9.5	15.9	17.4	10.8	8.9	7.5	10.4	9.8	9.8	100

Table 2. Mean hemoglobin concentrations among adolescent girls by age

Age	Number	Mean	SD
11	20	10.16	1.08
12	34	9.91	1.27
13	37	10.01	1.59
14	23	9.83	1.31
15	19	9.43	1.63
16	16	9.89	0.97
17	22	9.61	1.38
19	21	9.84	1.64
19	21	9.82	1.53
Total	213	9.85	1.4

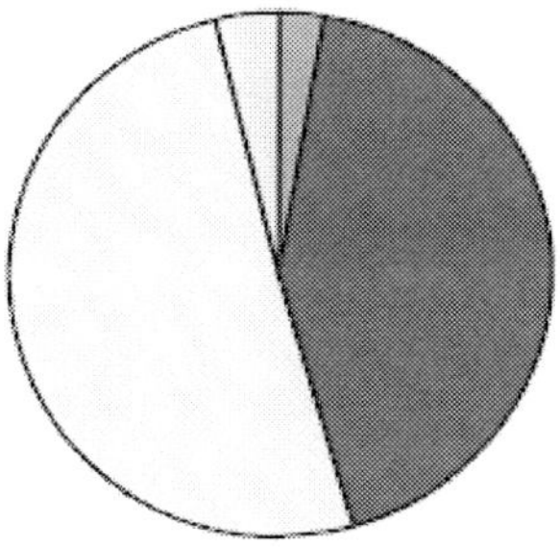

Figure 1. Prevalence of anemia.

Table 3. Prevalence of anemia among adolescent girls by age

Age (year)	Number	Hemoglobin concentration gm/dl			
		Normal (>12.0)	Mild (10-12)	Moderate (7-10)	Severe (<7.0)
11	20	5.0% (1)	45.0% (9)	50.0% (10)	0.0% (0)
12	34	2.9% (1)	41.1% (14)	53.9% (18)	2.9% (1)
13	37	5.4% (2)	51.3% (19)	37.8% (14)	5.4% (2)
14	23	0.0% (0)	43.4% (10)	56.5% (13)	0.0% (0)
15	19	0.0% (0)	31.6% (6)	57.9% (11)	10.5% (2)
16	16	0.0% (0)	43.7% (7)	56.2% (9)	0.0% (0)
17	22	0.0% (0)	36.3% (8)	59.0% (13)	4.5% (1)
18	21	5.0% (1)	33.3% (7)	57.1% (12)	4.8% (1)
19	21	5.0% (1)	52.3% (11)	38.0% (8)	4.8% (1)
Total	213	2.8% (6)	42.7% (91)	50.7% (108)	3.8% (8)

Prevalence of Anemia

The prevalence of anemia among adolescent girls is given in Table 3 and Figure 1. Overall, 97.2% of adolescent girls had some degree of anemia, of which 42.7% were mildly anemic, 50.7% were moderately anemic, and 3.8% were severely anemic, according to WHO criteria. Only about 3% of these girls had normal Hb concentration above 12.0 gm/dl. Severe cases of anemia were found more in 15 year-old group (10.5%) followed by 13, 18, 19, and 17 year-old groups. No any severe case was found in 11, 14, and 16 year-old groups. Moderate cases were found most in 17 year-old groups (59%) followed by 15, 18, 14, 16, 12, 11, 19, and 13 year-old groups, respectively.

Prevalence of Iron Deficiency

Out of 213 samples, only 43 samples were analyzed for serum ferritin. The mean SF concentration was found to be 53.99 ± 51.57 ng/ml. It ranged between 6 and 190 ng/ml. The wide range in SF observed among adolescent girls, were mainly due to infections and inflammatory disease.

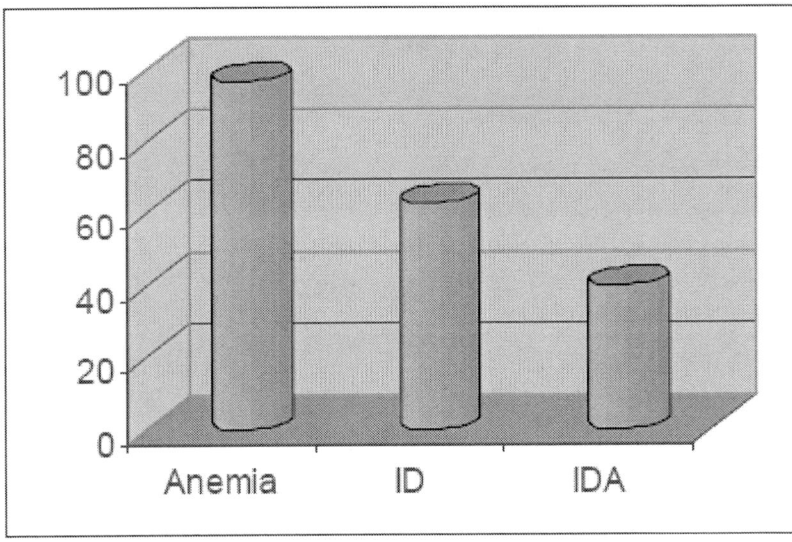

Figure 2. Prevalence of anemia, iron deficiency (ID), and IDA.

The proportion of adolescent girls with SF bellow 15 ng/ml, was 59.53% indicating depleted body iron stores (hypoferritin). 23.26% of them showed the SF value between 15-50 ng/ml, while 37.21% showed highest values (>50 ng/ml).

Iron Deficiency Anemia

Iron deficiency in term of low SF level, was found in 39.3% of adolescent girls who were anemic (Hb<12.0 gm/dl). According to the other studies in which the SF cut-off value was taken at 50.0 ng/ml, 62% of adolescent girls were having IDA. Iron deficiency anemia is depicted in Figure 2.

Discussion

This study revealed a considerably low Hb concentration among the adolescent girls in slum who resided on road side of Allahabad district, Uttar Pradesh. The overall prevalence of anemia was 97.2% in adolescent girls, which is higher than rural girls (80%) reported for the state, but it is lower than in tribal girls documented (100%) [10]. A multimeric study carried out in urban regions of Mumbai, Gujarat, and Delhi, found anemia in adolescent girls as 62-65%, 57-65%, and 48-50% respectively. In a Mother Care sponsored project, USAID, documented a prevalence of 85.3% (<11 gm/dl) in rural Haryana, while the same was even higher than 98%, when applied WHO cut-off value of 12 gm/dl [11]. Similarly, another survey carried out in the state of Madhya Pradesh, observed a wide prevalence of anemia (96.3%) in school girls aged 10-19 year-old [12]. Uttar Pradesh is a state, where malnutrition is significant, and studies carried out in the district of Varansi, Udhamsingh, and Lucknow, found that 45-46% adolescent girls are anemic [13,14]. Tribal adolescent population, in Karnal district survey for anemia, had 76.2% prevalence of anemia. It has also been shown that school-going girls are less anemic (83%) than non-school going girls (93%) [11]. The urban slum adolescent girl's population in Hyderabad, had 73-88% prevalence of anemia [15,16].

Adolescent girls, aged 11-14 year-old, were comparatively having higher Hb values as compared to those girls aged 15-19 year-old, but prevalence of anemia in younger age group (11-14) were little lower (96.5%) than the older age group (15-19) which having 98%. These results would suggest that the older adolescence was one in which girls were nutritionally vulnerable as evidenced by their low Hb concentrations. This finding is attributed to fact that iron requirements are considerable in adolescents, particularly during the period of rapid growth. The onset of puberty in girls is earlier than in boys, but menarche may be at 12-13 year-old with the growth spurt is still in its progress at the age of 14 year-old and linear growth is also continuing until the age of 18 year-old. With the onset of menarche (at puberty) and in the absence of adequate dietary intake, adolescent girls become highly susceptible to anemia, and this could have reflected in lower Hb levels.

Serum ferritin is an indicator of the relative extant of depletion of body iron stores. In present study, the mean ferritin concentration was found to be 54.0 ng/ml. The prevalence of adolescent girls with depleted iron store was 59.53% in the study area, based on the cut-off level of 15 ng/ml for serum ferritin [7]. This was only iron deficiency (ID), not iron deficiency anemia (IDA). The IDA prevalence was shown to be 39.3% in adolescent girls who were already anemic, in term of Hb level was found to be below 12.0 gm/dl. However, several other studies suggested that iron deficiency starts at ferritin concentration of less than 50 ng/ml. If using this criterion, 62% of the studied adolescent girls would have had IDA.

The causative factors of anemia, other than iron deficiency in the study population, include bioavailability of iron from diet. Their diet is deficient of fresh foods (meat, poultry, and fish), which are the excellent source of heme iron. Consumption of more leafy vegetable can decrease absorption of iron, because they contain fiber phytate and inosital phosphate.

This study indicates the importance of including adolescents in the risk group of population to improve their iron status and the need for planning intervention program that would increase the iron levels among adolescent girls through prophylaxis treatment, dietary modification, and helminthes control. Other than dietary, worm infestations and communicable diseases, like malaria, could have considerable impact on iron status of the population, which needs to be further investigated for controlling and preventing iron deficiency anemia among adolescent girls, who are the future mothers of our society.

References

[1] Brabin L, Brabin B. The cast of successful adolescent growth and development in girls in relation to iron and vitamin A status. *Am. J. of Clin. Nutr.* 1992; 55: 955-8.

[2] NCCNAI. Report of the National Consultation on Control of Nutritional Anemia in India. New Delhi: Ministry of Health and Family Welfare, and UNICEF; 1998.

[3] Seshadri S. Oral iron supplementation to control anemia in adolescent girls: Community trials of effectiveness of daily vs weekly supplementation. Baroda: UNICEF Project of Department of Food and Nutrition/WHO Collaborating Centre for Anemia Control, Maharaja Sivajirao University of Baroda; 1998: 22.

[4] Kanamal S. Strategy for combating anemia in adolescent girls. *Ind. J. Nutr. Diet* 1995; 62: 375-7.

[5] Bhaskaram P. Immunology of mild-moderate micronutrient deficiency. *Br. J. Nutr.* 2001; 75: 85-7.

[6] Logoff B, Jimenez E, Hagen J, et al. Poorer behavior and development outcomes more than 10 years after treatment for iron deficiency in infancy. *Pediatrics* 2000; 839-46.

[7] WHO. Iron Deficiency Anemia-Assessment, Prevention and control. A guide for programme manager. Geneva: Worlds Health Organization; 2002.

[8] Sharma A, Prasad K, Rao KV. Identification of an appropriate strategy to control anemia in adolescent girls of poor communities. *Ind. Pediatr* .2000; 37: 261-7.

[9] INACG. Measurement of iron status. A report of the international Nutritional Anemia Consultative Group (INAGC). Washington, DC: Nutrition Foundation, Inc; 1985: 4.

[10] NNMB. Prevalence of micronutrient deficiencies: NIN, Hyderabad, Technical Report No. 22. Hyderabad: National Nutrition Monitoring Bureau; 2003: 1-9.

[11] Raina N, Gupta A, Sharma M, et al. Prevention of anemia in pregnant women and adolescent girls in a rural area of Haryana, India: SWACH project report. Haryana: Survival for Women and Children (SWACH) foundation; 1999.

[12] Gawarikar RS, Gawarikar SB, Tripathy BC. Prevalence of anemia in adolescent girls of Ujjain in western Madhya Pradesh. *Ind. J. Nutr. Diet* 2005; 39: 493-9.

[13] Chaudhary S, Misra CP, Shukla KP. Nutritional status of adolescent girls in rural area of Varansi. *Ind. J. Soc Prev. Med.* 2003; 34: 53-61.

[14] Singh J, Singh JV, Srivastava AK. Health status of adolescent girls in slums of Lucknow. *Ind. J. Commun Med.* 2006; 31: 4-6.

[15] Saibaba A, Mohan Ram M, Raman Rao GV, et al. Nutritional status of adolescent girls of urban slums and the impact of ICE on their nutritional knowledge and practices. *Ind. J. Commun. Med.* 2002; 27: 10-2.

[16] Vasanthi G, Pawashe AB, Susie H, et al. Iron nutritional status of adolescent girls from rural area and urban slum. *Ind. Pediatr.* 2004; 31: 127-32.

In: Asia Pacific Life Sciences Compendium
Editor: Bandit Chumworathayi

ISBN: 978-1-62257-338-7

Chapter 10

Authorship and Disclosures in Scientific Papers - Implications for Novice Writers

Siew Yim Loh *
Department of Rehabilitation, Faculty of Medicine, University Malaya,
Kuala Lumpur, Malaysia

Abstract

In scientific publications, authorship and disclosures are as essential elements as the content of the manuscripts. There is a need for novice researchers and new writers to be fully informed regarding what constitutes authorships. A clear demarcation on who should be included in the authorship list and who should be excluded, with a guideline on who to acknowledge, conflict of interests and disclosures should be disseminated to new writers. Lastly, ghost writers and simultaneous submissions should not be encouraged by the scientific community. Awareness of these essential component in writing can promote integrity and good publishing ethics amongst novice writers.

Keywords: authorship, disclosure, novice researchers

Background

Scientific publications rely on authors to provide unbiased and complete reports of research. Therefore, the responsibility and integrity of the authors and coauthors are critical. As readers of health journals, we need to be concerned of the welfare of the patients and to be alert and confident of the results as well as the limitations and biases of the clinical trials or studies. Developed countries have long embarked on this race to publish but many developing countries in the Asia Pacific region is just beginning to join in the heat. The escalating pressure on academicians to 'publish or perish', first promoted amongst the medical

*Tel: +60379676612, Fax: 60379674766, E-mail: syloh@um.edu.my

communities in the UK in 1990s, has since shifted onto the lives of academicians in the developing world. The mounting pressure on Asian scientists to publish has led to a situation where novice writers struggle to get their paper printed in peer-reviewed journals which are predominantly westerned-governed, english-medium journals. Akins to the recent 2010 H1N1 influenza pandemic, unethical writing must be curbed so that it does not spread widely. In short, awareness and information dissemination on good ethical publishing practices in scientific writing is critical. Therefore, the aim of this paper is to highlight these essential definition and role of the author and coauthors in scientific papers.

WHAT CONSTITUES AUTHORSHIP?

The International Committee of Medical Journal Editors defines an author as " ... *someone who has made substantive intellectual contributions to a published study* ... ". [1] Specifically, authorship is taken as constituting the cumulative sums of all three criteria of - 1) substantial contributions to conception and design, or acquisition of data, or analysis and interpretation of data; 2) drafting the article or revising it critically for important intellectual content; and 3) final approval of the version to be published. Criteria 1 implies that the coauthors should participate actively during the research process, whilst criteria 2 emphasized co-drafting and co-revising in critical manners. This guide clearly excludes supervisors who provide scanty comments, or mere language or stylistic corrections without sufficient effort into the research process and scientific paper writing. In short, it is a good measure for journal to insist that the first author provides a list of the contributions of coauthors as rightly, the first author is the best person to judge who has contributed substantially (and in what) and who has not (and thus may be deserving of only an acknowledgement).

WHO ARE THOSE WHO DO NOT QUALIFY AUTHORSHIP?

A recent good practice by journal editorial board is the provision of a disclosure form to the first author to provide a list of people whose contributions to an article do not rise to the level of authorship, and thus will be inserted in the 'Acknowledgments' section. [2] This list includes medical writers hired to polish manuscripts, or an esteemed researcher' who did not contribute substantially to the paper, and it excludes ghost authors. A ghost author is a professional writer, paid to write the paper but which is then officially credited to another person. Often, the drug companies use this service by paying doctors or scientist to write favorable reports in journal on the outcomes of trials involving drugs or devices. Thus, the ghost author is typically not acknowledged in the scientific papers published in medical journals. The Citizen Commission on Human rights, [3] a watchdog nonprofit public benefit organization in USA, launched their campaign on the investigation and the psychiatric violations of human rights. In their documentary entitled "Making a killing---the untold story of psychotropic drug', the indiscriminate use of ghost writers and unethical publishing works were reported. This is a clear impactful example of violation of the use of scientific writing by pharmaceutical companies with clear conflicts of interest, and as such, this reason as to why the use of ghost writer should not be encouraged.

GUIDELINES ON AUTHORSHIP

In general, there are numerous guidelines on authorship, depending on which journal one subscribes to. The Uniform Requirements for Manuscripts Submitted to Biomedical Journals (ICMJE), where more than 500 biomedical journals subscribed to their guidelines, is therefore, a good start for novice writer to read about authorship. ICMJE outline that authorship credit should be based on three criteria - 1) substantial contributions to conception and design, or acquisition of data, or analysis and interpretation of data; 2) drafting the article or revising it critically for important intellectual content; and 3) final approval of the version to be published. However, many novice Asian authors may not be aware of it. Some common mis-beliefs that heads of department, or people who look for fundings are entitled to claim authorships will need to be curb from any greater disseminations. Katz et al [2] 's guide is a cautionary guide for novice writer and it reiterates that:-

- Individuals who do not meet criteria for authorship should not be included as authors .
- Others who participate in the research or in manuscript preparation, but do not meet authorship criteria, should be listed in the Acknowledgments.
- All authors are responsible for the accuracy of the data and reporting of the data.
- All individuals who meet criteria for authorship should be listed as authors.
- Editorial offices will not correspond with or accept correspondence from non-authors.

IS THERE A FIXED RULE ON NUMBER OF AUTHORS?

With regards to the number of authors in one paper, most journals do not stipulate a fixed rule for the number of authors to be listed for any particular work. Nevertheless, it is difficult to justify more than two authors for an editorial or commentary like this paper. Having said that, there are papers with a long list of names and it seemed odd and even questionable as it does not commensurate to the scale of the project being reported. Perhaps these are the "participating investigator" (to be listed under the acknowledgement section) which is typically associated with large multicentre studies where the number of collaborating doctors or scientists may run into dozens. Thus it is fair to insist that author must justify any paper with more than a dozen authors. [4] In brief, an average of five and not more than 10 authors could be a justifiable figure for any one paper, and depending on the depth/length of the paper.

WHAT IS THE ORDER OF AUTHORSHIP?

A review of several journals' instruction to author, does not show any consensus fixed rule on the order of authors. A widely-agreed formula is that the first author is the most important one and the other authors are generally listed in descending order of extend of their contributions. In some institution, the senior most 'contributing' academic author is the last author, viewed as being the most prestigious. Nevertheless, the order of authorship should be a joint decision of the study contributors.

GUIDELINE ON ACKNOWLEDGEMENT

How should authors acknowledge the contribution of those who do not deserved to be addressed as coauthors? The section to acknowledge these people is rightly, the 'acknowledgement' section in a scientific paper. This section is typically reported after the conclusion and just before the reference section in a scientific paper. Financial and material support, including grants are also to be acknowledged in this section. With reference to the International Committee of Medical Journal Editors, [5] this list may include those who provided purely technical help, contributes to writing assistance, or the consultant who provided general comments scantily, as well as the department head or chairman and even supervisor who provided only general support. In short, these are people who have contributed materially to the paper but whose contributions do not justify authorship (i.e. do not meet the three basic criteria describe earlier). They may be listed as "clinical investigators" or "participating investigators", with their contribution described. This may include "served as scientific advisors", "collected data", or "provided and cared for the patients". [5]

GUIDELINES ON CONFLICTS OF INTERESTS AND DISCLOSURE

All authors need to be particularly attentive to issues of conflict and disclosure. This is in relation to the pressure that arises from financial aids or incentive. In the interest of transparency, most journals will require that all manuscripts with commercial support include a statement in the Methods section describing these roles. The Journal of American Medical Association [6] proposed three guiding criteria which include---a) no relevant financial interests pertaining to the manuscript, b) any conflicts of interest, including specific financial interests and relationships and affiliations relevant to the subject matter or materials discussed in the manuscript are disclosed in an attachment., and c) all financial and material support (including those providing data or access to patients, interviewees, etc.) for this research and work are clearly identified in an acknowledgement/disclosure section to be published with the manuscript.

SIMULTANEOUS SUBMISSIONS

I recalled a conversation not too long ago with a senior academician in which she admitted to attempting, not just double submission but multiple submissions in order to meet the demands on her key performance index of published papers. As academicians, we live in an era where the pressure to publish can exert enormous strain and upset our otherwise good and sound judgment. The adage 'publish or perish' still stand true. Most if not all journals, prohibit simultaneous duplicate submission and its usually stipulated in their guidelines for authors, This is fair and need to convey to the novice writer so that they do not violate this rule out of ignorance and so that the others will stay clear of this violation in their pursuits for more papers.

Conclusion

Having an awareness about disclosure and authorship can definitely help novice writers to make decision on who to include as coauthors and this process of decision-making should be from the beginning of the manuscript writing (and not at the end). This awareness on authorship facilitates integration of good publication guidelines into practice; thus reflecting the age-old adage (from a preacher's sermon)---'**Truth** does not change behaviors, but its **truth-applied** that changes behaviors (of new writers). Integrity in general, and being informed helps to ensure the integrity of the papers published in the journal.

References

[1] Uniform Requirements for Manuscripts Submitted to *Biomedical Journals*: writing and editing for biomedical publication. Available at www.icmje.org Accessed on July 30, 2009.

[2] Katz, PP; Yelin, EH; Lockshin MD. Journal policies on authorship and disclosures. *Arthritis Care and Research* 2006; 55 (3); 345-7.

[3] Citizens Commission on Human rights. Investigating and exposing psychiatric violations of human rights. Available at www.cchr.org Accessed on July 10, 2009.

[4] Biros, MH; Advice to authors: getting published in Academic Emergency Medicine. Available at www.saem.org/inform/aempub.html Accessed on July 29, 2009.

[5] International Committee of Medical Journal Editors. Uniform requirements for manuscripts submitted to biomedical journals. Updated October 2008. Available at www.icjme.org Accessed on July 29, 2009.

[6] JAMA, HSR; Authorship Responsibility, Disclosures, and Acknowledgments---adapted from the Journal of American medical association 2006. Available at www.hsr.org/hsr/information/content/hsr-authors.doc Accessed on July 29, 2009.

In: Asia Pacific Life Sciences Compendium
Editor: Bandit Chumworathayi
ISBN: 978-1-62257-338-7

Chapter 11

ROLE OF HONEY, GHEE, AND CURCUMA LONGA IN DRESSING OF WOUNDS

Mohamad Gamil Abdel Monaem[1], Ragaa Hosny Mohamad[*,2] and Hussien Abdel-Rahman Al-Mehdar[3]
[1]Surgical Department, National Cancer Institute, Cairo University, Egypt
[2]Cancer Biology Department, National Cancer Institute, Cairo University, Egypt
[3]King Abdel Asis Faculty of Science, Jedda, Saudi Arabia

ABSTRACT

Background: The present study was designed to evaluate an ancient method of dressing using honey, ghee, and Curcuma longa, and to find the scientific background of using these materials in promoting tissue repair.

Method: In the present study honey, ghee and curcuma longa extracts were analyzed chemically. The honey compounds were separated and identified using Gas chromatography/Mass spectrometry (GC/MS). Honey phenolic and flavinoid compounds were analyzed by HPLC technique and the radical scavenging activity of honey sample extracts was measured using BrÜcker Elexsys 500 instrument operated at X-band Frequency. Fatty acids of ghee were assayed using a gas chromatograph equipped with a flame-ionization detector and the concentrations of vitamins C were measured spectrophotometrically. Analysis of Curcuma longa pure essential oils was carried out by gas chromatography (GC) and gas chromatography-mass spectroscopy (GC/MS) technique. The mixture of honey ghee and Curcuma longa dressings were mixed in clean vessels with help of an electric blender. The present study included thirty six patients in the surgical department at heliopolis hospital, Cairo, during the period between December, 2004, and December, 2007. The patients were divided in to two groups; the first group was treated with HGC-dressing, the second group was dressed with glycerin magnesia and/or povidine-iodine dressing.

Result: The chemical analysis of the mixture revealed that honey, ghee, and Curcuma longa extracts have many compounds which accelerate wound healing. The healing percentage is much better with H-G-C dressing ($p=0.001$).

* Corresponding Authour: Cairo University, Tel: +020101856864, Fax: +0223644720, Email: ragaatahtawy@gmail.com, hosny2@hotmail.com.

Conclusion: This method of wound dressing has proved to be an effective method of encouraging healing and controlling infection. It was proved that honey, ghee, and Curcuma longa, contain eicosapentanoic, linoleic, α-linolenic, -linolenic, arachidonic acids, and vitamin C, which are important in controlling infection and promoting wound healing.

Keywords: honey, ghee, curcuma longa, gas chromatography (GC), mass spectrometry (MS)

BACKGROUND

There is a rapidly increasing interest in the use of honey as a wound dressing, although common clinical opinion would appear to be that there is no evidence to support its use as a wound dressing. [1] Even where reviews of clinical evidence for the use of honey have been published, a negative impression is often obtained from consulting these, as the conclusions stated are that the evidence is of low quality and/or that there is a need for more evidence. [2-6] But the myriad of advertisements for modern wound dressings possibly blinds people to the fact that only small, poor-quality trials exist to support the use of these products. [7] For example, if the Medline database is searched for evidence to support the use of nanocrystalline silver dressings, it can be seen that there is in fact very little published evidence.

A recent systematic review of publications on the use of advanced dressings in the treatment of pressure ulcers has found that their generalized use in the treatment of pressure ulcers is not supported by high quality evidence. [8] In evidence-based medicine, decisions should be made on the basis of the available evidence: where randomized controlled trials of the highest quality have not been conducted, it is necessary to consider evidence of a lower quality.

Ibn Sina, in his book, “the law of medicine”, used honey, ghee, and curcumin, in treatment of wounds. As recently, there is evidence to support the use of honey as wound dressing and many of cases not responding to standard treatment for along time. So the present study was designed to evaluate Ibn Sina’s method of dressing wounds and to find the scientific back ground of using these materials in accelerating wound healing.

METHOD

Materials

1. Honey: Bee Honey Sider. Sider honey was obtained from Saudi Arabia.
2. Cow ghee obtained from cows lived in the Marsa Matrouh farm (Animal Production Institute).
3. Dried Curcuma longa was purchased from a local retail spice market in Mansoura, city, Egypt.

Chemical Analysis of Honey, Ghee and Curcumin

1. Analysis of Honey

Analysis of honey was performed at the Central Laboratory of Kato Aromatic Co., Cairo, Egypt. Accordingly, a study of the separation and identification of the honey compounds has been undertaken using Gas chromatography/Mass spectrometry (GC/MS). Steam distillation/solvent extraction was found to be the method of choice. The volatile was found to contain numerous organic compounds which belong to various classes namely: heterocyclic, and hydrocarbons. The linear retention indices (LRI) were determined by injection of a solution containing the homologous series of normal alkanes: FLUKA "Alkane standard solution C8-C20", Sigma-Aldrich, No. 04070.

Solid Phase Microextraction (SPME) Procedure

The SPME procedure was performed using a manual SPME holder (Supelco) equipped with a 65 μm polydimethylosiloxane/divinylbenzene (PDMS/DVB) (Supelco) coating. Before every use of the fiber, it was conditioned at the GC injection port, at a temperature of 250°C for 30 minutes. About 1-1.5 g of the investigated sample was placed into a 4 ml vial closed with a screw and heated to 40°C for 60 minutes and the fiber was then exposed to honey headspace. After 30 min, the SPME fiber was withdrawn from the vial and promptly introduced into the GC injector under the conditions reported below.

Gas Chromatography-mass Spectrometry (GC/MS) Procedure

GC/MS analyses were performed using Agilent 6890 GC coupled to an Agilent 5975 quadruple mass detector. The SPME fiber was desorbed in GC injector at 220°C for 5 minutes in splitless mode and chromatographic separation was carried out on a 30 m x 0.25 mm x 0.25 μm film thickness HP-5MS (5% Phenyl Methyl Siloxane) capillary column. The GC oven temperature was programmed from 40°C (held for 5 min.) to 250°C at a rate of 5°C. Helium was used as a carrier gas, at a constant flow of 0.9 ml/min. Mass spectra were recorded in EI mode at 70 eV, scanning the 20-550 m/z range. The interface and source temperature were 150 and 230°C, respectively.

The identification of the isolated volatile compounds was achieved by comparing the obtained mass spectra of unknown peaks with those stored in the NIST.02 (Central Laboratory of Kato Aromatic Co., Cairo, Egypt) and Wiley.7n. mass spectral electronic libraries. Identifications were confirmed where possible, by comparison with authentic substances used as references and by use of linear retention indices (LRI). Relative area values (as a percentage of total volatile composition) were directly obtained from total ion current (TIC). All analyses were carried out in duplicate.

HPLC Analysis of Honey Phenolic Compound

Prior to HPLC analysis, the dried phenolic extracts were redissolved in methanol and filtered through a 0.45 mm Millipore filter (Millipore Corp., Bedford, MA). [9] HPLC analysis was performed on an LC-10A series liquid chromatograph (Shimadzu, Japan), equipped with a C18 column (150 x 4.6 mm, 5-μm particle size) (Jones Chromatography, UK), thermo-stated at 35 ¼C. A gradient elution (Weston, et al., 2000) was modified to separate the extracted phenolics. Solvent (A) was 1.0% acetic acid in water, and solvent (B) was 1.0% acetic acid in methanol. Elution was performed at a solvent flow rate of 1.0 ml/min. The gradient profile of the system was 3% solvent B at the initial stage, 8% solvent B at 10 min, 31% solvent B at 20 min, 45% solvent B at 30 min, 50% solvent B at 50 min, 97% solvent B at 75 min, and again 3% solvent B at 81 min. The eluted phenolic compounds were monitored at 280 nm, and identified by comparing the chromatographic retention times with those of authentic standards.

HPLC Analysis of Honey Flavonoids

Honey flavonoid extraction for HPLC was performed by using 200 μg of honey sample passed through a column Lichrochart RP-18 (Merck, Germany) (25 x 0.4 cm), Supelco; pore size 9 nm, 5-μm particle size). Elution was with water: formic acid (19:1 v:v; solvent A) and acetonitrile (solvent B), and the flow rate was 1ml/min. Gradient elution started with 20% B, reaches 25 % B at 25 min and 30% B at 35 min, and then the system became isocratic until 50 min, reaches 50 % B at 60 min and 70 % B at 67 min. The flavonoids were detected with UV detector and the chromatograms were recorded at 290 nm. Flavonoid identification was carried out by direct HPLC comparison of authentic flavonoids and was based on co-chromatography in 290 nm. Response factors for the authentic flavonoids and the concentration of flavonoids in each honey sample were calculated according to Ogan & Katz [10] and Annual book of ASTM Standards. [11]

Determination of Free Radical-Scavenging Activity

For the measurement of the radical scavenging activity of honey sample extracts against stable 1,1-diphenyl-2-picrylhydrazyl (DPPH) was measured using BrÜcker Elexsys 500 instrument operated at X-band Frequency in National Research Center, Dokki, Egypt, according to the method described by Ramadan et al. [12] Sample was measured at room temprature in Ulta pure silica tube, microwave frequency was 9.79821 (Hz), microwave power (W) 0.002026, receiver gain 60 and Axis range was width 6000 and center 3480. Briefly, a mixture of 3 ml of honey methanolic extract at different levels (50–250 μg) + 6μml of 0.03 mg/ml DPPH (Sigma Aldrich, Germany) in methanol and shaken vigorously, incubated in the dark for 30 min, and the absorbance of the remaining DPPH was determined at 517 nm against a blank. Lower absorbance of the reaction mixture indicated higher free radical scavenging activity. Radical scavenging activity was calculated as follows: % Inhibition = [(Abs Blank–Abs Sample)/Abs Blank] x 100

Preparation of Honey, Ghee, and Curcuma Longa, Dressings

Two liters of pure liquid honey, one liter of melted ghee (clarified butter) and 10 gm clean sterile Curcuma longa powder were thoroughly mixed in clean vessels with help of electric blender. Then pre-cut sterilized surgical gauze pieces were soaked in the mixture, then separated and packed in clean sterile box and put in refrigerator until used. The honey is used raw, since unprocessed honey has been found to be more effective and 2-3 years old ghee is preferable to younger preparations.

2. Analysis of Ghee

Fatty acids of ghee were assayed using a gas chromatograph equipped with a flame-ionization detector according to the procedure reported by Gorban and Izzeldin. [13] The fatty acids were identified by comparison of retention time with known standards and were expressed as percentage of total fatty acids. The concentrations of vitamins C was measured spectrophotometrically according the method reported by Ghasemi and Nayebi. [14]

3. Analysis of Curcuma Longa (CLM)

Curcuma longa sample were obtained via steam distillation as pure essential oils from a commercial sources. The essential oil samples were stored in glass vials with teflon-sealed caps at -18±0.5°C in the absence of light.

A. Gas Chromatography

Analysis of Curcuma longa pure essential oils was carried out by gas chromatography (GC) and gas chromatography-mass spectrmscopy (GC/MS) technique at the Kato Aromatic Company, Cairo, Egypt. Essential oil samples were analyzed and the relative peak areas for individual constituents averaged. Quantification was computed as the percentage contribution of each compound to the total amount present. The relative percentages were determined using a Fisons (Rodano, Milano, Italy) 9130–9000 series gas-chromatograph equipped with a Fisons EL980 processor, a FID detector and a MEGA SE52 (Mega, Legnano,Italy) 5% poly diphenyl 95% dimethylsiloxane bonded phase column (i.d. = 0.32 mm, length 30 m, film thickness = 0.15 mm).

Operating conditions were as follows:

Injector temperature, 280°C; FID temperature, 280°C; carrier gas (Helium), flow rate 2 ml/min and split injection with split ratio 1:40. Oven temperature was initially 45°C and then raised to 100°C at a rate of 1°C/min, then raised to 250°C at a rate of 5°C/min and finally held at that temperature for 10 min. 1 of each sample, dissolved in CH2Cl2 (1:100 v/v), was injected. The percentage composition of the oils was computed by the normalization method from the GC peak areas, calculated by means of three injections from each oil, without using correction factors.

B. Gas Chromatography/Mass Spectrometry Analysis

Essential oil constituents were analyzed by a Hewlett Packard HP5890 series II plus gas chromatograph equipped with a HPMS 5989b mass spectrometer using electron impact. The gas-chromatographic (GC) conditions were the same as reported for GC analysis and the same column was used.

The mass spectrometry (MS) conditions were as follows:

Ionization voltage, 70 eV; emission current, 40 mA; scan rate, 1 scan/s; mass range, 35-300 Da; ion source temperature, 200°C. The MS fragmentation pattern was checked with those of other essential oils of known composition, with pure compounds and by matching the MS fragmentation patterns with NIST NBS75K mass spectra libraries and with those in the literature according to Adams. [15]

The relative amounts of the individual components were obtained from GC analysis, based on peak areas without FID factor correction. The constituents of the essential oils were also identified by comparing their GC retention indices. A mixture of aliphatic hydrocarbons (C8-C24) in hexane (Sigma-Aldrich, St. Louis, USA) was injected as under the above-mentioned temperature programmed to calculate the retention indices using the generalized equation of Deters et al. [16]

Statistical Analysis

Relevant statistical tests (Student-t, and ANOVA) were performed using SPSS (version 12) at $p<0.05$ as the significant level. [17] Quantitative variables were summarized using mean±SD, median, minimum, and maximum values. Differences were considered significant when p was <0.05 and highly significant when $p<0.01$.

Patients

The present study was performed in the surgical department at Heliopolis hospital, Cairo, during the period between 2004 and 2007. It included 20 patients with diabetic foot after surgical debridement, 15 patients with pilonidal sinus after excision and leaving the wound open and one patient with necrotic skin ulcer in left hand dorsum after extravasations of chemotherapeutic agent.

The patients under investigation were divided in to two groups: The first group was treated with H-G-C dressing including 12 patients with diabetic foot, 8 patients with pilonidal sinus and the patients with necrotic ulcer. The second group (8 with diabetic foot and seven with pilonidal sinus) were dressed with glycerin magnesia and/or povidine-iodine dressing. Samples were taken from the wound for culture and sensitivity after 7, 10, and 14 days.

Area Measurement

Accurate blind surface measurement of the wounds by transparency tracing was used for assessment of the wound healing at 1st, 7th, 14th, 21st, 28th, and 35th days after operation. The following equation was used for determining the percentage of wound:

Wound Percentage = (Area of wound on day x / Area of wound on day 0) X 100

In this equation, day x was the day of wound area measurement (days 7, 14, 21, 35, after wounding) and day 0 was the day wounding had done. On the other hand, the percentage of healing was measured as: 100-Wound Percentage = Healing.

Dressing Technique

Wound areas were cleaned with normal saline and any necrotic areas removed. A piece of gauze impregnated with honey, ghee, and Curcuma longa, was applied directly to the wound, held in place by cotton dressing and a bandage or adhesive plaster. This dressing was renewed every 48 hours. The second group was dressed with glycerin magnesia and/or povidine-iodine dressing on daily basis.

Table 1. Volatile compounds related to wound healing identified in investigated honey samples assessed by GC/MS of ether extracts expressed as percent of total chromatogram area

	Volatile compounds	*Sider*
1	1- (I H-pyrrol-2-yl) ethanone.	0.80
2	Formic acid (Methanoic acid)	0.23
3	Acetic acid (Ethanoic acid)	1.10
4	2-methyl cyclopentanone	0.44
5	Propanoic acid, 2-methyl- (Isobutyric acid)	0.28
6	Butanoic acid (Butyric acid)	0.48
7	Butanoic acid, 3-methyl- (Isovaleric acid)	1.82
8	Butanoic acid, 2-methyl- (Butyric acid, 2-methyl)	1.10
9	Pentanoic acid (Valeric acid)	1.40
10	2,3- dihydrobenzofurane	1.59
11	Hexanoic acid (Caproic acid)	0.20
12	4-oxo-1H-quinoline-2-carboxylic acid	0.95
13	Octanoic Acid (Caprylic acid)	0.53
14	Benzeneacetic acid (Acetic acid, phenyl)	3.34
15	Nonanoic acid (Pelargonic acid)	0.55
16	Decanoic acid (Capric acid)	0.06

The ion current generated depends on the characteristics of the compound concerned and it is not a true quantization.

RESULT

The identified volatile compounds of honey are presented in Table 1. Of these compound organic acids: e.g. formic acid, acetic acid, hydrocarbons, aromatic hydrocarbons, esters are related to the wound healing and Kynurenic acid has analgesic activity. Nine phenolic acids were identified in Sider honey extract: malic acid, palmitic acid, benzoic acid, gallic acid, cinnamic acid, caffeic acid, kaempferol ferulic acid, Vanillic, Ellagic, p-Coumaric, and Homogentisic. (Table 2) The flavonoids present in the honey sample analyzed by HPLC are shown in Table 3.

In the present data, oleic, linoleic, α-linolenic (n-3), -linolenic (n-6), and α-Tocopherol (μmol/l), Eicosapentanoic (n-3), linoleic (n-6), α-linolenic (n-3), -linolenic (n-6), and arachidonic (n-6) acid, are among the fatty acid content of ghee. (Table 4)

The analysis of Curcuma longa showed 48 compounds (Table 5) in which ar-tumerone (27.78%), tumerone (17.16%), Camphene (13.82%), ar-curcumene (7.0%), and dehydro-curcumene (0.08), 2-methyl-5-(1-methylethyl) phenol (Carvacorol) (5.28%), ar-curcumene (3.29%), and Curlone (13.82%), are of importance to wound healing.

Table 2. Phenolic compounds identified in honeys using HPLC

Compound	**Structure**	**R.T.(min)**	**Quantity (µg/100g of Sider honey**
Gallic acid	3,4,5-ris(benzyloxy)benzoic Acid	6.04	330.4 ± 12.7
Caffeic acid	3,4-Dihydroxy-Trans-Cinnmate	22.13	69.023 ± 6.84
Ferulic acid	Trans-4-hydroxy-3-methox ycinnamic acid	27.71	53.3 ± 4.72
Benzoic acid	C_6H_5COOH	28.4	79.7 ± 4.31
Cinnamic acid	3-Phenylacrylic acid; Phenylacrylic acid	35.81	19.0 ± 2.5
Vanillic	4-hydroxy-3-methoxybenzoic acid	17.9	13.0 ± 1.7
Ascorbic acid(C)	*R*)-3,4-dihydroxy-5-((*S*)- 1,2-dihydroxyethyl)furan-2(5*H*)-one	1.2	2.2±2.5
Ellagic	3, 4, 5-Trihydroxybenzoic acid	14.2	36.3 ± 1.60
p-Coumaric	4-hydroxycinnamic acid, β-(4-hydroxyphenyl)acrylic acid	13.4	31.5 ± 2.35
Homogentisic Melanic acid	2-(2,5-dihydroxyphenyl)acetic acid	3.7	27.4 ± 5.32

Detection at 280 nm, data are presented as mean±standard deviations (n = 5).

Table 3. Flavonoid compounds identified in honeys using HPLC technique (ug/100g honey

No	Flavon	Structure	Rt %
1	Apigenin	5,7,4'-trihydroxyflavone	31.1
2	Chrisin	5,7-dihydroxy flavone	0.93
3	Hesperetin	5,7,3'- trihydroxy-4'-methoxyflavanone	32.1
4	Kaempferol	3,5,7,4'- tetrahydroxyflavone	31.9
5	Myricetin	3,5,7,3',4',5'-hexahydroxyflavone	19.6
6	Naringenin	5,7,4'-trihydroxyflavanone	30.8
No	Flavon	Structure	Rt %
7	Pinocembrin	7-Dihydroxyflavanone; 5,7-dihydroxy-2-phenyl-2,3-dihydro-4H-chromen-4-one;	41.2
8	Quercetin	3,5,7,3',4'-pentahydroxyflavone	27.7
9	Formonontin	7-hydroxy-4'-methoxyisoflavone	12.43
10	Eriodictyol	5,7,3',4'-tetrahydroxyflavanone	21.15
11	Galangine	3, 5,7-trihydroxy flavone	0.09

In H-G-C group, patients with diabetic foot (n=12), 83.3% of wounds were sterile by day 7th to 10th (vs 70% of patients with diabetic foot in the second group), sloughs separated completely and painlessly in day 9 compared with 17th day in patients treated with glycerine magnesia and/or PI. With H-G-C, granulations appeared by day 8th (second group day 13th); and 88% of H-G-C wounds (compared to only 41% in the second group) were healed by day 35th and healing percentage was 99% by day 42nd. No overgrowth of granulation tissue was noticed with H-G-C dressing.

Table 4. The fatty acids related to wound healing and vitamin C content of cow ghee

Proportion Name of Fatty Acids	Cow
Monounsat. FA, g	
Oleic	0.84±0.21
Polyunsat. FA, g	
Eicosenoic	1.7±0.07
Linoleic	0.08±0.02
Linolenic	0.05±0.01
Arachidonic	0.03±0.01
α- Tocopherol μmol	17.42 ±1.23**
L-ascorbic acid	19.9 ± 1.8*

Cow fatty acids were expressed as mg/100gm, α-Tocopherol (μmol/l of unsaturated to saturated acids data expressed as mean±SD.

Table 5. Constituents of Volatile oil *C. longa* (CLV) rhizome related to wound healing

Sl.No	Compounds	Relative (%)
1	α-Terpinene	2..35
2	m-cymene	0.21
3	Limonene	0.17
4	Desmethoxycurcumin	0.17
5	p-Cymen-8-ol 2-Hydroxy-(p-cymene)8-phenol	0.09
6	Bisdemethoxycurcumin	0.63
7	Caryophyllene	0.98
8	Ar-tumerone	27.78
9	Ar-Curcumene	3.29
10	α - Zingiberene	4.37
11	diferuloylmethane,	0.59
12	Dihydrocurcumene	2.08
13	Tumerone	17.16
14	Caryophyllene oxide	0.14
15	α-turmerone	7.4
16	2-methyl-5-(1-methylethyl)phenol(Carvacorol)	3.28
17	Curlone	13.82
18	γ-Gurjunen epoxide-(1)	0.17
19	Acoradiene	0.42

In patients with pilonidal sinus wounds treated by H-G-C (n=8), the wounds were sterile and 7 out of 8 (87.5%) became painless by day 5th vs only 2 of 7 in the second group (28.5%). With H-G-C granulations appeared by day 3 (100%) and healing percentage was 97.5 by day 42nd (second group = 58%). The healing percentage is much more better with H-G-C dressing (p=0.001). (Table 6)

Table 6. Wounds healing percentage of cases treated with with H-G-C versus routine dressing

Type of Dressing	Number of Cases		Outcome and Percentage of wound Healing					
			7-day Wound healing	21-day Wound Healing	28-day Wound healing	35-day Wound healing	42-day Wound Healing	P-value
H-G-C Dressing	DF	12	25.8%	44.5%	72.5%	88%	99%	0.001*
	PS	8	13.5	30%	63%	75.3%	97.5%	
PI-Dressing	DF	8	8.4%	16%	30%	41%	60%	0.001*
	PS	7	6.7%	21%	35%	43%	58%	

DF = Diabetic Food, PS = Pilonidal sinus, * = Highly significant.

DISCUSSION

There is a rapidly increasing interest in the use of honey as a wound dressing, although common clinical opinion would appear to be that there is no evidence to support its use as a wound dressing. But the myriad of advertisements for modern wound dressings possibly blinds people to the fact that only small, poor-quality trials exist to support the use of these products. [7]

Gadhvi [18] reported that honey seems to contain proteolytic enzymes that digest necrotic tissue to separate slough from living tissue. Honey appears to inhibit growth of pathogens whilst separating the slough and cleaning the wound. Ghee appears to promote healthy granulation, fills wound cavities and epithelializes them. Honey alone keeps the wound clean but cannot promotes healing, and while ghee alone promotes granulation and epithelization in clean healthy wounds, it cannot seperates sloughs and control infection. The two together, however, work as the most ideal dressing. Also, honey alone can cause pain but does not do so when combined with ghee. There is strong evidence of presence of a growth factor in H-G. Healthy granulation and epithelization appear simultaneously and rapidly. There is visible healing with every change of dressing. We find the same observations in our work. In the present series we had a 42 year-old female patient with a wound for 5 months resulted from sloughing the skin (4x4 cm) after extravasation of a chemotherapeutic drug which did not respond to conventional treatment. With H-G-C, the wound healed within seven weeks.

The use of honey as wound dressing comes from laboratory studies that have clearly demonstrated that honey has bioactivities that would be beneficial in wound care. In work with cultures of leukocytes, honey has been shown to stimulate cytokine production by monocytes. [19,20] The release of cytokines is what initiates the tissue repair process as well as the immune response to infection. Also, stimulation by honey of other aspects of the immune response, the proliferation of B- and T-lymphocytes and the activity of phagocytes, has been shown by Abuharfei et al. [21] Additional to this work with cells in culture, it has been demonstrated that honey stimulates the production of antibodies in mice in response to antigens from Escherichia coli. [22] These findings suggest that part of the effectiveness of honey in clearing and preventing infection in wounds that is so widely seen in the clinical evidence may be due to enhancement of the body's own immunity as well as being due to the antibacterial activity of honey. The number of publications on laboratory studies showing that honey has antibacterial activity with a very broad spectrum is very large [23]

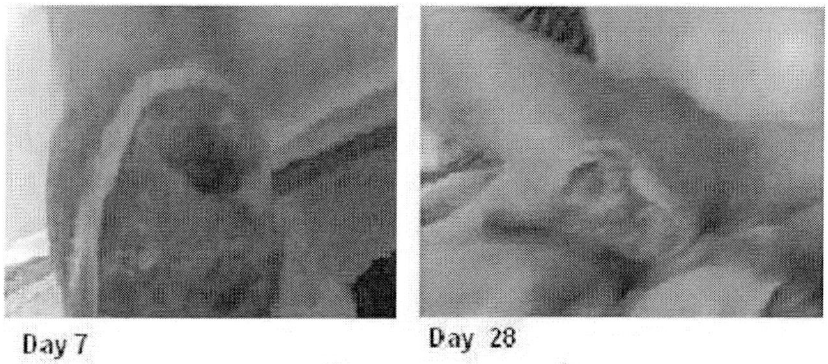

Case 1. Female diabetic patient 50 years old suffered from gangerene of the heal of left foot, excision of the gangerenous patch and debridement was performed. Dressing was done using H-G-C.

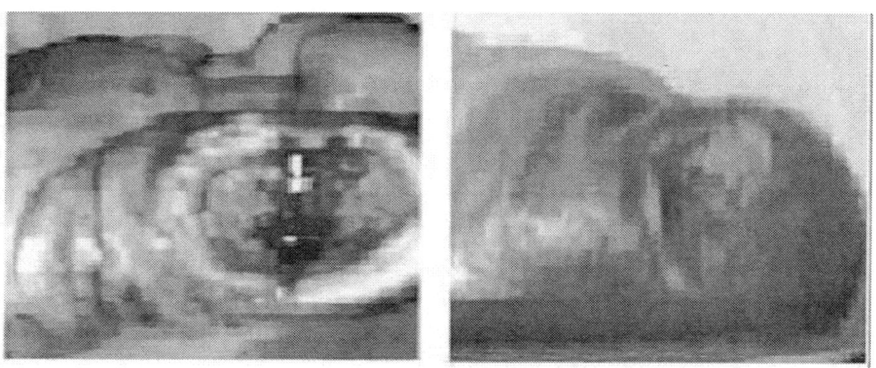

Case 2. Femal diabetic patient 57 years old suffered from trophic ulcer in the ventral aspect of left big toe, excision of the ulcer and debridement of the underlining tissue and bone was performed. Dressing was done using H-G-C.

Honey has an anti-inflammatory activity as it has been noted that if sufficient honey is kept in place, by applying it by way of impregnated dressings and changing these frequently enough, then its anti-inflammatory activity will reduce the amount of exudate and thus remove the need for frequent dressing changes. [24] This coincides with our findings that we need only dressing every 48 hours.

There is a very large amount of evidence for honey having significant anti-inflammatory activity. There is evidence from histological observation of biopsy samples taken in a clinical trial of honey on burns [25] and from biochemical assays of indicators of inflammation in other clinical trials on burns. [26,27] One of these biochemical studies was in the form of a randomized controlled trial with 60 patients, comparing honey with silver sulfadiazine, and it was demonstrated that honey decreased oxidative stress by mopping up the free radicals arising from burns. [27]

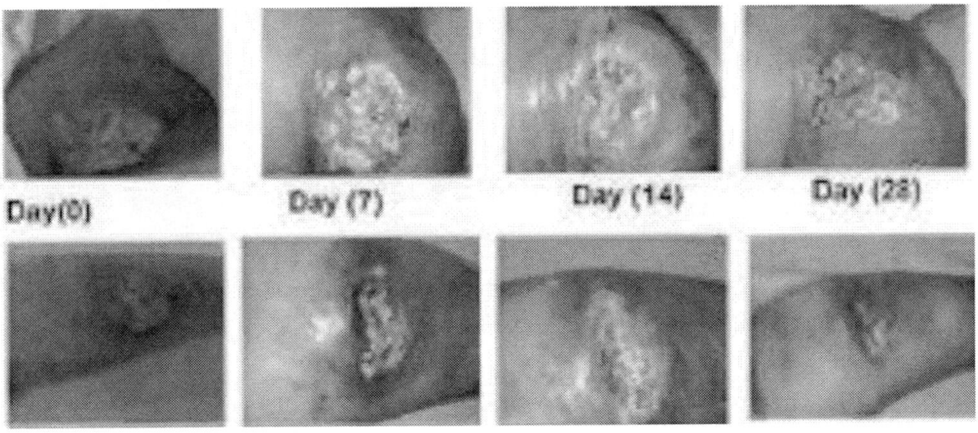

Case 3. Female diabetic patient 70 years old suffered from two patches of gangrene at the ventral surface of the right thigh and leg. Debridement was performed. Dressing was done using H-G-C.

There is also histological evidence for the anti-inflammatory activity of honey from some of the studies on experimental animals. [28] This indicates that the anti-inflammatory activity of honey is a direct action. [29] Molan [30] stated that the bacterial activity of honey clears infection and protects wounds from becoming infected, and thus it provides a moist healing environment without the risk of bacterial growth occurring. It also rapidly derbies wounds and removes malodor. Its anti-inflammatory activity reduces edema and exudates and prevents or minimizes hypertrophic scarring. It also stimulates the growth of granulation tissue and epithelial tissue so that healing is hastened. Furthermore, it creates a non-adherent interface between the wound and the dressing so that dressings may be easily removed without pain or damage to newly grown tissue. Our results supported these data.

Honey is a supersaturated solution of sugars of which fructose (38%) and glucose (31%) are the main carbohydrates. [31] It was produced by honey bees from the nectar of plants. Some of the components of the honey are added by the bees during the maturation process or are derived from the plants. [32]

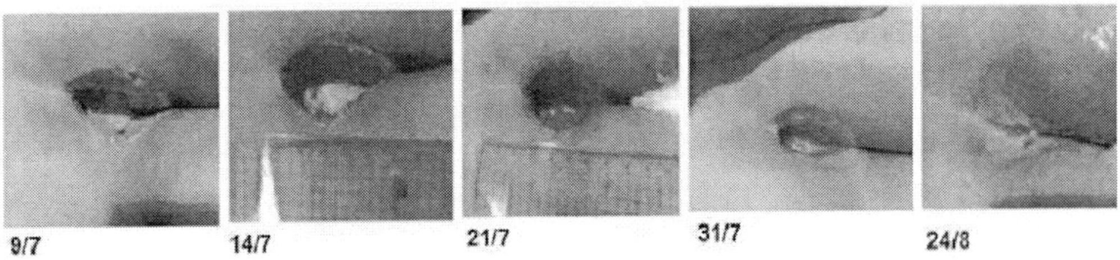

Case 4. Male patient 21 years old suffered from pilonidal sinus. It was excised using the open method. Dressing was done using H-G-C.

A wide range of minor constituents is also present in honey, many of which are known to have antioxidant properties. These include phenolic acids and flavonoids, [33] certain enzymes (glucose oxidase, catalase), [34] organic acids, [35] ascorbic acid, [34] carotenoid-like substances, [36] and amino acids, [37] Frankel [38] also found that honey had significant antioxidant activity. To illustrate the differences in honey phenolics due to the geographical origin (propolis-derived phenolics) and the similarities between floral derived phenolics of monofloral honey samples, it is clear that the profiles are quite different. [33]

The GC/MS and HPLC analysis of Egyptian and Sider honey revealed the similarities in hydroxyacetic acid, 3-hydroxypropanoic acid, malic acid, palmitic acid, 4-hydroxybenzoic acid, 3,4-dimethoxy-cinnamic acid, Caffeic acid, glycerol and phosphoric acid. As well, the two types of honey are similarities in 4-OH benzoic, vanillic, caffeic, p-coumaric acids, and quercitin, quercitin-3-methyl ether. [39] The GC/MS and HPLC analysis of Sider honey contains hydroxyacetic acid, 3-hydroxypropanoic acid, malic acid, palmitic acid, 4-hydroxybenzoic acid, 3,4-dimethoxy-cinnamic acid, Caffeic acid, glycerol and phosphoric acid which have anti-inflammatory effect. [40]

Caffeic acid and its derivatives are used as a component of antimitogenic, anticarcinogenic, anti-inflamatory, and immunomodulator, in biological chemistry because exogenous application of benzoic acid causes salicylic acid accumulation. [41] Salicylic acid-induced pathway is characterized by the production of a cascade of pathogenesis related proteins where a number of biochemical and physiological changes has been associated with pathogen infection. These include the production of antifungal chitinases, glucanases and thaumatins, and oxidative enzymes such as peroxidases, polyphenol oxidases and lipoxygenases. [42] β-pinene was found in significant concentrations in citrus oils (lemon, citron, sweet lime) with antimicrobial activity against Saccharomyces cerevisiae. [43] This terpene was one of the most important components of a honey EO, able to reduce the growth kinetics of many microorganisms. [44] The presence of KYNA in honey may explain its pain-relieving effects reported in the literature in contrast to that reported by Ghadive. [18] The substance, acting in concert with honey flavonoids (COX-2 inhibitors), by antagonizing the NMDA receptor may contribute to the antinociceptive effect of honey. Moreover, kynureninates, owing to their antimicrobial properties, can favour the successful outcome of wounds and burns. [45]

In the present study old ghee was used. The acidity of the whey should be about 0.8 to 0.9 per cent lactic acid in old rather than fresh whey, which will be of less uniform acidity, should be used to give a process which is easier to control and a product which is more uniform in texture. Control of the whey by measurement of the acidity by titration is recommended. [46] Ghee is rich in oleic, linoleic, α-linolenic (n-3), -linolenic (n-6), and α-Tocopherol. Eicosapentanoic (n-3), linoleic (n-6), α-linolenic (n-3), -linolenic (n-6) luric, merystic, palmitic, steirc and arachidonic (n-6) acids, significantly enhances cellular migration in the IEC-6 model of wound healing. [47]

Eicosapentanoate, linoleate, -linolenate, -linolenate and arachidonate, are all capable of improving reconstitution of epithelial integrity following mucosal injury, inhibition of eicosanoid synthesis reduced the enhancement of restitution by n-6 fatty acids back to control levels. [47] Vitamin C and beta-carotine (ascorbic acid, ascorbate) has been shown to enhance neurite outgrowth, promote fibroblast adhesion during wound healing, and reduce xenobiotic-induced leukocyte hyperactivity and inflammatory damage. [48] The present results showed that ghee is rich in vitamin C and beta-carotine.

The wound-healing activity of turmeric has been widely studied and it has been seen that its local application is effective. In Chinese medicine, it has been used for this purpose since ancient times. [49] It has been reported to possess antibacterial, anti-fungal, anti-inflammatory, [50,51] and analgesic activities. [52] The reparation action mechanism of curcumin extracts involved an increase in the levels of beta transforming growth factor plus an increase in the activity of the enzyme nitric oxide synthase. [53] The anti-inflammatory property and the presence of vitamin A & proteins in turmeric result in the early synthesis of collagen fibers by mimicking fibroblastic activity. [54] Curcumin stabilizes lysosomal membranes and causes uncoupling of oxidative phosphorylation. It also possesses strong oxygen radical scavenging activity, which confers anti-inflammatory properties. [55-58]

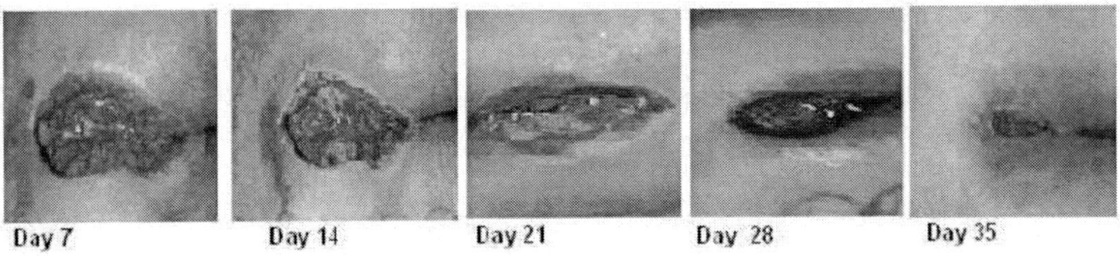

Case 5. Male patient, 18 year-old, suffered from pilonidal sinus. It was excised using the open method. Dressing was done using H-G-C.

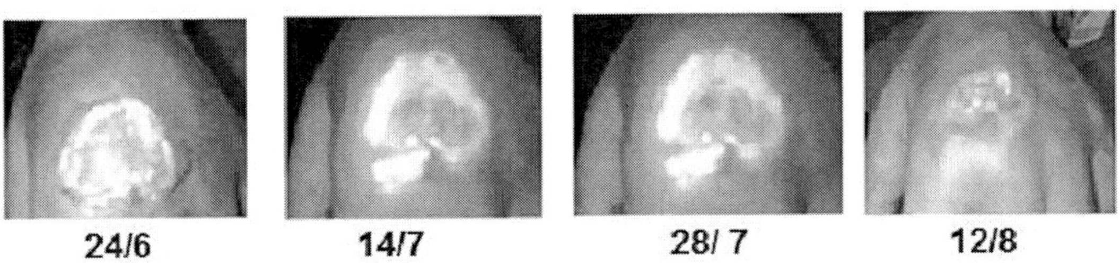

Case 6. 42 year-old female patient with a wound for 5 months resulted from sloughing the skin (4x4 cm) after extravasation of a chemotherapeutic drug which did not respond to conventional treatment. With H-G, the wound healed within 7 weeks.

Curcumin is non-toxic even at high dosages, and has been classified as 'generally recognized as safe' (GRAS) by the National Cancer Institute. [59] Carvacrol is also a powerful antibacterial or antifungal agent based on it biological activity [60] together with p-cymene. [61] Probably aromatic turmerone alone, or in synergy with turmerone, curlone, is responsible for the highest antioxidant and antimutagenic activities of curcuma longa. [62]

Also, the present data agree with the previous studies that established the ability of curcumin and related compounds that inhibit free radical generation and act as free radical scavengers and antioxidants, [63] inhibiting lipid peroxidation and oxidative DNA damage. [64] Curcuminoids induce glutathione S-transferase and are potent inhibitors of cytochrome P450. [65] The best-known antioxidant mechanism of curcuma and its components is their capacity to eliminate reactive oxygen species such as the hydroxyl radical, [63] superoxide radical, [66] and singlet oxygen. [67]

CONCLUSION

In conclusion, this method of wound dressing has proved to be an effective method encouraging healing. Its anti-microbial effect clears infection and protects wounds from becoming infected, and thus provides a moist healing environment without the risk of bacterial growth. It rapidly debrides wounds and remove malodour. Its anti-inflammatory activity reduces edema and exudates and minimizes hypertrophic scarring. It also stimulates the growth of granulation tissue and epithelial tissue so that healing is hastened. From this study, the data support the concept that honey is bioavailable and it can accelerate the wound healing together with ghee and Curcuma longa.

Acknowledgment

Thanks are due to Dr Mahmoud Mohamad El-Merzabani, Prof of Pharmacology, National Cancer Institute, for continuous support.

References

[1] Gethin, G. Is there enough clinical evidence to use honey to manage wounds? *J Wound Care* 2004;13(7):275-8.
[2] Roth, S. *Psychology: the art of wooing nature*. Jason Aronson, Northwale, N.J., 1990.
[3] Office of Complementary Medicines. *Honey scientific report* [Online]; 1998. Retrieved from http://wwwtgagovau/docs/pdf/cmec/ honeysrpdf.
[4] Fox, C. Honey as a dressing for chronic wounds in adults. *Br J Community Nurs* 2002;7(10):530-4.
[5] Moore, OA; Smith, LA; Campbell, F. Systematic review of the use of honey as a wound dressing. *BMC Complement Altern Med* 2001;1(1):2.
[6] Mwipatayi, BP; Angel, D; Norrish, J. The use of honey in chronic leg ulcers: a literature review. *Primary Intention* 2004;12(3):107-12.
[7] Vermeulen, H; Ubbink, DT; Goossens A. :Systematic review of dressings and topical agents for surgical wounds healing by secondary intention. *Br J Surg* 2005; 92 (6):665-72.)(7).
[8] Bouza, C; Saz, Z; Muñoz, A et al. Efficacy of advanced dressings in the treatment of pressure ulcers: a systematic review. *J Wound Care* 2005;14(5):193-9.
[9] Ferreres, F; Garcia-Viguera, C; Tomás-Lorente, F; Tomás-Barberán, FA. Hespertin, a marker of the floral origin of citrus honey. *J. Sci. Food Agric.,* (1993) 61, 121-123.
[10] Ogan, K. and Katz E. Liquid chromatographic separation of alkyl phenols with fluorescence and ultra violet detection. *Anal. Chem.* (1981): 53: 160-163.
[11] Annual book of ASTM Standards, part 11.03" *Atmospheric analysis" American Scosity for testing and materials,* Philadelphia, Pennsylvania(1983).
[12] Ramadan, MF; Kroh, LW and Moersel, JT. Radical scavenging activity of black cumin (Nigella sativa L.), coriander (Coriandrum sativum L.) and niger (Guizotia abyssinica cass.) crude seed oils and fractions. *J. of Agric. and Food Chem.* 2003, 51 (24): 6961-6969.
[13] Gorban, A. M. S., and O. M. Izzeldin. 2001. Fatty acids and lipids of camel milk and colostrum. *Int. J. Food Sci. Nutr.* 52:283–287..
[14] Ghasemi J. and Nayebi S. 2006: Application and comparison of two-rate and differential kinetic methods in simultaneous spectrophotometric determination of

ascorbic acid and L-cysteine in pharmaceutical samples Khimiko-Farmatsevticheskii Zhurnal, Vol. 40, No. 1, pp. 41–48.

[15] Adams, RP. (2001). Identification of essential oil components by gaschromatography/quadrupole mass spectrometry. Carol Stream IL, USA: Allured. USA. *Chem.*, 34: 823-826.

[16] Deters, M; Siegers, C; H.nsel, W; Schneider, K and Hennighausen, G. Influence of curcumin on cyclosporin-induced reduction of biliary bilirubin and cholesterol excretion and on biliary excretion of cyclosporin and and its Metabolites *Planta Med* 2000; 66: 429-434.

[17] Saunders DB and Trapp GR. *Basic and clinical biostatistics*, 3nd edition, Connecticut, Appleton & Lang(2001).

[18] Gadhvi NP. *An ancient dressing revisited. Digest* 1995,Vol21,No9 pg53-55.

[19] Tonks A, Cooper RA, Price AJ, et al. Stimulation of TNF-a release in monocytes by honey. *Cytokine* 2001;14(4):240-2.

[20] Tonks AJ, Cooper RA, Jones KP, et al. Honey stimulates inflammatory cytokine production from monocytes. *Cytokine* 2003;21(5):242-7.

[21] Abuharfeil N, Al-Oran R, Abo-Shehada M. The effect of bee honey on proliferative activity of human B- and T-lymphocytes and the activity of phagocytes. *Food Agric Immunol* 1999;11:169-77.

[22] Al-Waili NS, Haq A. Effect of honey on antibody production against thymus-dependent and thymus-independent antigens in primary and secondary immune responses. *J Med Food 2004*; 7(4):491-4.

[23] Molan PC. The antibacterial activity of honey: 1. The nature of the antibacterial activity. *Bee World* 1992;73(1):5-28.

[24] Molan PC, Betts J. Using honey dressings: the practical considerations.*Nurs Times* 2000;96(49):36-7.

[25] Subrahmanyam M. A prospective randomised clinical and histological study of superficial burnwound healingwith honey and silver sulfadiazine. *Burns 1998;* 24(2): 157-61.

[26] Subrahmanyam M, Sahapure AG, Nagane NS, et al. Effects of topical application of honey on burn wound healing. *Ann Burns Fire Disasters* 2001;14(3):143-5.

[27] Subrahmanyam M, Shahapure AG, Nagane NS, et al. Free radical control—the main mechanism of the action of honey in burns. *Ann Burns Fire Disasters* 2003;16(3):135-8.

[28] Bilsel Y; Bugra D; Yamaner; S, et al. Could honey have a place in colitis therapy? Effects of honey, prednisolone, and disulfiram on inflammation, nitric oxide, and free radical formation. *Dig Surg* 2002;19:306-12.

[29] Church; J, Honey as a source of the anti-stiffness factor. *Fed Proc Am Physiol Soc* 1954;13(1):26.

[30] Molan: Honey As *A Wound Dressing The International Journal of Lower Extremity Wounds2006*,5(1)p.40-54.

[31] Tomás-Barberán FA; Martos, I; Ferreres, F; Radovic, BS; Anklam, E)" HPLC flavonoid profile as markers for the botanical origin of European unifloral honeys."" *J.Sci..Food Agric.,* . 2001,81, 485-496.

[32] Lusby, PE; Coombes, A; Wilkinson, JM. Honey: a potent agent for wound healing? *J Wound Ostomy Continence Nurs* 2002; 29: 295-300

[33] Regnault-Roger C; Hadidan, R; Biard, JF and Boukef K. (1987) " High performance liquid and thin-layer chromatographic determination of phenolic acids in palm(Phoenix dactilifera) products." *Food Chemistry*, 25(1), 61-71. (36) and

[34] Cherchi, A; Spanedda, L; Tuberoso, C; Cabras, P. (1994)" Solid-phase extraction and HPLC determination of organic acids in honey."*J. Chromatogr.*, 669, 59-64. (33),

[35] White. JW (1975): *Composition of honey. A Comprehensive survey*; Crane, E. Ed. Carne, Russak & Company; New York, pp 157-206 .)(32),

[36] Tan, ST; Wilkins, AL; Holland, PT; and McGhie,TK.(1989): Extractives from New Zealand unifloral honeys. 2. Degraded carotenoids and other substances from Heather Honey. *J. Agric. Food Chem.* 37, 1217-1221

[37] 3White, JW; Rudyj, ON. (1978) " The protein content of honey. *J. Apic. Res.* 17,234-238.)(34).

[38] Frankel, S; Robinson, G.E; Berenbaum, M.R;." Antioxidant content and correlated characteristics of 14 monofloral honeys." *J. Apic. Res.*, 1998 37, 27-31.

[39] Ziouti, A; El-Bboustani, E; Macheix JJ; Geibel M; Treutter, D. and Feuch W " Cell wall- bound phenols in date palm leaves and roots." *Acta- Horticulturae,* 1994,381, 276-279.

[40] Saija; A, Tomaino; A, Trombetta; D, In vitro and in vivo evaluation of caffeic and ferulic acids as topical photoprotective agents. *Int J Pharm* 2000; 199: 39-47.

[41] Yalpani, N; Leon, J; Lawton, M; Raskin, I. Pathway of salicylic acid biosynthesis in healthy and virus-inoculated tobacco. *Plant Physiol.*, 1993, 103, 315-321.

[42] Soylu, S, Bennett, MH; Mansfiels, J.W.. Induction of phytoalexin accumulation in broad bean (Vicia faba L.) cotyledons following treatments with biotic and abiotic elicitors. *Turk. J. Agric.*, 2002,26, 343-348.

[43] Belletti N, Ndagijimana M, Sisto C, Guerzoni ME, Lanciotti R, Gardini F Evaluation of the antimicrobial activity of citrus essences on Saccharomyces cerevisiae. *J. Agricult. Food Chem.* 2004. 52: 6932-6938.

[44] Tepe B; Daferera D, Sokmen, A, Sokmen M, Polissiou M Antimicrobial and antioxidant activities of the essential oil and various extracts of Salvia tomentosa Miller (Lamiaceae). *Food Chem.* 2005, 90:333-340.

[45] Beretta Giangiacomo; Caneva Enrico; Facino Roberto Maffei Kynurenic acid in honey from arboreal plants: MS and NMR evidence. *Planta medica* 2007; 73(15): 1592-5.

[46] Nembang, L.B.: A report on traditional milk products in the Himalayan area. *FAO Manuscript.* (1989).

[47] Derek J; Ruthig and Kelly A. Meckling-Gill: Both (n-3) and (n-6) Fatty Acids Stimulate Wound Healing in the Rat Intestinal Epithelial Cell Line, IEC-6. *Journal of Nutrition.* 1999; 129: 1791-1798.

[48] Benjamin Weeks, Pedro Perez: A novel vitamin C preparation enhances neurite formation and fibroblast adhesion and reduces xenobiotic-induced T-cell hyperactivation. *Med Sci Monit.* 2007 26;13 (3).

[49] Srimal RC. Turmeric: a brief review of medicinal properties. *Fitoterapia,* 1997; 68 (6): 483-94 (ref.2503).

[50] Chopra, RN; Nayar, SL; Chopra, IC (1986). Glossary of Indian Medicinal Plants. *CSIR*, New Delhi.

[51] Rao, SGV; Selvaraj, J; Senthil, R;Radhakrishnan, RN; Murali, Manokar B (2003). Efficacy of some indigenous medicines in wound healing in rats. *Indian Journal of Animal Sciences,* 73: 652-653.

[52] Srimal, RC; Khanna, NM; Dhawan, BN (1971) A preliminary report on ntiinflammatory activity of Cucurmin. *Indian J. Pharmacol*, 3: 10.

[53] Alonso, Jorge. Tratado de Fitofármacos y Nutracéuticos. *Barcelona: Corpus*, 2004; p: 395-403 (633.8 ALO).

[54] Kumar, AS; Singh, HP; Prem Parkash; Singh,SP. Efficacy of some indigenous drugs in tissue repair in buffaloes. *Indian Vet. J*, 1993; 70:42-44

[55] Kohli K, Ali J, Ansari MJ, Raheman Z: Curcumin: a natural antiinflammatory agent. *Indian J Pharmacol* 2005, 37:141-147.

[56] Kunchandy E, Rao M: Oxygen radical scavenging activity of curcumin. *Int J Pharm* 1990, 58:237-240.

[57] Toennesen HH: Studies on curcumin and curcuminoids. XIII. Catalytic effect of curcumin on the peroxidation of linoleic acid by 15-lipoxygenase. *Int J Pharm* 1989, 50:67-69.

[58] Ahsan H, Parveen N, Khan NU, Hadi SM: Pro-oxidant, anti-oxidant and cleavage activities on DNA of curcumin and its derivatives demethoxycurcumin and bisdemethoxycurcumin. *Chem Biol Interact* 1999, 121:161-175.

[59] Osawa T, Sugiyama Y, Inayoshi M, Kawakishi S: Antioxidative activity of tetrahydrocurcuminoids. *Biosci Biotechnol Biochem* 1995, 59:1609-1612.

[60] Knowles JR, Roller S, Murray DB; Naidu AS. Antimicrobial Action of Carvacrol at Different Stages of Dual- Species Biofilm Development by Staphylococcus aureus and Salmonella enterica Serovar Typhimurium *Appl environ microbiol.* 2005; 71(2): 797-803.

[61] Gabriella Kiskó and Sibel Roller :Carvacrol and p-cymene inactivate Escherichia coli O157:H7 in apple juice. *BMC Microbiology* 2005, 5:36doi:10.1186/1471-2180-5-36.

[62] Guddadarangavvanahally K Jayaprakasha; Bhabani S Jena; Pradeep S Negi and Kunnumpurath K Sakariah. Evaluation of Antioxidant Activities and Antimutagenicity of Turmeric Oil: A Byproduct from Curcumin Production. Z. *Naturforsch.* 57c, 828D835 (2002).

[63] Sreejayan N; Rao MN. Free radical scavenging activity of curcuminoids. *Arzneimittelforschung*, 1996, 46(2): 169-171

[64] Sreejayan; MN Rao: Curcuminoids as potent inhibitors of lipid peroxidation. The *Journal of pharmacy and pharmacology* Volume: 46 ISSN: 0022-3573 1995-11-15.

[65] Shukla PK, Khanna VK, Khan MY, Srimal RC.Protective effect of curcumin against lead neurotoxicity in rat. *Hum Exp Toxicol.* 2003 Dec;22(12):653-658.

[66] Rao CV, Rivenson A, Simi B, Reddy BS: Chemoprevention of colon carcinogenesis by dietary curcumin, a naturally occurring plant phenolic compound. *Cancer Res,* 1995, 55(2):259-266.

[67] Kumuda C Das, and Chandan K Das. Curcumin (diferuloylmethane), a singlet oxygen (1O2) quencher *Biochemical and Biophysical Research Communications* Vol. 295, Issue 1, 2002, Pages 62-66.

In: Asia Pacific Life Sciences Compendium
Editor: Bandit Chumworathayi

ISBN: 978-1-62257-338-7

Chapter 12

NEURO-ACTIVITY EFFECTS OF PHOENIX DACTYLIFERA FRUIT AND PITS EXTRACT IN CHRONIC STRESSED RAT MODELS

Ragaa Hosny Mohamad,[1,*] Amal Mohamad El-Bastawesy,[2] Mohamad Gamil Abdel Monaem,[3] Zekry Khalid Zekry,[4] Hussien Abdel-Rahman Al-Mehdar,[5] Amany A El-Bassmy,[6] Magdy Al-Masry,[7] Sabry Mohamad Sharawy[1] and Mahmoud Mohamad El-Merzabani[8]

[1]Cancer Biology Department, National Cancer Institute, Cairo University, Egypt
[2]Food Technology Research Institute, Cairo University, Egypt
[3]Surgical Oncology Department, National Cancer Institute, Cairo University, Egypt
[4]Medical Oncology Department, National Cancer Institute, Cairo University, Egypt
[5]Faculty of Science, King Abdul Aziz University, Saudi Arabia
[6]Biostatistician, National Cancer Institute
[7]Psychiatry Department, Faculty of Medicine, Cairo University, Egypt
[8]Pharmacist, National Cancer Institute, Egypt

ABSTRACT

Objective: The present study aimed to investigate the neuroactivity effects of the methanolic extract of Phoenix dactylifera (MEPC) fruits and pits compared with fluoxetine. Method: MEPC of PC fruits and pits functional constituents was assayed. The levels of IL4, TNFα, IFNγ, lipid profile, and MDA in serum and 5HT, 5HIAA, IL1β, TNFα, IFNγ, were assayed in brain tissues of rats, the animal behavior, NK and LAK cell count were studied. Result: 5HT and 5HIAA levels in brain tissues were diminished in MEPC of fruit treated group than the other groups. The levels of the IL4, TNFα, IFNγ, in serum and IL1β, TNFα, in brain tissues were improved in MEPC fruit treated groups than the other groups. Moreover, the sucrose consumption was improved and SOD activities

[*]Tel: +020101856864, Fax: +0223644720, Email: ragaatahtawy@gmail.com

in liver tissues were significantly increased in group treated with PC fruit extract than other groups, and decreased MDA and NO levels lead to repair liver injury. Conclusion: MEPC of fruit and pits have effectiveness on inhibition of the selective serotonin reuptake, prevent the hyperactivity of the immune system, oxidative stress profile, lipid profile, and improved the sucrose consumption, which play a role in improving the etiology of depression.

Keywords: Phoenix dactylifera, pits, fluoxetine, chronic mild stress, depression

INTRODUCTION

In The Middle East and India, there is a traditional belief that Phoenix dactylifera (PC) fruits and pits have been traditionally claimed for use in treatment of on the depression and stress, also, believed to aid in control of serotonin pathway. Carlini [1] performed a study focused to use herbal medicines which recognized as active on the central nervous system, and they have at least a hypothetical potential to affect chronic conditions such as anxiety, depression, headaches or epilepsy, that do not respond well to conventional treatments. The Phoenix dactylifera is a member of the family Arecaceae. [2] Its fruit is a nutritious source of sugar, minerals and vitamins, phenolic compound, and flavonoid. [3] The various parts of this plant are widely used in traditional medicine for the treatment of various disorders, which include memory disturbances, fever, loss of consciousness, nervous disorders. [4,5] Al-Qarawi et al [6] stated that the Phoenix dactylifera extract have nephro-protective activity. Acute, as well as long term exposure to noise, can produce excessive free radicals such as superoxide dismutase. [7] The oxidative mechanisms may play a role in psychiatric disorder. [8] Bilici et al [9] reported that the major depression, especially melancholia, is associated with elevated antioxidative enzyme activities and lipid peroxidation. Karolewicz et al [10] demonstrated that NOS inhibitors could alter the neurotransmitter utilization in vivo and have antidepressant and anxiolytic-like activities in rodents. These impairments could be reversed by chronic administration of antidepressants such as selective serotonin reuptake inhibitors and monoamine oxidase inhibitors. [11] Breggin [12] stated that the serotonin (5-hydroxytryptamine, 5-HT) is an important brain chemical that acts as a neurotransmitter to communicate information among nerve cells, 5-HT are rapidly metabolized and inactivated by monoamine oxidase and aldehyde dehydrogenase to the major metabolite by the liver, and excreted by the kidneys. In addition, Griebel et al [13] proved that the 5-hydroxyindoleacetic acid (5-HIAA) is commonly decreased in animals and in patients experiencing stress and depression, suggesting a dysfunction of the 5-hydroxytryptaminergic system. Dysregulation in the levels of lymphocyte activities and its immune mediators, such as IL1β, IL6, IL12, IL6R, IL2, IL2R, IL1Ra, and IFNγ, leads to imbalance between individual pro- and anti-inflammatory cytokines also imbalance in the Th1 and Th2 cause major depression. [14] Depression is associated with down regulated in the cytokines levels, natural killer cells and lymphokine-activated killer cells activities [15] with present changes in serum lipid profile, including A high triglyceride level and decreased HDLc serum levels were associated with depression. [16] Moreover, major depression patients who have lowered 5-HT level were also associated with lowered levels in lipid profile. [17] In the present study, we therefore decided to evaluate the methanolic extract of Phoenix dactylifera fruits (MEPDF) and its pits

(MEPDP) from type name called Agwa, it is cultivate in Al-madenah Al-monawarh- Saudi Arabia to protect neurons against oxidative injury and prevent the depressive that do not respond well to conventional treatments for verification of a traditional ethnomedical practice.

METHOD

Drugs and Chemicals

Fluoxetine (Prozac), 20 mg capsule, is a serotonin reuptake inhibitor (SSRI) that helps patients with depression. It is purchased from Eli Lilly Company, USA. The o-phthalaldehyde (OPT), disodium ethylenediamine tetraacetate dihydrate (EDTA), sodium lactate, lymphocyte separation medium, Dulbecco's Modified Eagle Medium (DMEM Medium), Cat. No. 12-604, and fetal bovine serum (FBS), were obtained from Sigma Co., USA. Nonidet P40 (NP-40) was purchased from Fluka (Switzerland). Nitro-blue tetrazolium chloride (NBT) was obtained from Fluka, Switzerland. Nicotinamide adenine dinucleotide (NAD+) and phenazine methosulfate (PMS) were from Sigma Co., USA. Lactate dehydrogenase (LDH) substrate includes the following compositions: nitroblue tetrazolium (NBT) reduction, nicotinamide adenine dinucleotide (NAD) catalyst, phenazine methosulfate (PMS), and sodium lactate. 5-HT, 5-HIAA MP in brain tissues were measured using an ELISA method following the manufacturer's instructions (Biomedicals, USA). IL-1β, TNFα IL-4, and IFNγ were also measured using an ELISA method following another manufacturer's instructions (RandD System, UK). Serum concentrations were expressed as pg/g protein. The assay kits for MDA, NOS, and cholesterol, LDL, HDL, and triglyceride lipid profile, were purchased from Biosources Company. All other reagents used were of analytical grade.

Plant Materials

Fresh fruits of Phoenix dactylifera (Agwa) and its pits, at the "rutabe stage", were donated from Saudi Arabia. The extract from fruit and pits, were separated from 50 g of date fruit and its pits, the pits were soaked in water, washed to free them of any adhering date flesh, air dried and then dried (12 h) at about 50°C. Date pits, from the same date type was separately milled in a heavy-duty grinder to pass 1-2 mm screens and then kept at -20°C. The methanolic extract of dried fruit and its pits was prepared in the approved laboratories of National Research Center, Cairo, Egypt, using the procedure mentioned below.

1. Lipid analysis in extract: Lipid extraction was carried out as described by Besbes et al [18], with an SER 148 Solvent Extractor (Sigma Co., USA) equipped with six Soxhlet posts.
2. Evaluation of phenolic compounds: Qualitative evaluation of phenolic compounds was carried out by using a Jasco PU 980 HPLC system (Jasco Ltd., Essex, UK) equipped with a carbon chain grafted silica gel C18 column (25±0.46 cm i.d., particle size 5 mm; Waters S.A.S., St. Quentin en-Y velines, France). The total phenolics,

expressed as caffeic acid equivalents (mg/kg of oil), were determined at 725 nm by using Folin-Ciocalteau reagent.

3. Analysis of sterols: Sterols were evaluated in PC Fruit and pits according to the AOCS [19] method. The solvent used for sterols elution was hexane/diethyl ether (70:30, v/v). After drying, the spots were visualized by using 2,7-dichlorofluorescein (0.2% in pure ethanol). Sterol band was identified under UV light by comparing its Rf value with that of authentic cholesterol.
4. Determination of tocopherols: Tocopherols were evaluated in PC fruit and pits according to the AOCS [19] method. The solvent used for tocopherol elution was hexane/2-propanol (98.5:1.5, v/v) with a constant flow rate of 1 mL/min.
5. Analysis of Amino acid: Five milligram of the powdered PC pit and fruit sample was weighed and placed in 2ml ampoules, to which the internal standard (norleucine) and 0.45 ml of 6N HCl were added. Norleucine was used as internal standard because it is an amino acid not commonly found in proteins. The ampoules were evacuated, sealed and placed in an oven for 24 hrs at 110 °C. Amino acid composition was analyzed by ion exchange chromatography following the method decided by Lucas and Sotelo[20], PC pit and fruit samples were hydrolyzed with HCl (6.0 mol/l) at 145°C for 4h. At the end of hydrolysis, the hydrolysate was adjusted to pH 6.8. According to Lucas and Sotelo [20], 1ml hydrolysate was diluted with a 1ml buffer pH 1.3, and from this dilution a 100μl aliquot was injected on a Technicon amino acid analyzer model NC2P (Technicon International, Geneva). Tryptophan was isolated by enzymatic hydrolysis with pepsin and panchreatine reaction for 24h. A hydrolysate aliquot was made to react with p-dimethylamino –benzldehyde and the product was quantified with a spectrophotometer at 590 nm and standard Tryptophan curve (Lucas and Sotelo, [21]; Ras et al., [22]).

Animals

320 albino rats weighing 80-120 gm were used. They were purchased from the Animal House in National Cancer Institute, Cairo University. They were caged in a room under standard laboratory conditions (temperature 25 °C, relative humidity 55±5% and lighting 6 hours).

The animals were fed on standard diet and water ad libitum. They were maintained at least 12 h light/dark cycle and standard condition before the start of the experiment, and were tested in the light phase. All studies were carried out in accordance with the guidelines provided by The National Cancer Institute, Cairo University, Egypt, and the Animal Ethical Committee approved for the purpose of control and supervision of experiments on animals.

Experimental Protocols for Preparation Depression Model

Preparation of the rat depression model treated with PC fruit and pits extract, saline, or flextime, was done. A rat model of chronic stress-induced depression was established according to methods described by Grønli et al [23] and Willner. [24] The rats received a variety of stressors for 21 day, described in Figure 1.

Sucrose Intake Procedure

The sucrose intake procedure performed here mainly referred to Muscat and Willner [25] with some modifications. After stresses procedure all stressed rat was housed in one cage and fasted before sucrose intake testing, after which sucrose 1% 10 g/L sucrose solution were consummated in 48 hours the exposure to 1% (w/v) sucrose 1 hour at base-line tests (twice weekly, for 21 day, following 23 hour period of food and water deprivation). The intake sucrose consumption was monitored by weighing pre-using bottles at the end of the test at twice weekly intervals for the next 6 weeks.

Day	food deprivation	water deprivation	cage tilt(45°)	Paired Hosing	intermittent illumination	soiled cage	low intensity stroboscopic	Whit Noise
Mon							8:00↓ 10:00	
Tuesday								8:00 ↓ 10:00
Wesday	08:00 ↓ 15:00	08:00 ↓ 15:00						
Tuesday			08:00 10:00 ↓					
Friday				08:00↓ 10:00				
Saturday					20:00 7:00 ↓			
Sunday						17:00 ↓ 4:00		

Figure 1. Procedure of chronic mild stress.

Drug Treatment Protocol

The stressed rats: The animals were randomly divided into four groups (40 rats per group): model control, saline+chronic stress-induced, PC pits extract orally (500 mg/kg)+chronic stress-induced, PC fruit extract orally (500 mg/ kg)+chronic stress-induced, fluoxetine 10 mg/kg twice daily orally+chronic stress-induced.

Control healthy rats: divided in to 4 groups, vehicle (0.9% saline solution), PC pits extract (500 mg/kg), PC fruit extract (500 mg/ kg), fluoxetine (20mg/kg) twice daily. Drugs and extract (0.9% saline) were given orally once a day at 8:00-9:00 a.m. from the beginning of stress regime.

Experimental design

The Animals were divided in to eight groups 40 animals for each it's described below:

1. 40 rats used as a control group received 0.9% saline solution (vehicle).
2. 40 rats used as a control group which received 500 mg/kg PC pits rutabe extract..
3. 40 rats used as a control group which received 500 mg/kg PC fruit rutabe extract
4. 40 rats as a control group which received 10 mg/kg fluoxetine twice daily.
5. 40 rats used as a stressed group which received 0.9% saline solution.

6. 40 rats used as a stressed group which received500 mg/kg PC pits rutabe extract.
7. 40 rats used as a stressed group which received 500 mg/kg PC fruit rutabe extract.
8. 40 rats used as a stressed group which received 20 mg/kg fluoxetine twice daily.

Each group included 40 rats. Drugs and extract (including distilled water) were given orally once a day at 8:00–9:00 a.m. from the beginning of stress regime. All drugs were either dissolved or suspended in a 0.9% saline solution.

Blood Sampling and Tissue Collection

After the end of stress procedure and the last sucrose intake test, all animals were left without any treatment until the following morning, the rats were decapitated. Blood was sampled, centrifuged (860 g, 20 min) at 4°C and then stored at -80°C for IL-4, TNF-α, IFN-γ, lipid profile, and MDA assays. Whole spleen, brain and liver were quickly removed on an ice-plate, and washed, with cold 0.9% saline solution. The liver was frozen in liquid nitrogen and stored at -80oC for subsequent SOD and NOS activity assays. The prefrontal cortex, hippocampus, hypothalamus were carefully dissected from brain for 5-HT and 5-HIAA, IL-1β and TNF-α assays. The spleen was quickly removed under sterile conditions immediately for the assays of NK cell and LAK cell activities. 5-HT and 5-HIAA assays: Brain region tissue was homogenized in 20 volumes of 1 N formic acid/acetone (15/85, v/v) according to the method of Smith et al. [26] The samples were reconstituted in the enzyme immunoassay buffer phosphate buffer with BSA and sodium azide for analyses of 5-HT levels. The samples were mixed with 4 ml of n-heptane and 5 ml of 0.1 M HC for analyses of 5-HIAA, according to Curzon and Green's method [27] Following the protocol provided with the ELISA kit experiments were done in duplicate. NK cell and LAK cell activities: Splenic NK cell and LAK cell activities were measured by the LDH release assay as previously described by Korzeniewski and Callewaert, [28] and Wu et al. [29] The NK cell and LAK cell activities were expressed as percentage specific lysis calculated by the following formula:

%Cytotoxicity = ¼ ExMxSx100

where E is the experimental release of LDH from the target cells incubated in the presence of spleen cells. M is the maximum release of LDH determined by lysing the target cells with 1% NP-40. S is the spontaneous release of LDH from the target cells incubated in the absence of spleen cells.

Cytokine analysis in serum and brain regions-linked immuno-sorbent assay: The serum samples cytokine IL-4, TNFα, and IFNγ levels were analyzed by using an ELISA method. The brain regions samples were thawed in an ice bath and sonicated for 20 seconds in cold sterile potassium persulfate (PBS) (pH 7.4-7.6) with protease inhibitor cocktail (Roche). The sonicated samples were centrifuged and the supernatant diluted with sterile PBS according to the methods described by Cai et al [30] and O'Connor et al [31] then the cytokine IL-1β, TNFα concentration were estimated in all brain regions samples and were expressed as pg/g protein and the optical density value was read at 450 nm in ELISA plate reader. SOD and NOS assays: Rat livers were homogenized in 9 volumes of ice-cold 0.9% saline solution using a motor-driven Teflon glass homogenizer. The homogenate centrifuged for 10 min at

4000 rpm at 4oC. The supernatant obtained was used for assays of SOD and NOS activities using the pyrogallol autoxidation method and a commercial NOS kit, respectively. The optical density was read on spectrophotometer (Model U-3000). The amount of SOD that inhibits 50% the rate of reduction under the specified conditions was regarded as one enzyme unit. [32] MDA assay: The serum MDA level was spectrophotometrically measured using a commercial NOS kit by the absorbance of a red-colored product with thiobarbituric acid method described by Bilici et al. [9] Lipid profile assay: TC and TG concentrations were measured enzymatically using commercially available kits. The assays were performed in accordance with the manufacturer's instructions. HDLc was measured according to a previously described method by Allain et al. [33] LDLc was determinated according to a previously described method by Assmann et al. [34]

Statistical Analyses

Values of different parameters were expressed as the mean±standard deviation (SD). The significance of the difference was statistically evaluated using one-way analysis of variance (ANOVA), followed by Bonferroni alpha test to assess the significance of differences between individual groups. P>0.05 was considered in-significance. [35]

Result

Table 1 revealed that the separation of components of unsaponifiable matter of Phoenix dactylifera (PC) fruit and pits extracted by TLC. The sterol profile fraction of the unsaponifiable matter was studied. This has been considered as the major unsaponifiable fraction in many oils. The order under GC conditions used was as follows: cholesterol, campesterol, stigmasterol, b-sitosterol, D5-avenasterol, D5.2,4-stigmastadienol.

Table 1. Separation of components of unsaponifiable matter of Phoenix dactylifera L by TLC

Types	No. of Fractions	Rf. Values	Components
Date fruit	1	0.08	Unidentified
	2	0.15	Unidentified
	3	0.28	Cholesterol
	4	0.39	B-sitosterol
	5	0.53	Tocopherols
	6	0.62	Unidentified
	7	0.96	Non of sterols and triterpenes
Date pits	2	0.29	Cholesterol
	3	0.40	B-sitosterol
	4	0.57	Tocopherols
	5	0.99	Non of sterols and triterpenes

Sterol composition in PC fruit and pits extract oils are presented in Table 2. High levels of sterols were present in both oils, which made up 3500 mg/kg date extract and 3000 mg/kg date seed oil. The latter was characterized by slightly lower amounts of unsaponifiables (7.76 g/kg of total lipids against 8.92 g/kg total lipids). For both oils, sterol fraction exceeded 38% of total unsaponifiables. In PC pits extract, β-sitosterol and campesterol were among the major components, together constituting about 90% of total sterols. In the two studied date seed oils, the sterol marker was b-sitosterol, which constituted about 83.31 and 78.66% of the total sterol content in PC fruit and its pits extract, respectively. In date seed oils, cholesterol was found at about 0.90% of the total sterols for PC fruit, and about 0.50% for PC pits extracts variety. The next major component, i.e., campesterol, constituted 9.10-10.19% of total sterols in date seed oil.

Table 2. Separation of (%) of Phenolic, Sterol, and Tocopherol Composition of Phoenix dactylifera L Fruit and Pits

Peak	Component	Fruit Rutab*	Pits Rutabe*
Phenol Composition (%)			
1	Cholesterol	0.96 a A ± 0.08	0.58a B ± 0.05
2	Campesterol	9.10b A ± 0.20	10.19b B ± 0.15
3	Stigmasterol	2.42c A ± 0.05	2.29c B ± 0.03
4	b-sitosterol	83.31 d A±1.38	78.66 d B ± 1.16
5	D5-avenasterol	0.45 e A ± .15	4.50 e B ± 0.08
6	D5.2,4-stigmastadienol	0.23 e A ±0.01	0.41 f B ± 0.01
	Total sterols†	3.50 A ± 0.60	3.00 A ± 0.20
	Total unsaponifiables†	8.92 A ± 0.39	7.76 B ± 0.34
Sterol Composition (%)			
1	NI	12.50± 1.22	8.60 ± 0.82
2	Hydroxytyrosol	10.22 ± 0.41	6.94 ± 0.23
3	Gallic acid	4.11 ± 0.59	2.48 ± 0.35
4	Protocatechuic acid	9.62± 1.32	4.26 ± 0.64
5	3,4-dihydroxyphenylacetic acid	–	1.56 ± 0.06
6	Tyrosol	4.50 ± 0.16	8.10 ± 0.50
7	Caffeic acid	1.30 ± 0.11	4.95 ± 0.08
8	NI	43.16 ± 0.54	54.80 ± 0.82
9	p-coumaric acid	0.26 ± 0.08	0.22 ± 0.06
10	Oleuropein	0.18± 0.02	0.11 ± 0.03
	Total phenols†	520.81± 10.22	220.32 ± 6.30
Tocopherol Composition (%)			
1	α-tocopherol	24.97aA ± 1.40	38.85aB ± 1.80
2	γ-tocopherol	3.76bA ± 0.12	5.40bB ± 0.10
3	δ-tocopherol	1.22cA ± 0.04	2.40cB ± 0.06
4	Total unidentified peaks	70.05A ± 1.56	53.35B ± 1.96

Moreover, the relative percentage of D5-avenasterol was higher in pits extracts than in PC fruit extract (4.50 against 0.45% of total sterols). PC fruit and pits, contain a high level of β-sitosterol (2360-2916 mg/kg), compared to olive oil, and have a high level of b-sitosterol

(683-2610 mg/kg), D5-avenasterol (34-266 mg/kg), D7-stigmastanol (150-500 mg/kg). HPLC chromatogram for phenolic compound separation from PC fruit and pits extract, and the present data in Table 2, revealed that there are seven phenolic compounds identified in PC fruit extract. On the other hand, eight phenolic compounds were identified in PC pits extract.

The same phenolic compounds were found in both PC fruit and pits extract, except for 3,4-dihydroxyphenylacetic acid, which was found only in PC pits extract. For PC fruit extract, phenolics identified were hydroxytyrosol, with the highest relative content, followed by protocatechuic acid, tyrosol, gallic acid, caffeic acid, to a lesser extent, p-coumaric acid and oleuropein. However, for pits extract, tyrosol was present at the highest percentage, followed by hydroxytyrosol, caffeic acid, protocatechuic acid, gallic acid, 3,4-dihydroxyphenylacetic acid and, to a lesser extent, p-coumaric acid and oleuropein.

All values given are mean of triplicate results ± standard deviation. NI; not identified. *Means in each column sharing a common lowercase letter are not significantly different at $P<0.05$. Means within the same row sharing a common capital letter are not significantly ($P<0.05$) different. † in mg/kg of pits seed oil expressed as caffeic acid in Phenol Composition (%), g/kg of Pits oil in Sterol Composition (%) and Tocopherol Composition in g/kg of Fruit and Pits oil.

Tocopherol Profile

The tocopherol chromatograms for both PC pits extract cultivars were qualitatively similar, but showing only quantitative differences. For the two date seed oils, three peaks corresponding to α-, γ and δ-tocopherols were identified, and these accounted for 29.95 and 46.65% of the total tocopherols for its seeds oils, respectively. (Table 2) In the Phoenix dactylifera (PC) fruit and pits oil examined, α-tocopherol was the main component and constituted about 24.97 and 38.85% of the total tocopherols for date fruits and date pits oil, respectively. (Table 2)

Meanwhile, γ -tocopherol accounted for ~3.76% of PC fruit extract and ~5.40% of PC pits extract whereas the content of d-tocopherol ranged from 1.22% for PC fruit extract to 2.40% for PC pits extract. Total unidentified peaks ranged from 53.35% for date seed oils to 70.05% for date and date seed oils. These results indicate that date pits oil could be used in cosmetics, pharmaceuticals, and food. On the other hand, there are no reports available on the unsaponifiable matter of date seeds oil; hence, this study may be important in terms of potential nutritional, functional and economic utility of Phoenix dactylifera date seeds as a new source of oil.

It is evident that date seeds do not provide a high yield of oil, but may serve as a potential source of natural antioxidants such as phenols, tocopherols, and sterols. Table 3 revealed that there are ten essential amino acids in Phoenix dactylifera (PC) fruit and pits extracted contained tyrosine and serine ammonia, lysine, Tyrosine, Serine, Phenylalanine, Threonine in a low content. Valine, histidine, praline, arginine and leucine, are present in date extract with median levels but Phoenix dactylifera (PC) fruit tryptophan are contains a high percentages ryptophan.

Table 3. Amino Acid contents of Phoenix dactylifera for Fruit and Pits extract

Amino Acid	Amino acid in PC Fruit Conc. (u g/ml)	Amino acid in PC Pits Conc. (u g/ml)	Amino Acid	Amino acid in PC Fruit Conc. (u g/ml)	Amino acid in PC Pits Conc. (u g/ml
Tryptophan	293.51±23.4	93.43±23.4	Ammonia	14.53±6.94	8.36±2.41
Leucine	57.84±14.3	22.73±14.3	Lysine	33.68±9.32	18.54±3.33
Phenylalanine	30.05±9.72	51.32±9.72	Histidine	210.78±26.7	64.54±3.41
Proline	91.78±16.43	40.65±18.65	Valine	66. 25 ± 24.51	37. 62 ± 6.32
Arginine	64.48±17.41	23.37±15.43	Tyrosine	17. 10±9.83	9. 43± 53.71
Threonine	21.34±14.93	8.84 ±9.43	Serine	27.76± 17.31	11.42± 6.54

Effects on Sucrose Intake

Table 4. showed the difference in sucrose intake between the control and stressed groups. The present data showed that the chronic mild stress was gradually reduced the consumption of the sucrose solution.

As compared to the 15.4 g intake in the baseline test, 4 weeks later, the sucrose intake was reduced to 5.6 g in the stressed animals. Table 4 also showed that the unstressed rats treated with PC fruit and pits extract and fluoxetine presented no significant variation in their preference for sucrose consumption.

But in the stressed animals, when compared to the sucrose intake at week 0, treatment with PC fruit and pits extract caused a gradual recovery of the sucrose intake more than fluoxetine. Stressed animals with PC fruit and pits extract and fluoxetine treatment for six weeks showed significant improvement from their own predrug baseline, and there was no significant difference between the drug treated stressed and vehicle control animals.

The results demonstrated that six weeks of fluoxetine treatment needed more time for a complete recovery.

Table 4. Sucrose intake test of rats treated with Phoenix dactylifera L Fruit and Pits extracts and fluoxetine (FL)

Groups	Duration of drug treatment (weeks)			
	0	1	2	3
Vehicle control	36.19 ± 10.2	33.4 ±10.8	34.5 ±11.9	35.1 ± 10.9
Stress + vehicle	8.6 ± 1.1**	6.8 ±1.8*	8.95 ± 7.39**	9.9 ± 9.5**
Groups	Duration of drug treatment (weeks)			
	0	1	2	3
PC Pits Extract control	30 ±1.5	31.6 ±0.9	33.4 ±1.8	32.1 ±1.4
PC pits Extract+Stress	22.9 ±0.8**	25.3 ±0.8**	26.6 ±1.7 ##	28.1 ±8.5#
PC fruit extract control	36.2 ±1.8	34.0 ±1.4	35.9 ±15.8	36.18 ± 9.5
PC fruit extract + Stress	26.5 ±1.0**	28.0 ±1.1**	29.9 ± 0.5**	34.6 ±11.3#
FL control	18.7 ±8.6	18.9 ±7.9	19.2 ±5.2	20.0 ±10.2
Stress + FL	13.10 ± 6.43d **	15.6 ±3.4**	5.6 ± 3.7*	7.4 ±0.7**

In contrast, the gradual onset of action of PC fruit and pits extract appeared to exert significant antidepressant effects after a shorter time. Data are expressed as mean ± SD. (n =10).*P < 0.01 as compared to vehicle-treated control at each time-point.**P < 0.001 as compared to vehicle-treated control at each time-point.# P < 0.01 as compared to drug-treated stressed animals at Week 0. ## P < 0.001 as compared to drug-treated stressed animals at Week 0.+P < 0.01 as compared to FL-treated stressed animals at each time-point.+ + P < 0.001 as compared to FL-treated stressed animals at each time-point Rats were dosed with the date and seeds as described in the Materials and methods section. Two other groups of rats were treated with fluoxetine (FL) in parallel. The values indicate the amount of sucrose taken (g).For the PC fruit extract treated stressed group, some apparent effects appeared as early as after three weeks of drug treatment, at the end of six weeks and thereafter the amounts of sucrose solution taken by the stressed animals receiving Phoenix dactylifera fruits extract were consistently and more effectively than fluoxetine in recovering the sucrose intake in stressed animals.

Effects on 5-HT and 5-HIAA Levels

Table 5 revealed that the group treated with PC fruit and pits extract had effected on the levels of 5-HT and 5-HIAA in the brain parts of stressed animal model more than the group treated with fluoxetine. In the chronic mild stressed, the group showed significantly reduced in the 5-HT and 5-HIAA concentrations in all the examined brain parts compared to the vehicle unstressed animals with the exception of 5-HT in the striatum where the observed decrease did not satisfy our strict requirement for statistical significance.

Table5. 5-HT and 5-HIAA levels in brain regions tissues treated with PC fruit and pits extracts and fluoxetine

Parameters	Prefrontal Cortex		Hippocampus		Hypothalamus	
Groups	5-HT	5-HIAA	5-HT	5-HIAA	5-HT	5-HIAA
Vehicle control	591.1 ±126.8	1050.5 ± 211.3	1473.5 ± 242.7	1704.7 ±192.0	1435.1 ±340.8	792.9 ±103.7
Stress + vehicle	735.6 ±113.5*	476.5 ±97.5**	893.9 ±82.9**	938.7 ±87.4**	928.9 ±141.7*	356.1 ±71.6**
PC pits Extract	580.0 ± 21.4	943.5 ±135.3	1318.9 ±165.8	1497.8 ±136.7	1326.9 ±129.4	921.7 ±233.6
PC pits extract + Stress	526.1 ± 40.2*	693.1 ±117.9	1205.6 ± 353.1	789.7 ±9.5**	1508.9 ± 292.8##	746.9 ±46.1*
PC fruit extract	548.1 ± 79.6	1329.6 ±157.0	1343.9 ±369.9	1657.4 ±178.4	1426.9 ±404.5	832.5 ±271.4
PC fruit extract+ Stress	496.7 ± 114.7*	995.6 ±138.4##	13)6.3 ± 390.1	907.8 ±104.6**	1395.3 ±325.8 #	538.3 ±97.2*
FL control	771.2 ± 117.3	1016.4 ± 167.7	1462.1 ±385.5	773.4 ±180.2	1577.7 ±449.2	901.8 ±148.4
Stress + FL	758.3 ±107.2 #	794.8 ±55.9##	1587.9 ±341.5#	812.4 ±134.6 ##	1682.9 ±244.3 ##	809.4 ±56.9##

Phoenix dactylifera (PC) fruit and pits extract was capable of reversing the effect of chronic mild stress on 5-HT (p<0.01) in the prefrontal cortex, but more than fluoxetine. However, both fluoxetine and PC fruit and pits extract were effective in reversing the effect of chronic mild stress on 5-HIAA. In the hippocampus, PC fruit and pits extract were able to reverse the chronic mild stress effect in both 5-HT and 5-HIAA levels, whereas fluoxetine was only capable of reversing 5-HT levels. In the striatum, PC fruit and pits extract were effective in reversing the both the 5-HT and 5-HIAA levels while fluoxetine was effective in reversing 5- HT but not HIAA levels. The situation in the hypothalamus was more or less the same as in the hippocampus in that PC fruit and pits extract were effective in reversing both parameters but fluoxetine was effective in reversing the 5-HT levels only. Data are expressed as mean±SD (n =10). *P<0.01 as compared to vehicle-treated control. **P<0.001 as compared to vehicle-treated control. #P<0.01 as compared to vehicle-treated stressed animals. ##P<0.001 as compared to vehicle-treated stressed animals. Rats were dosed with the date and seeds as described in the Method section. Two other groups of rats were treated with fluoxetine (FL) in parallel. The values indicate the levels of 5-HT of 5-HIAA in ng/g of tissue.

Effects on Splenic NK Cell and LAK Cell Activities

Table 6 illustrated that the PC fruit and pits extract were significantly elevated the splenic NK cell and LAK cell activities more than the group treated with fluoxetine in comparison with stressed animal model. It is worth noting that, after PC fruit extract treatment, the NK cell activity in the stressed animals was higher than that of the stressed controls, and PC fruit extract enhanced LAK cell activity in stressed rats to exceed the activity in the unstressed controls. In stressed animal model, fluoxetine slightly increased the NK cell activity.

Effects on Serum IL-4, IFN-γ and TNF-α Levels

Table 6 showed that the vehicle-stressed animals had a significantly higher serum IL-4 and TNF-α (p<0.01) and decrease IFN-γ levels (p<0.01) in stressed group than normal control, but administration of PC fruit and pits extract to the stressed animals could effectively reverse the stress-induced decrease in IL-4, TNF-α (p<0.01), and increase in IFN-γ levels (p<0.01), compared to the stressed control animals than stressed with fluoxetine group. While the stressed with PC fruit and pits extract group, the levels of IL-4, TNF-α, and IFN-γ levels, were improved than all other groups. Data are expressed as mean ± SD (n =40).*P < 0.01 as compared to vehicle-treated control..**P < 0.001 as compared to vehicle-treated control. ##P < 0.001 as compared to vehicle-treated stressed animals. Rats were dosed with the date and seeds for 6 weeks as described in the Materials and methods section. Two other groups of rats were treated with fluoxetine (FL) in parallel. The date extract treatment showed significant effects on (1) IL2 concentration (p=0.019), (2) TNFα and IFNγ (pg/l) concentrations in stress and non stress (p=0.0001 and, p=0.0001, respectively), cut off of value of IL-4 , IFNγ, TNF-α are 123.2 pg/ml and32.6 pg/ml 85.6 pg/ml respectively

Table 6. Descriptive data on cytokines concentrations in different brain regions and NK cell and LAK cell activities at different type of treatment

Parameters Groups	Activity (%) In serum			Splenic Tissues	
	TNFα (pgl/ml) (mean ± SD)	IL-4 (pg/ml) (mean ± SD)	IFNγ (pg/ml) (mean ± D)	NK (mean ± SD)	LAK (mean ± SD)
Vehicle control	35.57 ±14.69	125.57 ±0.69	88.84 ± 18.96	27.29 ± 1.34	55.86 ±3.46
Vehicle+Stress	1899.9 ± 147.01**	180.54 ±1.48*	41.25± 22.32**	9.49 ±0.85**	75.07 ±3.46**
PC pits Extract	792±42.2**	136.86 ±1.22**	84.4.13 ± 46.59*	25.58 ±5.92	72.31 ±2.93
PC pits Extract + Stress +	779.4 ±199.1 **	143.43 ±0.65**	56.15 ± 87.72 ##	20.25 ±4.88 ##	67.14 ±2.56**
PC fruit extract	6992.84 ± 40.45 **	127.24 ±1.50*	86.97± 52.97*	26.04 ±6.83	73.28 ±1.99
PC fruit extract +Stress	678.7 ± 1914.7##	130.04 1.19**	77.14 ± 61.59 ##	25.80 1.83**	79.91 ±3.30 ##
FL control	848.69 ± 124.78 **	137.78 ±1.11**	65.91 ± 38.31 **	19.93 ±4.12*	74.71±2.91
Stress + FL	958.7 ± 1914.7	131.62 1.58 ##	36.12 ± 68.43 ##	29.26 1.43##	73.21 ±1.49##

Effects on IL-1β, TNF-α, and IFN-γ Levels in Brain Region Tissues

Showed in Table 7, IL-1β, and TNF-α levels in brain region tissues were elevated significantly ($p<0.01$) in stressed group than normal control, but administration of PC fruit and pits extract to the stressed animals could effectively reverse the stress-induced decrease in IL-1β, TNF-alpha ($p<0.01$), and increase in IFN-γ level ($p<0.01$) compared to the stressed control animals than stressed with fluoxetine group. In the stressed with PC fruit and pits extract group, the levels of IL-1β and TNF-alpha levels were improved significantly ($p<0.01$) more than all other groups under treatment.

Effects on Serum Lipid Profile

It was found that the stressed rats exhibited significantly lower values of TC, HDLc, and LDLc, and a higher value of TG ($p<0.01$) as compared to the unstressed animals (Table 8). Date seeds extract and date extract and fluoxetine, significantly reversed the decreased TC and HDLc values to normal.

However, only PC fruit and pits extract were capable of reversing the LDLc levels to normal ($p<0.01$). For TG, all the three drugs were capable of decreasing the elevated levels to values below the normal value. Judging from the numerical values of the reversed levels, PC fruit extract appeared to be more efficacious than PC pits extract and fluoxetine.

Interestingly, PC fruit and pits extract was observed to produce some effects on the HDLc and TG levels of the unstressed controls, but fluoxetine did not affect the unstressed controls in any of the parameters. Data are expressed as mean±SD (n =10). *P<0.01 as compared to vehicle-treated control. Rats were dosed with the date and seeds for 6 weeks as described in the Method section. Two other groups of rats were treated with fluoxetine (FL) in parallel Cut of value of TNFα =73.6 pg/ml and IL-1 β=64.3 (pg/g).

Table 7. Descriptive data on cytokines concentrations in different brain regions tissues at different type of treatment

Parameters Groups	Prefrontal Cortex		Hippocampus		Hypothalamus	
	IL-1 β ((pg/g)	TNFα (pgl/g)	IL-1β ((pg/g)	TNFα (pgl/g)	IL-1β ((pg/g)	TNFα (pgl/g)
Vehicle control	65.51 ± 555.2	76.2 ± 121.2	76.5 ± 555.2	35.31 ± 106.3	57.78 ± 22.19	75.9± 112
Stress alone	94.32 ± 14.52*	97.4 ±140.2*	94.32 ± 14.52*	87.6 ± 12.2*	284.32 ± 14.52*	182.9±153. 7*
PC pits Extract + Stress	75.77 ± 10.27*	80.21± 141.1*	71.6 ± 653.9*	48.6 ± 1413.9*	104.98 ± 115.89*	136.1±122. 4*
PC Fruit Extract+ Stress	63.77 ± 28.62*	78.2 ± 127.3*	60.8 ± 724.5*	45.6 ± 104.1*	150.95 ± 77.9*	130.2 ± 137.2*
Stress + FL	80.47 ± 41.07	87.4 ±86.3*	85.85 ± 51.78	57.6 ± 131.2*	230.62 ± 173.09*	148.9±153. 7*

Table 8. Descriptive data on serum lipid profile concentrations in different type of treatment

Group	Lipid profile in Serum			
	TC (mg/dl)	HDLc (mg/dl)	LDLc (mg/dl)	TG (mg/dl)
Vehicle control	168.21 ±0.14	50.5 ±3.02	83.2 ±6.9	110.6 ±7.9
Vehicle +Stress	185 ±0.07**	82 ±0.02*	116.3 ±5.5**	134 ±0.09*
PC pits Extract	80.2 ±17.4	56.0 ±7.03	86.4 ±0.06	114.5 ±31.10
PC pits Extract+ Stress +	91.6 ±0.10 ##	62.2 ±9.38 #	90.2 ±0.10#	129 ±0.04**
PC fruit extract	79.16 ±0.14	52.5 ±4.53	84.8 ±0.10	115.7 ±30.9
PC fruit extract +Stress	86.11 ±0.08 ##	54.1 ±5.71**	87 ±0.03 ##	124 ±0.07**
FL control	171.71 ±30.09	63 ±0.02**	97±30.22	131.00 ± 36.63
Stress + FL	192.22 ±35.07#	76.7 ±0.04##	107.4 ±0.10**	171 ±0.05**

Effects on SOD Activity

It was found that vehicle-treated stressed rats showed significantly higher SOD activity ($p<0.01$) as compared to the vehicle-treated unstressed animals (Table 9). The stressed animals, which received PC fruit and pits extract and fluoxetine respectively, demonstrated significantly reduced SOD activity within the normal range compared to group received fluoxetin. Data are expressed as mean ± S..M. ($n = 10$).*$P < 0.01$ as compared to vehicle-treated control.**$P < 0.001$ as compared to vehicle-treated control.# $P < 0.01$ as compared to vehicle-treated stressed animals.## $P < 0.001$ as compared to vehicle-treated stressed animals. Rats were dosed with the date and seeds for 6 weeks as described in the Materials and methods section. Two other groups of rats were treated with fluoxetine (FL) in parallel.

Data are expressed as mean±SD (n =10). *P<0.01 as compared to vehicle-treated control. **P<0.001 as compared to vehicle-treated control. ## P<0.0001 as compared to vehicle-treated stressed animals. Rats were dosed with the PC fruit and pits extract for 6 weeks in parallel with the two other groups of rats were treated with fluoxetine (FL).

Effects on NOS Activity

The present data in Table 9 showed a decreased liver NOS activity (p<0.01) in the groups treated with PC fruit and pits extract, more than group treated with fluoxetin, as well as untreated stressed rats.

Effects on Serum MDA Levels

Serum MDA levels were significantly increased (p<0.01) in the vehicle-treated stressed animals, as compared with the vehicle-treated unstressed animals (Table 9). Group treated with PC fruit and pits extract showed reduction in the MDA level more than group treaded with fluoxetin alone.

Table 9. Descriptive data on liver SOD and NOS activities and serum MDA concentrations in different type of treatment

Groups	Enzyme activity (U/mg protein) in liver tissue		Serum
	SOD (U/mg protein)	NOS (U/mg protein)	MDA level (nmol/mgtissue)
Vehicle control	193.21 ± 46.38	44.93±5.98	1.64 ± 1.05
Vehicle +Stress	84.58 ± 23.83**	179.08 ± 26.16**	6.02 ± 4.50**
PC pits Extract	176.60 ±0.18*	50.03±5.10**	1.74±0.34**
PC pits Extract+ Stress +	128.12 ± 54.60 ##	119.08 ±24.21** ##	2.84 ± 0.83**
PC fruit extract	187.21 58 ±0.10	45.93±4.35**	1.69±0.34**
PC fruit extract +Stress	145.21 ± 35.56 ##	69.08±11.53 **	1.97 ± 0.98**,##
FL control	1694. 94 ±0.12**	79.03±23.14**	2.38 ±0.32**
Stress + FL	84. 76 ±0.16 ##	145.13 ± 19.23**	4.63 ±0.72**,##

DISCUSSION

PC fruit and its pits methanolic extracts posses a wide spectrum of CNS activity due to many reasons. Firstly, the date fruits and pits have a high total polyphenols [36] and phenol compounds, this explains that the phenolics antioxidants have a beneficial effect by preventing oxidative stress in coronary heart disease. [37] Most data in the present study revealed that the fraction of the unsaponifiable matter having a high content from sterols in PC fruit and pits. In the literature, the sterols have been associated with antioxidant effects. [38] Furthermore, the present results agree with several studies performed by Vyawahare et al [39] who stated that the oral administration of PC extract demonstrated an anxiolytic-like effect in mice, as it significantly increased the time spent in open arms and decreased the time spent in closed arms.

The present data showed the PC fruit and its pits highly contain vitamin E family (tocols). The present data agree with study performed by Aparicio and Aparicio-Ruiz [40] where they proved that the vitamin E family consists of eight "tocols" α,β,γ and δ-tocopherol and α,β,γ and δ-tocotrienol and γ-tocopherol comprise more that 60% of total vitamin E content and it agree with Ramadan et al [38] where they proved that PC seed oil had a high amount of α-tocopherol (~30%) and tocotrienols (~60%). Data in the present study interpreted the antioxidant properties of PC fruit and its pits, and agree with study performed by Sokol [41] who proved that the tocopherols and tocotrienols possess antioxidant properties and are active as vitamins E, as well as vitamin E deficiency affects nervous system development in children. [42] Tryptophan is essential amino acids and present in the PC fruit and its pits extracts are a precursor for the synthesis of the neurotransmitter serotonin and Tryptamine, as well as for the synthesis of the antipellagra vitamin nicotinic acid and the epiphyseal hormone melatonin as well as tryptophan and its metabolites regulate neurobehavioral effects such as appetite, sleeping-walking-rhythm and pain perception. Although the essential amino acids content which were found in PC fruit more than that essential amino acid present in PC pits . [43] PC fruit and its pits extract gradually reversed the chronic mild stress induced reductions in sucrose consumption but PC fruit extract more effective than fluoxetine which time required triggering this onset in behavioral change. So, the PC fruit extract reduced the antidepressant and chronic stress in the immobility behavioral in animal model. The present study agreed with the previous study performed by Guo et al [44] where they proved that the ginger as an herbal medicine is active in pre-clinical mice models of behavioral depression with effects comparable to known antidepressants.

The present data of the biochemical markers relevant for antidepressant activity in animal model revealed that the enhancement and alterations in brain 5-HT and 5-HIAA concentrations were observed in several brain regions. Rats exposed to chronic mild stress showed significantly decreased 5-HT and 5-HIAA in most of the brain parts relative to the unstressed animals. This 5-HT depletion may have a causative role in producing increased a hedonic behavior in the stressed rats. Administration of PC fruit and pits extract and fluoxetine elicited significant increases in 5- HT and 5-HIAA levels in various brain areas in stressed rats. Furthermore, the action of PC fruit and pits extract on the 5-HT level appeared to be more potent than that of fluoxetine in the cortex region of the brain. This agreed with Huang and Williams [45] who suggests that serotonin 5-HT1B antagonists alone or combined with selective serotonin re-uptake inhibitors might hold unique promise as efficacious antidepressants.

The present data showed that the chronic mild stress suppressed splenic NK cell and LAK cell activities. Such decreased activities were reversed by the administering animal model with PC fruit and pits extract and fluoxetine, while PC fruit extract decreased activities more than other groups. In the present data, stressed animals dosed with date extract, in fact, gave an even higher NK cell activity than the unstressed animals. In stressed animals group, not dosed with PC fruit and pits extract and fluoxetine, gave elevated LAK cell activity above the normal level in unstressed animals. The present data confirmed the study performed by Morimoto et al [46] where they stated that the increasing in the serotonergic activity in the brain correlates with an increase in NK cell activity. Based on the fact that 5-HT modulates constituents of the immune system and that the 5-HT1A receptor is involved in depression, it is possible that the activity of NK cells is regulated by 5-HT via the 5-HT1A receptor. [47]

The present data revealed that the extract of the PC fruit decreased the brain tissues IL-1β and serum levels of IL-4, TNF-α (in serum and tissues), and IFNγ, increased the levels than the stressed and other treated with fluoxetine. These data confirmed the study performed by Maes [48] who hypothesizes that inflammatory, oxidative, and nitrosative (IOandNS) pathways, are causally related to depression following external (psychological) and internal (organic) stressors and that IOandNS pathways are novel targets for antidepressant and he review that depression is accompanied by an inflammatory reaction as indicated by an increased production of pro-inflammatory cytokines, such as interleukin-1β (IL-1β), tumor necrosis factor-α (TNF-α) and interferon-γ (IFN)-γ. These cytokines are stress-sensitive and may cause depressive behaviors.

The latter may be induced by an increased catabolism of tryptophan, the precursor of serotonin, to neurotoxic (tryptophan catabolites along the indoleamine oxidase pathway). Inflammatory biomarkers are detected in animal models of depression development. Nevertheless, in the present study, a significant decreased in Th1 cytokine IFNγ before treatment with fluoxetine and extracts of the PC fruit of and pits extracts, respectively, elevated the IFN-γ levels after treatment. Therefore, our finding also is in agreement with the hypothesis that the balance in the Th1 and Th2 cytokines could be important in the path-physiology of depression. [14]

The high content of monounsaturated and polyunsaturated fatty acids, as well as natural sterols and low amount of cholesterol in extract of PC fruit and its pits contribute to lowering of rat serum cholesterol levels and LDL cholesterol this lead to inhibit cholesterol absorption from the intestine then protect against cardiovascular disorder, the present data agree with study performed by Tapiero et al [50] and Parceresa et al. [51] In addition, finally, the fatty acids in the adipose tissue are better predictors of depression than those of serum cholesteryl esters. This indicates that depression relates more strongly to long-term than to short-term fatty acid intake. So, the reason for this may be the reported slow rate of deposition of dietary polyunsaturated fatty acid to the brain. [52] The present finding concluded that the PC fruit and pits extracts which contain sterols may be used as new therapeutic agents for treatment of hypercholesterolemia. El-Melegy et al [53] concluded that there was a strong relationship between the oxidant/antioxidant status and dyslipidemia leading to significant lower of the serum levels of glutathione peroxidase activity, HDL, and cholesterol.

CONCLUSION

The present study concluded that the PC fruits and pits have a high levels of sterols, total polyphenols, phenol compounds, and vitamin E family [α-tocopherol (~30%), tocotrienols (~60%)] and high levels of sterols. This explains that the PC fruit and its pits possess antioxidant properties, have a beneficial effect by preventing oxidative stress increase scavenge the free radicals, increase levels of 5-HT and 5-HIAA (serotonergic activity) in most of the brain parts, which correlates with an increase in NK and LAK cell activity, as well as cytokines and reversed the chronic mild stress induced reductions in sucrose consumption. PC fruit and pits extracts, which contain sterols, may be used as new therapeutic agents for treatment of hypercholesterolemia.

ACKNOWLEDGMENT

Thanks are due to Mr. Abdul Latif yossif and Abdul Latif yossif Pharm- El-Oshaira Al-Madena Al-Monawara Saudi Arabia for funding and continuous support. As well as King Abdul Aziz University, in Saudi Arabia for their supporting me all scientific services that we needed to complete that research

REFERENCES

[1] Carlini, EA. Plants and the central nervous system. *Pharmacol. Biochem. Behav.* 2003; 3: 501-12.

[2] Phillipson JD. Phytochemistry and medicinal plants. *Phytochemistry*, 2001; 56: 237-43.

[3] Chandra, A; Chandra, Anju; Gupta, IC. Date palm research in Thar Desert. 1 st ed. Jodhpur India: Scientific Publishers; 1992.

[4] Anonymus. The wealth of India (A dictionary of Indian raw materials and industrial products) Raw materials. 1st ed. Vol. 3. New Delhi: Council of Scientific and Industrial Research; 1985.

[5] Nadkarni, KM. editor. Indian Materia Medica Vol. 1. Mumbai: Bombay Popular Prakashan Pvt. Ltd; 1976.

[6] Al-Qarawi, AA; Abdel-Rahman, H; Mousa, HM; Ali, BH; El-Mougy, SA. Nephroprotective action of Phoenix dactylifera. in gentamicin-induced nephrotoxicity. *Pharmaceutical. Biol.* 2008; 4: 227-30.

[7] Endo T, Nakagawa T, Iguchi F. Elevation of superoxide dismutase increases acoustic trauma from noise exposure. *Free Radical. Bio. Med.* 2005; 38: 492-8.

[8] Akyol, O; Zoroğlu, SS; Armutcu, F; Sahin, S; Gurel, A. Nitric oxide as a physiopathological factor in neuropsychiatric disorders. *In Vivo*, 2004, 18 (3): 377-90.

[9] Bilici, M, Efe Hb, Köroglu MAa, Uydu HAb, Bekaroglua M, Deger O. bAntioxidative enzyme activities and lipid peroxidation in major depression: alterations by antidepressant treatments. *Journal of Affective Disorders*, V64, Issue 1, 2001, P 43-51.

[10] Karolewicz, B; Paul, IA; Antkiewicz-Michaluk, L. Effect of NOS inhibitor on forced swim test and neurotransmitters turnover in the mouse brain. *Polish Journal of Pharmacology* 2001, 53 (6), 587–596.

[11] Gambarana, C; Scheggi, S; Tagliamonte, A; Tolu, P; De Montis, M.G. Animal models for the study of antidepressant activity. *Brain Res. Brain Res. Protoc.* 2001; 7 (1), 11–20.

[12] Breggin, PR. Medication madness: A psychiatrist exposes the dangers of mood-altering medications. New York: St. Martin's Press; 2008.

[13] Griebel, G; Rodgers, RJ; Perrault, G; Sanger, DJ. The effects of compounds varying in selectivity as 5-HT1A receptor antagonists in three rat models of anxiety. *Neuropharmacology* 2000; 39 (10): 1848-57.

[14] Smith, RS. The macrophage theory of depression. *Med. Hypotheses,* 1991; 35 (4): 298-306.

[15] Holden, R; Pakula, IS; Mooney, PA. An immunological model connecting the pathogenesis of stress, depression and carcinoma. *Medical Hypotheses*. 1998, 51 (4), 309–314.

[16] Steegmans, PH; Hoes, AW; Bak, AA; van der Does, E; Grobbee, DE. Higher prevalence of depressive symptoms in middle-aged men with low serum cholesterol levels. *Psychosomatic Medicine* 2000; 62 (2), 205–211.

[17] Almeida-Montes, LG; Valles-Sanchez, V; Moreno-Aguilar, J; Chavez-Balderas, RA; Garcia-Marin, JA; Cortes Sotres, JF; Hheinze-Martin, G. Relation of serum cholesterol, lipid, serotonin and tryptophan levels to severity of depression and to suicide attempts. *Journal of Psychiatry and Neuroscience* 2000, 25 (4), 371–77.

[18] Besbes, S; Blecker, C; Deroanne, C; Drira, NE; Attia, H. Date seeds: chemical composition and characteristic profiles of the lipid fraction. *Food Chem.* 2004, 84, 577–84.

[19] AOCS. Official Methods and Recommended Practices of the American Oil Chemists' Society, 5th Ed, AOCS Press, Champaign, IL. 1997.

[20] Lucas, B; Sotelo, A. (1982) Amino acid determination in pure proteins, foods and feeds using two different hydrolysis methods. *Anal. Biochem*. 123: 349-356.

[21] Lucas, B; Sotelo, A. (1980) Effect of different alkalies, temperatures and hydrolysis times on tryptophan determination in pure proteins and foods. *Anal. Biochem* 109: 192-197.

[22] Ras, RMV; Tara, MR; Krishnan, CK (1974) Colorimetric estimation of tryptophane content of pulses. *J. Food Sci. Technol.* 11: 213-216.

[23] Grønli, J; Murison, R; Fiske, E; Bjorvatn, B; Sørensen, E; Portas, CM. et al. Effects of chronic mild stress on sexual behavior, locomotor activity and consumption of sucrose and saccharine solutions. *Physiol. Behav.* 2005; 84: 571-7.

[24] Willner, P. Validity, reliability and utility of the chronic mild stress model of depression: a 10-year review and evaluation. *Psychopharmacology* 1997; 134: 319–29.

[25] Muscat, R; Willner, W. Effects of dopamine receptor antagonists on sucrose consumption and preference. *Psychopharmacology* 1989; 99: 98-102.

[26] Smith, JE; Lane, JD; Shea, PA; McBride, WJ; Aprison, MH. A method of concurrent measurement of picomole quantities of acetylcholine, choline, dopamine, norepinephrine, serotonin, 5-hydroxytryptophan, 5-hydroxyindolacetic acid, tryptophan, tyrosine, glycine, aspartate, glutamate, alanine, and gamma-aminobutyric acid in single tissue samples from different areas of rat central nervous system. *Anal. Biochem*. 1975; 64:149–169.

[27] Curzon, Gp; Green, AR. Rapid method for the determination of 5-hydroxytryptamine and 5-hydroxyindoleacetic acid in small regions of rat brain. *British Journal of Pharmacology* 1970; 39 (3), 653–5.

[28] Korzeniewski, C; Callewaert, DM. An enzyme-release assay for nature cytotoxicity. *Journal of Immunological Methods* 1983; 64 (3), 313–20.

[29] Wu, R; Coleman, N; Stanley, M. Different susceptibity of cervical keratinocytes containing human papillomavirus to cellmediated cytotoxicity. *Chinical Medicine Journal* 1996; 109 (11), 854–8.

[30] Cai, Z ; Pang, Y ; Lin, S ; Rhodes, PG. Differential roles of tumor necrosis factor-alpha and interleukin-1 beta in lipopolysaccharide-induced brain injury in the neonatal rat. *Brain Res*. 2003; 975 (1-2): 37-47.

[31] O'Connor, KA; Johnson, JD; Hansen, MK; Wieseler Frank, JL; Maksimova, E; Watkins, LR; Maier, SF. Peripheral and central proinflammatory cytokine response to a severe acute stressor. *Brain Res.* 2003; 991: 123-32.

[32] Fairbanks, V; Klee, GG. Biochemical Aspects of Haematology, Tietz Textbook of Clinical Chemistry. Philadelphia: Saunders WB; 1994.

[33] Allain, CC; Poon, LS; Chan, CS; Richmond, W; Fu, PC. Enzymatic Determination of Total Serum Cholesterol. *Clinical Chemistry*, 1974, 20 (4), 470–5.

[34] Assmann, G; Jabs, HU; Kohnert, U; Nolte, W; Schriewer, H. Low density lipoprotein determination in blood serum following precipitation of low density lipoprotein with polyvinyl sulfate. *Clinical Chemistry Acta*, 1984; 140 (1), 77–83.

[35] Saunders, DB; Trapp, GR. Basic and clinical biostatistics, 3nd edition, Connecticut, Appleton and Lang.2001.

[36] Papadopoulus, G; Boskow, D. Antioxidant effect of natural phenols on olive oil. *J. Am. Oil Chem. Soc.* 1991; 68, 669-71.

[37] Tuck, KL; Hayball, PJ. Major phenolic compounds in olive oil: Metabolism and health effects. *J. Nutr. Biochem.* 2002; 13: 636-44.

[38] Cercaci, L; Rodriguez-Estrada, MT; Lercker, G. Solid phase extraction–thin layer chromatography-gas chromatography method for the detection of Hazelnut oil in olive oil by determination of esterified sterols. *J. Chromatogr.* 2003; A985, 211–20.

[39] Vyawahare, NS; Pujari, RR; Rajendran, R; Khsirsagar, AD; Ingawale, DK; Patil, MN. Neurobehavioral effects of Phoenix dactylifera in mice. *J Young Pharmacists*, 2009; 1: 225-32.

[40] Aparicio, R; Aparicio-Ruiz, R. Authentication of vegetable oils by chromatographic techniques. *J. Chromatogr.* 2000; A881, 93–104.

[41] Ramadan, MF; Mörsel, JT. Recovered lipids from prickly pear [Opuntia ficus indica (L.) mill] peel: A good source of polyunsaturated fatty acids, natural antioxidant, vitamins and sterols. *Food Chem.* 2003; 83, 447–56.

[42] Sokol, RJ; Vitamin, E. In Present Knowledge in Nutrition (E.E.Ziegler and L.J. Filer, eds.) 1996; pp. 130–136, ILSI Press, Washington, DC.

[43] Heine, W; Radke, M; Wutzke, KD (1995). The significance of tryptophan in human nutrition. *Amino acids*, 9: 191-205.

[44] Rimm, EB; Stampfer, MJ; Aschero, A; Giovannucci, E; Golditz, GA; Willet, WC. Vitamin E consumption and the risk of coronary heart disease in men. *N. Engl. J. Med.* 1993; 328, 1450–56.

[45] Guo, YS; Kong, LD; Wang, YM; Huang, ZQ. Antidepressant evaluation of polysaccharides from a Chinese herbal medicine Banxia-houpu decoction. Phytotherapy Research (MS No. PhyRes/2001/000148, in press).

[46] Huang, Y; Williams, WA. Enhanced selective serotonin re-uptake inhibitors as antidepressants. Expert Opinion on Therapeutic Patents. 2007, Vol. 17, No. 8, Pages 889-907.

[47] Morimoto, K; Takeshita, T; Inoue-Sakurai, C; Maruyama, S. Lifestyles and mental health status are associated with natural killer cell and lymphokine-activated killer cell activities. *The Science of the Total Environment* 2001; 270 (1–3), 3–11.

[48] Frank, MG; Johnson, DR; Hendricks, SE; Frank, JL. Monocyte 5-HT1A receptors mediate pindobind suppression of natural killer cell activity: modulation by catalase. *International Immunopharmacolog* 2001; 1 (2), 247–53.

[49] Maes, M. The cytokine hypothesis of depression: inflammation, oxidative and nitrosative stress (IOandNS) and leaky gut as new targets for adjunctive treatments in depression. *Neuroendocrinology Letters* 2008, Vol. 29 No. 3; 25; 29 (3).

[50] Tapiero, H; Townsend, DM; Tew, KD. Phytosterols in the prevention of human pathologies. *Biomed. Pharmacother.* 2003; 57, 321–5.

[51] Parceresa, J; Richardson, DG; Rafeacas, M; Codony, R; Boatella, J. Fatty acids, tocopherol and sterol content of some hazelnut varieties (Conylus avellana L.) harvested in Oregon (USA). *J. Chromatogr.* 1998; A805, 259–68.
[52] Mamalakis, G; Jansen, E; Cremers, H; Kiriakakis, M; Tsibinos, G; Kafatos, A. Depression and adipose and serum cholesteryl ester polyunsaturated fatty acids in the survivors of the seven countries study population of Crete. *Eur. J. Clin. Nutr.* 2006; 60 (8): 1016-23.
[53] El-Melegy, NT; Mohamed, NA; Sayed, MM. Oxidative Modification of Low-Density Lipoprotein in Relation to Dyslipidemia and Oxidant Status in Children with Steroid Sensitive Nephrotic Syndrome. *Pediatr. Res.* 2008; 63 (4): 404-9.

In: Asia Pacific Life Sciences Compendium
Editor: Bandit Chumworathayi
ISBN: 978-1-62257-338-7

Chapter 13

Additive Effect of Natural Product Together with Chemotherapy on Biochemical and Molecular Changes in Egyptian Hepatocellular Carcinoma Patients

Ragaa Hosny Mohamad,[1,*] Zekry Khalid Zekry,[2] Mohamad Gamil A El-Siad,[3] Samia A Akl,[4] Hussien Abdel-Rahman Al-Mehdar,[5] Siad Ebrahim El-Shaieb,[6] Amany A El-Bassmy,[7] Sabry Mohamad Sharawy[1] and Mahmoud Mohamad El-Marzbani[8]

[1]Medical Biochemistry, National Cancer Institute, Cairo University, Egypt
[2]Clinical Oncology, National Cancer Institute, Cairo University, Egypt
[3]Surgical Oncology, National Cancer Institute, Cairo University, Egypt
[4]Clinical Pathology, National Cancer Institute, Cairo University, Egypt
[5]Immunology Department, King Abdul Aziz University, Saudi Arabia
[6]Internal Medicine Department, Menofia University, Egypt
[7]Cancer Epidemiology and Biostatistics National Cancer Institute, Cairo University, Egypt
[8]Drug Research and Experimental Oncology, Cancer Biology Department, National Cancer Institute, Cairo University, Egypt

Abstract

The present work aimed to evaluate the therapeutic effect of oral co-administration of Nigella sativa (NS) oil, fish oil (FO), evening primrose oil (EPO) and vitamin E (Vit E) with chemotherapeutic agents in treatment of hepatocellular carcinoma (HCC) patients

* Tel +020101856864, Fax: +0223644720, Email: ragaatahtawy@gmail.com

compared to control group and chemotherapy only group. 80 HCC male patients and 20 controls, the patients were divided into four groups G1 received only chemotherapy, G2, G3, G4 received an additional NS, EPO, and FO dose, respectively. VEGF, VEGF per platelet (plt) count, basic (b) FGF, TGF-β, IFN-γ-R1-mRNA, and prostaglandin E, were measured in studied groups, before and after treatment.

After receiving different types of treatment, serum VEGF and b-FGF levels were decreased. Mean of VEGF/plt count was also decreased significantly in all groups. There is a significant correlation between serum VEGF level and platelet count in different groups of study. Serum levels of TGF-β, which were high before treatment, showed a significant reduction reaching normal levels in patients receiving NS.

Moreover, basal levels of IFN-γ-R1-mRNA were lower in HCC patients compared to control, and its level was significantly increased in patients receiving NS and FO. The natural oil especially, Nigella sativa, can act as antiangiogenic agent and improve the antitumor effect and the performance status of chemotherapeutic agents in HCC.

Keywords: 5-Flurouracil, tamoxifen, nigella sativa, evening primrose oil, fish oil

BACKGROUND

Hepatocellular carcinoma (HCC) represents 80-90% of primary liver malignancies and is the third cause of cancer deaths worldwide. [1] It is the fifth most common cancer in men and the ninth in women. [2] It carries a grave prognosis, most patients die within 1 year after diagnosis and five-year survival is between 2% and 6%. This is dependent on tumor size and associated diseases at the time of diagnosis.

Hence, patients with cirrhosis have a shorter survival. [3] Currently, no effective systemic therapy exists for HCC. Recombinant [alpha]-interferon (IFN) has been suggested to have some antitumor efficacy in this illness. Although the synergistic antitumor activity of IFN with 5-fluorouracil (5-FU) has been reported in several gastrointestinal malignancies, it showed no significant benefit in treatment of HCC patients. [4] The volatile oils from black seeds and evening primrose contain γ-linolenic acid (GLA), a member of the ω-6 family of polyunsaturated fatty acids which exerts selective cytotoxic effects on cancer cells without affecting normal cells. [5]

Also, evening primrose oil is used in traditional medicine to treat chronic diseases. It has been shown that exogenous supplementation with GLA sensitizes breast cancer cells to antimitotic drugs and endocrine therapies. [6,7] The whole powdered seeds as well as successive alcoholic extracts of Nigella sativa (NS) L. family Ranunculaceae, revealed a remarkable anti-inflammatory activity in rats. [8]

Fixed oil of NS and derived thymoquinone has been shown to inhibit eicosanoid generation in leukocytes and membrane lipid peroxidation. [9] HCC is a highly vascular tumor, thus, angiogenesis is considered to be important for its progression. [10]

Vascular endothelial growth factor (VEGF) is a well known, potent angiogenic factor which enhances vascular permeability, promoting the extravasation of protein to form a stromal matrix and tumor invasion. [11] The VEGF content of platelets is several-fold higher in cancer patients than in healthy subjects. [12,13]

Platelets are also known to endocytose and concentrate plasma proteins such as VEGF, and later transport them into their granules. [14] Previous studies suggested that VEGF secreted from tumor cells could be stored and transported by platelets in the bloodstream, and that this reservoir of VEGF might have a role in tumor angiogenesis and invasion. [15]

Based on this hypothesis, VEGF load in platelets may indirectly show tumor-originated VEGF and predict tumor angiogenic activity better than serum VEGF. Other study used serum VEGF per platelet count to correct variation of serum VEGF levels in patients with different platelet counts. [16]

The induction of tumor anti-angiogenesis can control tumor expansion and suppresses metastasis of tumor cells to distant sites. [17] It has been suggested that angiogenesis is an early event in tumorigenesis. [18] Tumor angiogenesis may be regulated by angiogenic factors such as VEGF, [19] transforming growth factor-ß, [20] and tumor necrosis factor. [21]

The aim of the present work is to investigate the therapeutic effect of oral co-administration of natural oils (NS oil, FO, and EPO) or vit E, with chemotherapeutic agents on HCC patients, in comparison to therapeutic effect of chemotherapy alone.

METHOD

Materials

1. Fluorouracil: Obtained from Roche Company (USA).
2. Tamoxafin: Obtained from Novartis Company (USA).
3. Lecovrerin: Obtained from Lederle Company (Hong Kong).
4. Vitamin E: Obtained from Pharco Pharmaceuticals, Alexandria-Egypt.
5. Nigella sativa oil (Nigella sativa L. family Ranunculaceae): Commercial name; Nigellar capsule, each containing 500 mg Nigella sativa oil, obtained from Kahira Pharmaceuticals and Chemical Industries Company, Cairo, Egypt.
6. Fish oil: Commercial name; EPA, each capsule contained 129 mg fish oil, obtained from Incromega Croda Chemicals Europe Ltd, UK).
7. Evening primrose seeds oil (Oenothera Biennis L. family Oenotheraceae): Commercial name; Primaleve, each EPO capsule contained 500 mg evening primrose oil, obtained from Efamol Ltd.

Patients, Samples, and Treatments

The study included 80 male patients aged 25-75 years with inoperable (HCC). All patients were presented to the specialized liver clinic of National Cancer Institute (NCI), Cairo University, between September 2006 and April 2008.

The study also included 20, age- and sex-matched, healthy control subjects. HCC patients were diagnosed by CT guided biopsy. Tumors were staged according to UICC classification and the severity of liver diseases classified according to the Child-Paugh classification.

Patients suffering from autoimmune disease or metabolic disorders affecting the liver or previously received any therapy were excluded from this study.

The study had been approved by Ethical Committee of the National Cancer Institution, and an informed consent was obtained from every case.

Pretreatment

At start, blood samples were taken from patients to examine serum markers (VEGF, VEGF per platelet count, basic FGF, TGF-beta, IFN-γ-R1-mRNA, and prostaglandin E).

Treatment Protocol

The study was carried out between September 2006 and April 2008. Patients were divided into four groups 20 patients each. Treatment was given according to the following protocol:

All patients received 5-FU (500 mg/m2/2 weeks for 6 cycles) and simultaneously received;

Group 1: Tamoxafin 20 mg/day only.
Group 2: Tamoxafin 20 mg/day+2gm Nigella Sativa, two times daily.
Group 3: Tamoxafin 20 mg/day+2gm of Fish oil capsule, two times daily.
Group 4: Tamoxafin 20 mg/day+2gm of Evening primrose oil, two times daily.
Group 5: Fifteen healthy control subjects, collectively from the staff working in NCI, Cairo University.

Patients were given NS, FO, or EPO, as oral doses, one gram twice daily, thirty minutes before meals. Capsules were refrigerated or stored at room temperature in a dry, dark place.

Biochemical Analysis of Serum

VEGF, VEGF per platelet count, basic FGF, TGF-β, IFN-γ-R1-mRNA, prostaglandin E, zinc, SOD, vitamin C, and vitamin E levels, were estimated before and after treatment and compared to healthy control group.

All serum parameters were determined by commercially available human enzyme-linked immunosorbent (ELISA) assay kits obtained from RandD Systems, Minneapolis, for TGF-β, BioSource international, USA, for basic-FGF, Amersham Pharmacia Biotech, for VEGF and Abbott Laboratories, North Chicago, IL, USA, for AFP. Serum Prostaglandin E2 (PGE2) concentrations were determined using a competitive enzyme-linked immunoassay for a stable PGE2 derivative (Bicyclo Prostaglandin E2 Designs Company USA Kit). All assays were performed according to the manufacturer's instructions. Furthermore, vascular endothelial growth factor per platelet count (VEGF/plt) was determined.

Determination of IFNγR1-mRNA in Peripheral Blood Mononuclear Cells (PBMC)

RNA was extracted from PBMC (10^6-10^7 cells) using the TRIzol Reagent (Life Technologies, Gaithersburg, MD, USA). cDNA was produced from 3 μg of total RNA by

RT-enzyme using SuperScript RNase H-RT (Life Technologies). The PCR preparations contained 3 μl cDNA, 0.1 μM specific primers (27), 200 μM dNTPs, and 2.5 U Taq DNA polymerase (Amersham Pharmacia Biotech, Sweden). Amplification was carried out over 42 (IFNγR1) and 25 (ß-actin) cycles at 95°C for 1 min, 52°C for 2 min, and 72°C for 2 min, followed by 72°C for 10 min for final extension. PCR products were separated by 1.0% agarose gel electrophoresis and analyzed by absorbance using a GS-700 Image Analyzer (Bio-Rad, Hercules, CA, USA). IFNAR1-mRNA level in PBMC was assessed as the IFNAR1-mRNA/ß-actin-mRNA ratio. The results reported for each RNA sample is the mean of at least three RT-PCR assays.

Statistical Analysis

Statistical Package for Social Sciences (SPSS), version 12, was used. Quantitative variables were summarized using mean and SD, median, minimum, and maximum values. Qualitative data were summarized using frequencies and percentage. Comparison between different measurement values before and after treatment within each group were analyzed using paired-t test. Differences were considered significant when $p<0.05$, and highly significant when $p<0.01$.

RESULT

Table 1. shows the change in studied parameters, before and after treatment. Superoxide dismutase activities SOD (U/g Hb) were significantly reduced in G2 and G3, compared to healthy control subjects, G5. Also, the present study revealed that abnormal serum AFP in HCC patients at the time of diagnosis was significantly improved after administration of FO and EPO ($p<0.001$) reaching nearly the normal level after treatment with NS, G2. Serum AFP was also significantly reduced in G1 patients but still 40 times higher than normal ($p<0.001$). It was shown that all patients had higher basal levels of TGF-β (pg/ml) compared to healthy controls. These levels were significantly reduced in the first three groups reaching the normal levels in patients receiving NS and FO. Moreover, the group receiving NS showed marked improvement in the clinical picture, and marked reduction in TGF-β. Basal levels of IFN-γ-R1-mRNA were significantly lower in HCC patients compared to healthy controls. However, after treatment, there was an increase in IFN-γ-R1-mRNA levels among patients receiving NS and FO in comparison to G1 who received only chemotherapy. In addition, it was found that, after treatment, serum VEGF were significantly decreased in all groups. VEGF per platelet count was calculated by dividing serum VEGF concentration by the platelet count, in the investigated patient groups. Pretreatment patients showed a high VEGF level with mean VEGF/plt ratios of 21.32±0.45, 15.80±0.10, 15.5±1.21, 10.86±1.23 pg/10^3, in G1, G2, G3, G4, respectively, compared to only 0.020±0.009 in healthy control, but after treatment, serum VEGF and VEGF per platelet count, were significantly reduced in investigated patients, showing a lower mean of VEGF/plt ratio of 0.77±0.48, 0.12±0.10, 0.24±1.21, 0.33±0.13 pg/103, in G1, G2, G3, G4, respectively.

Table 1. Comparison of some molecular findings among the studied groups before and after treatment

Group Parameter	G1		G2		G3		G4		G5
	Before	After	Before	After	Before	After	Before	After	Control
VEGF (pg/mL)	2034.3±53.2	931.3±13.2	1962.9±13.1	389 ±13.1	1879.3±12.3	267.3±34.3	2171.7±22.4	445. 9±13.1	18.5 ±0.17
VEGF/plt (pg/10^3)	21.32±0.45	0.77±0.48	15.80 ±0.10	0.12 ±0.10	15.5 ±1.21	0.24 ±1.21	10.86 ±1.23	0.33 ±0.13	0.020±0.009
b-FGF (0-13.5 pg/ml)	1107±758	1098.6±10.2	1293±537	68.2±75.5	1371±381	80.8±92.9	937±555	325.3±30.6	14.3±20.7
IFNγR1-mRNA	0.12±0.03	0.15±0.15	0.14±0.01	0.25±0.10	0.12±0.03	0.23 ±0.10	0.18±0.22	0. 20±0.06	0.30±0.05
TGF-β (pg/ml)	963.5±10.1	431.2±4.7	1088.9±29.9	210.8±5.8	1184.9±65.6	239.0±24.3	919.5±73.8	897.7±28.6	189.8±12.6
AFP (Up to 10ng/dl)	5245±8316	441.7±44.9	7740±1209	10.5±6.9	3713±7205	22.2±14.8	687±1362	313.5±10.5	3.87±0.35
SOD(U/g Hb)	2773±265	2621±93.5	2898±915	2332±149	2922±1130	2421±83.5	2848±908	2737±110.1	2220±121.7
PGE2 (39.1-2,500 pg/mL)	2739±2338	562±24	3035±85	165±148	3034±67	199±34	3040±199	426±20	184.3±10.4

G1=5-Fu+Tamox+Vit E (n=15), G2=5-Fu+Tamox+Nigella sativa oil (n=15), G3=5-Fu+Tamox+Fish oil (n=15), G4=5-Fu+Tamox+Primrose oil (n=15), G5=Control (n=10), p<0.01*, p=<0.001**.

Table 2. Correlation between platelet and serum VEGF in different groups of study>

Time of Treatment	Studied Groups	Parameter assayed	R	P-Value
Before Treatment	5-Fu+Tamox+Vit E (n=15)	VEGF (15.6-1000 pg/mL) VS Platelets (140-420x10³/ uL)	0.94	0.279*
	5-Fu+Tamox+Nigella sativa oil (n=15)		0.98	<0.001**
	5-Fu+Tamox+Fish oil(n=15)		0.98	<0.001**
	5-Fu+Tamox+Primrose oil (n=15)		0.97	0.254*
After Treatment	5-Fu+Tamox+Vit E (n=15)	VEGF (15.6-1000 pg/mL) VS #Platelets (140-420x10³/ uL)	0.98	0.070*
	5-Fu+Tamox+Nigella sativa oil (n=15)		0.98	<0.001**
	5-Fu+Tamox+Fish oil (n=15)		0.98	<0.001**
	5-Fu+Tamox+Primrose oil (n=15)		0.98	<0.09*

A statistically significant correlation was found between serum VEGF level and #platelet count.

DISCUSSION

Platelets may undergo endocytosis and store various angiogenic factors, such as VEGF and platelet-derived growth factor (PDGF), in their granules; these molecules are then secreted immediately after platelet activation. [22] Accordingly, VEGF/plt ratio was detected to correct the variation in serum VEGF levels by different plt count. [23,24] VEGF/plt ratios of 1.4 pg/10^6 and higher were reported to be a significant risk factor for hepatic vein thrombosis and mortality in patients with hepatic cell carcinoma. [23]

Results of the present study showed that, before treatment, serum VEGF and b-FGF levels were significantly higher in all groups of patients, when compared to healthy control. However, after treatment, serum VEGF levels and VEGF per platelet count were significantly decreased, especially in NS treated group.

Also, Kim et al [23] reported that serum VEGF per platelet count in HCC patients was significantly higher than in liver cirrhosis patients and healthy controls. In addition, serum VEGF per platelet count was higher in patients with advanced stage and portal vein thrombosis ($p<0.01$). Moreover, the present study agreed with a study performed by Gautam et al [25] who proved that the thymoquinon, one of the Nigella sativa constituents, down-regulated the expression of angiogenic VEGF gene products, which result in potentiation of apoptosis induced by tumor necrosis factor and chemotherapeutic agent.

Also, in the present study, there was a statistically significant correlation between serum VEGF level per platelet count, which agreed with the previous study performed by Salgado et al [26] and George et al. [24] These facts would imply the abundance of tumor-originated VEGF in platelets of the HCC patients. These results were consistent with previous studies which proved that there was a direct correlation between platelet counts and serum VEGF. Platelets may transport VEGF to its target cells, restricting the angiogenic activity of circulating VEGF to sites where coagulation takes place, such as a healing wound. [27] The scavenger function of platelets may explain the lack of apparent angiogenesis observed outside the tumor in patients with high amounts of circulating VEGF. This view is supported by the fact that platelet aggregation with cancer cells is commonly observed during the process of angiogenesis and metastasis. [28]

Enomoto et al [29] also proved that the methanol soluble fractions of black cumin oil, which is prepared by compression of seeds of Nigella sativa L, showed inhibitory effects on arachidonic acid (AA)-induced platelet aggregation and blood coagulation, and it had more potent activity than aspirin, which is well known as a remedy for thrombosis.

Hepatocytes contain about 10% of total body pool of GSH. A normal homeostasis of this tripeptide is necessary for the activity of the cytochrome P450 enzyme system and of mitochondria, for the "traffic" from cytoplasm and nucleus and finally for the maintenance of a normal cellular redox status. GSH also affects the transcription of proinflammatory/anti-inflammatory cytokine genes, liver regeneration through cytokine-mediated nuclear factor-binding (NF-B) induction, NO bioavailability, and energy metabolism. [30]

The present study showed that the levels of non-enzymatic antioxidant parameters in serum (glutathione, vitamin C, and vitamin E) were significantly reduced compared with healthy control subjects. Also, the antioxidant activity of superoxide dismutase a SOD (U/g Hb) was significantly reduced in Nigella sativa and fish oil groups, when compared to healthy controls. This result agrees with a study performed by De La Cruz et al [31] who reported that dietary supplementation with Nigella sativa oil, which activates glutathione reductase, decreased tissue oxidative stress through reduction of lipid peroxidation. This may be due to increase of reduced glutathione in all tissues and decrease in oxidized glutathione.

The inhibition of oxidative stress can be due to its contents of thymoquinone, polyphenol, and extra tocopherol, those were reported to be associated with Nigella sativa oil. [32] Also, it was reported that ingestion of ω-3 polyunsaturated fatty acids, without adequate antioxidant protection, could result in increased free radical formation and lipid peroxidation. [33] This can be explained by the antioxidant effect of vitamin E that reduce the oxidative and nitrosative stress, increase GSH pool, and, therefore, counteract the liver disease in liver patients. [34,35] Moreover, the present study agree with study performed by Skrzydlewska et al [36], on 55 cases of a colorectal cancer patients tissues, who reported that the activity of superoxide dismutase, glutathione peroxidase, and glutathione reductase, was significantly increased whereas that of catalase was significantly decreased in tumor tissues collected from colorectal cancer patients receiving vitamin E.

Eicosanoids generated from arachidonic acid (AA), such as prostaglandin E2 (PGE2), leukotriene B4, thromboxane A2, and 12-hydroxyeicosatetraenoic acid, have been positively linked to carcinogenesis. [37] For example, PGE2 promotes tumor cell survival and is found at higher concentrations in cancer cells than in normal cells. [38] The mechanisms, whereby PGE2 promotes tumor survival, include inhibition of apoptosis and stimulation of cell proliferation. [39-41] It has also been reported that PGE2 increases tumor progression by promoting tumor angiogenesis. [42,43] Accordingly, dietary supplementation of long chains of ω-3 fatty acids, like eicosapentaenoic acid and docosahexaenoic acid, competitively inhibits the oxygenation of arachidonic acid by cyclooxygenase.

It has been shown that both the fixed oil of N. sativa, as well as thymoquinone (the main compound of the essential oil), inhibit non-enzymatic lipid perioxidation in liposomes. [44] It has also been shown that compounds isolated from N. sativa have appreciable free radical scavenging properties. [45] The antioxidant property of N. sativa would explain its action against CCl4 hepatotoxicity, [46] liver fibrosis, and cirrhosis, [47] and hepatic damage induced by Schistosoma mansoni infection. [48] The present study revealed that the serum AFP of HCC patients, who showed abnormal serum AFP at time of diagnosis, became significantly better after administration of oil capsules as well as vitamin E approaching to normal levels after treatment with NS, G2.

Cancer cells express tumor-specific antigens that normally would be recognized by the immune system and lead to destruction of the cancer cell; during tumorigenesis, most cancer cells acquire the ability to evade this immunosurveillance.

Although there are multiple mechanisms by which cancer cells evade an immune response, a major mechanism is active cancer cell-mediated immunosuppression via secretion of TGF-β, which is a potent immunosuppressive cytokine. [49] A role for TGF-β in cancer cell-mediated immunosuppression is supported by the following observations: cancer cells are able to produce and secrete TGF-β, [50] which is able to immunosuppress cancer patients in the absence of cytotoxic treatment. [51] TGF-β1, a member of a family of peptide factors regulating cell growth and differentiation, [52] had also been shown to exert potent immunosuppressive effects. [53,54] The immunosuppressive properties of TGF-β1 were shown to modulate the immune response by affecting proliferation, the activation state, and differentiation, of natural killer (NK) cells and cytotoxic T cells. [55,56]

TGF-β1 is expressed in most common forms of cancer. It has been found that elevated serum levels of TGF-β1 were a more sensitive indicator of small hepatocellular carcinomas, than AFP levels, suggesting that measuring serum TGF-β1 levels may be clinically useful for diagnosing these tumors. [57] These findings coincide with the results of the present study. Therefore, in HCC patients, the suppression of NK cell activity, which is believed to play an important role in host antitumor defense mechanisms, appear to be associated with the progression of HCC. [58] Taken together, it is possible that the TGF-β1, produced by HCC cells, may exert negative effect on antitumor immunity against HCC. The results of present study showed marked reduction in the levels of TGF-β1 approaching normal levels after NS oil treatment, which abolishes its negative effect on immunity.

Conclusion

The natural oils, especially Nigella sativa, can improve the antitumor effect of chemotherapeutic agents in HCC, together with the performance status, and act as an angiogenic agent.

Acknowledgment

Thanks are due to Dr Tark Ab-zaid and his El-Capten Pharm Company Cairo –Egypt for their supporting me all scientific services that we needed to complete that research

References

[1] Parkin, DM; Bray, F; Ferlay, J; Pisani, P. Global Cancer Statistics, 2002. *CA Cancer J. Clin.* 2005; 55: 74-108.

[2] El-Serag, HB. Hepatocellular carcinoma and hepatitis C in the United States. *Hepatology* 2002; 36 (Suppl 1): S74-S83.

[3] Arii, S; Yamaoka, Y; Futagawa, S; Inoue, K; Kobayashi, K; Kojiro, M; Makuuchi, M; Nakamura, Y; Okita, K; and Yamada R. Results of surgical and nonsurgical treatment for small-sized hepatocellular carcinomas: A retrospective and nationwide survey in Japan. *Hepatology* 2000; 32: 1224-29.

[4] Sherlock, S; Docley, J. Malignant liver tumors. In: Disease of the liver and biliary system, 11th edition. Oxford: *Blackwell Publishing* Co; 2002: 537-62.

[5] Colombo, M. Malignant neoplasms of the liver. In: Schiff ER, Sorrell MF, Maddrey WC, editors. Schiff's disease of the liver. Philadelphia: *Lippincott Williams and Wilkins*; 2003: 1377-1403.

[6] Bosch, FX; Ribes, J; Borras, J. Epidemiology of primary liver cancer. *Semin Liver Dis.* 1999; 19: 271-85.

[7] Benvegna, L, Gios, M; Boccato, S; Alberti, A. Natural history of compensated viral cirrhosis: a prospective study on the incidence and hierarchy of major complications. *Gut* 2004; 53: 744-9.

[8] Stuart, K; Tessitore, J; Huberman, M. 5-Fluorouracil and Alpha-Interferon in Hepatocellular Carcinoma. *Am. J. Clin. Oncol.* 1996; 19 (2): 136-9.

[9] Begin, ME; Ells, G; Horrobin, DF. Polyunsaturated fatty acid-induced cytotoxicity against tumor cells and its relationship to lipid peroxidation. *J. Natl. Cancer Inst.* 1998; 80: 188-94.

[10] Kenny, FS; Pinder, SE; Ellis, IO; Gee, JM; Nicholson, RI; Bryce, RP. Gamma linolenic acid with tamoxifen as primary therapy in breast cancer. *Int. J. Cancer,* 2000; 85: 643-8.

[11] Harris, NM; Crook, TJ; Dyer, JP; Solomon, LZ; Bass, P; Cooper, AJ; Birch, BR. Intravesical meglumine gamma-linolenic acid in superficial bladder cancer: an efficacy study. *Eur. Urol.* 2002; 42: 39-42.

[12] Belch, JJF; Hill, A. Evening primrose oil and borage oil in rheumatologic conditions. *Am. J. Clin. Nutr.* 2000; 71: 352S-6S.

[13] Javier, A; Menendez Luciano, V; Ramon, C; Ruth, L. Effect of γ-Linolenic Acid on the Transcriptional Activity of the Her-2/neu (erbB-2) *Oncogene Journal of the National Cancer Institute* 2005; 97 (21): 1611-5.

[14] Fortin, PR; Lew, RA; Liang, MH. Validation of a metaanalysis: the effects of fish oil in rheumatoi arthritis. *J. Clin. Epidemiol.* 1995; 48: 379-90.

[15] DeLuca, P; Rothman, D; Zurier, RB. Marine and botanical lipids as immunomodulatory and therapeutic agents in the treatment of rheumatoid arthritis. *Rheum. Dis. Clin. N Am.* 1995; 21: 759-77.

[16] Manku, MS; Morse, N; Belch, JJF. Effects of gammalinolenic acid supplementation on plasma essential fatty acids. *Prog. Lipid. Res.* 1986; 25: 469-73.

[17] Al-Okbi, SY; Soroor, KH; Abdel Samed, AM; Rashed, MM; Mohamed, AD. Nutritional status and blood antioxidant levels of Egyptian rheumatoid arthritis patients. *Kaser El-Aini Medical Journal* 2000; 6: 43-64.

[18] Houghton, PJ; Zarka, R; de las Heras, B; Hoult, JR. Fixed oil of Nigella sativa and derived thymoquinone inhibit eicosanoid generation in leukocytes and membrane lipid peroxidation. *Planta Med.* 1995, 61: 33-6.

[19] Chakravarty, N. Inhibition of histamin release from mast cells by nigellone. *Annals of Allergy* 1993; 70: 237-42.

[20] Li, XM; Tang, ZY; Zhou, G; Lui, YK; Ye, SL. Significance of vascular endothelial growth factor mRNA expression in invasion and metastasis in hepatocellular carcinoma. *J. Exp. Clin. Cancer Res.* 1998; 17: 13-7.

[21] Dvoraket, HF; Brown, IF; Detmar, M; Dvorak, AM. Vascular permeability factor/vascular endothelial growth factor, microvascular hypermeability, and angiogenesis. *Am. J. Pathol.* 1995; 146: 1029-39.

[22] Harriso,n P; Wilbourn, B; Debili, N; Vainchenker, W; Breton-Gorius, J; Lawrie, AS; Masse, JM; Savidge, GF; Cramer, EM. Uptake of plasma fibrinogen into αgranules of human megakaryocytes and platelets. *J. Clin. Invest.* 1989; 84: 1320-4.

[23] Kim, SJ; Choi, IK; Park, KH; Yoon, SY; Oh, SC; Seo, JH; Choi, CW; Kim, BS; Shin, SW; Kim, YH; Kim, JS. Serum vascular endothelial growth factor per platelet count in hepatocellular carcinoma: correlations with clinical parameters and survival. *Jpn. J. Clin. Oncol. 2004;* 34 (4): 184-90.

[24] George, ML; Eccles, SA; Tutton, MG; Abulafi, AM; Swift, RI. Correlation of plasma and serum vascular endothelial growth factor levels with platelet count in colorectal cancer: Clinical evidence of platelet scavenging? *Clin. Cancer Res.* 2000; 6: 3147-52.
[25] Gautam, S; Kwang, S; Bharat, B. Targeting Nuclear Factor-K: Activation Pathway by Thymoquinone: Role in Suppression of Antiapoptotic Gene Products and Enhancement of Apoptosis. *Mol Cancer Res.* 2008; 6 (6): 1059-70.
[26] Salgado, R; Vermeulen, PB; Benoy, I; Weytjens, R; Huget, P; Van Marck, E. Platelet number and interleukin-6 correlate with VEGF but not with bFGF serum levels of advanced cancer patients. *Br. J. Cancer* 1999; 80: 892-7.
[27] Maloney, JP; Silliman, CC; Ambruso, DR; Wang, J; Tuder, RM; Voelkel, NF; et al. In vitro release of vascular endothelial growth factor during platelet aggregation. *Am. J. Physiol.* 1995; 275: 1054-61.
[28] Trikha, M; Nakada, MT. Platelets and cancer: Implications for antiangiogenic therapy. *Semin Thromb Hemos.* 2002; 28: 39-44.
[29] Enomoto, S; Asano, R; Iwahori, Y; Narui, T; Okada, Y; Singab, NAB; Okuyama, T. Hematological Studies on Black Cumin Oil from the Seeds of Nigella sativa L. *Biol. Pharm. Bull.* 2001; 24 (3): 307-10.
[30] Song, BC; Chung, YH; Kim, JA; et al. Transforming growth factor-beta1 as a useful serologic marker of small hepatocellular carcinoma. *Cancer* 2002; 94: 175-80.
[31] De La Cruz, JP; Quintro, L; Villalobos, MA; De, La; Cuesta, FS. Lipid peroxidation and glutathione system in hyperlipemic rabbits: Influence of olive oil administration. *Biochem. Biophys. Acta* 2000; 1485: 36-44.
[32] Nergiz, C; Ötles, S. Chemical composition of Nigella sativa L. seeds. *Food Chem.* 1993; 48: 259-61.
[33] Wu, D; Meydani, SN. n-3 Polyunsaturated fatty acid and immune Function. *Proc. Nutr. Soc.* 1998; 57: 503-9.
[34] Corrales, FJ; Ruiz, F; Mato, JM. In vivo regulation by glutathione of methionine adenosyltransferase S-nitrosylation in rat liver. *J. Hepatol.* 1999; 31: 887-94.
[35] Von Herbay, A; Stahl, W; Niederau, C; Sies, H. Vitamin E improves the aminotransferase status of patients suffering from viral hepatitis C: a randomized, double-blind, placebo-controlled study. *Free Radic. Res.* 1997; 27: 599-605.
[36] Skrzydlewska, E; Sulkowski, S; Koda, M; Zalewski, B; Kanczuga-Koda, L; Sulkowska, M. Lipid peroxidation and antioxidant status in colorectal cancer. *World J. Gastroenterol.* 2005; 11 (3): 403-6.
[37] Rose, DP; Connolly, JM. Omega-3 fatty acids as cancer chemopreventive agents. *Pharmacol. Ther.* 1999; 83: 217-44.
[38] Chulada, PC; Thompson, MB; Mahle, JF; et al. Genetic disruption of Ptgs-1, as well as Ptgs-2, reduces intestinal tumorigenesis in Min mice. *Cancer Res.* 2000; 60: 4705-8.
[39] Cheuk, BL; Chew, SB; Fiscus, RR; Wong, PY. Cyclooxygenase-2 regulates apoptosis in rat epididymis through prostaglandin D2. *Biol. Reprod.* 2000; 66: 374-80.
[40] Tsutsumi, S; Haruna, R; Tomisato, W; et al. Effects of prostaglandins on spontaneous apoptosis in gastric mucosal cells. *Dig. Dis. Sci.* 2002; 47: 84-9.
[41] Leahy, KM; Ornberg, RL; Wang, Y; Zweifel, BS; Koki, AT; Masferrer, JL. Cyclooxygenase-2 inhibition by celecoxib reduces proliferation and induces apoptosis in angiogenic endothelial cells in vivo. *Cancer Res.* 2002; 62: 625-31.

[42] Rose, DP; Connolly, JM. Antiangiogenicity of docosahexaenoic acid and its role in the suppression of breast cancer cell growth in nude mice. *Int. J. Oncol.* 1999; 15: 1011-5.

[43] Cianchi, F; Cortesini, C; Bechi, P; et al. Up-regulation of cyclooxygenase 2 gene expression correlates with tumor angiogenesis in human colorectal cancer. *Gastroenterology* 2001; 121: 1339-47.

[44] Mitsuyoshi, H; Nakashima, T; Sumida, Y; Yoh, T; Nakajima, Y; Iscikawa, H; Inaba, K; Sakamoto, Y; Okanoue, T; Kashima, K. Ursodeoxycholic acid protects hepatocytes against oxidative injury via induction of antioxidants. *Biochem. Biophys. Res. Commun.* 1999; 263: 537-42.

[45] Houghton, PJ; Zarka, R; de las Heras, B; Hoult, JRS. Fixed oil of Nigella sativa and derived thymoquinone inhibit eicosanoid generation in leukocytes and membrane lipid peroxidation. *Planta Med.* 1995; 61: 33-6.

[46] Burits, M; Bucar, F. Antioxidant activity of Nigella sativa essential oil. *Phytotherapy Research* 2000; 14: 323-8.

[47] Nagi, MN; Alam, K; Badary, OA; Al-Shabanah, OA; Al-Sawaf, HA; Al-Bekairi, AM. Thymoquinone protects against carbon tetrachloride hepatotoxicity in mice via an antioxidant mechanism. *Biochem. Mol. Biol. Int.* 1999; 47: 153-9.

[48] Türkdogan, MK; Agaoglu, Z; Yener, Z; Sekeroglu, R; Akkan, HA; Avci, ME. The role of antioxidant vitamins (C and E), selenium and Nigella sativa in the prevention of liver fibrosis and cirrhosis in rabbits, new hopes. *Dtscch. Tierarzt Wschr.* 2000; 108: 71-3.

[49] Kawanaka, H; Tarnawski, AS. PGE (2) stimulates VEGF expression in endothelial cells via ERK2/JNK1 signaling pathways. *Biochem Biophys Res. Commun.* 2001; 286: 923-8.

[50] Kirkbride,KC; Blobe, GC. Inhibiting the TGF-beta signalling pathway as a means of cancer immunotherapy. *Expert Opin. Biol. Ther.* 2003; 3: 251-61.

[51] Bodmer, S; Strommer, K; Frei, K; et al. Immunosuppression and transforming growth factor-beta in glioblastoma: Preferential production of transforming growth factor-beta 2. *J. Immunol.* 1999; 143: 3222-9.

[52] Tsuchiyama, T; Nakamoto, Y; Sakai, Y; Marukawa, Y; Kitahara, M; Mukaida, N ; Kaneko, S. Hepatocellular Carcinoma Chemoattractant Protein-1 against Therapy Combined with Monocyte Antitumor Effects of Suicide Gene Prolonged, NK Cell-Mediate. *J. Immunol.* 2007; 178; 574-83.

[53] Blobe, GC; Schiemann, WP; Lodish, HF. Role of transforming growth factor-b in human disease. *N. Engl. J. Med.* 2000; 342: 1350-8.

[54] Nakamura, K; Kitani, A; Strober ,W. Cell contact-dependent immunosuppression by CD4(+)CD25(+) regulatory T cells is mediated by cell surface-bound transforming growth factor beta. *J. Exp. Med.* 2001; 194: 629-44.

[55] Brandes, ME; Wakefield, LM; Wahl, SM. Modulation of monocyte type I transforming growth factor-beta receptors by inflammatory stimuli. *J. Biol. Chem.* 1991; 266: 19697-703.

[56] Gorelik, L; Flavell, RA. Transforming growth factor-beta in T-cell biology. *Nat. Rev. Immunol.* 2002; 2: 46-53.

[57] Von Bernstorff, W; Voss, M; Freichel, S; et al. Systemic and local immunosuppression in pancreatic cancer patients. *Clin. Cancer Res.* 2001; 7: 925s-32s.

[58] Mahmoud, MR; El-Abhar, HS; Saleh, S. The effects of Nigella sativa oil against the liver damage induced by Schistosoma mansoni in mice. *J. Ethnopharmacol.* 2002; 79: 1-11.

In: Asia Pacific Life Sciences Compendium
Editor: Bandit Chumworathayi

ISBN: 978-1-62257-338-7

Chapter 14

EVALUATION OF IMMUNE STATUS IN PATIENTS WITH BREAST CANCER WHO RECEIVED SOME NATURAL SUPPLEMENTS AS ADJUVANT THERAPY AFTER COMPLETION OF ANTI-CANCER TREATMENTS

Ragaa Hosny Mohamad,[1*]
Mohamad Gamil A Abdel-Monaem Al-Siad,[2] Zekry Khalid Zekry,[3]
Mohamad Reda Hamza,[3] Mostafa Shawky El-Haddad,[4]
Siad Ebrahim El-Shaieb,[5]
Hussain Abdel Al-Rahman Al-Mehdar,[6] Sabry Mohamad Sharawy[1]
and Mahmoud Mohamad El-Marzbani[1]

[1]Department of Medical Biochemistry and Cancer Biology, National Cancer Institute, Egypt
[2]Department of Surgical Oncology, National Cancer Institute, Egypt
[3]Department of Medical Oncology, National Cancer Institute, Egypt
[4]Department of Radiation Oncology, Kasr Al Ainy Hospital, Cairo University, Egypt
[5]Department of Internal Medicine, Menofia University Hospital, Egypt
[6]Department Microbiology, Faculty of Science, King Abdul Aziz University, Saudi Arabia

ABSTRACT

Objective: The aim of the present study is to investigate the chemopreventive efficacy of nigella sativa, Bio Bran, and vitamin E, in the breast cancer patient.

Method: One hundred and sixty eight diagnosed breast cancer patients treated with six cycles of standard chemotherapy (CMF) followed by radiotherapy and Tamoxifen for at least one year. Subgroups of 42 patients each received in addition the above

[*] Corresponding author: Ragaa Hosny Mohamad. Cairo University, Egypt, Tel: +0201001856864, Fax: +0223642077, E-mail: ragaatahtawy@gmail.com.

mentioned, treated with either Nigella sativa, Bio Bran, or vitamin E. 20 healthy subjects, served as a control group, were included in this study. The following parameters were investigated; IL-18, CEA, CA15-3, sFas, sFas-L, VEGF, IL-2, IL-6, IFN-γ. Result: The data so far obtained revealed that different supplements improve all parameters under investigated, especially in group treated with Nigella sativa.

Conclusion: Supplementation with Nigella sativa in the breast cancer patients is proved to be useful in reducing recurrence and improve disease free survival.

Keywords: nigella sativa, bio bran, Vitamin E, breast cancer

BACKGROUND

Breast cancer is the most common malignancy among women in the world, with nearly a million new cases each year. It accounts for nearly 21% of all cancers among women worldwide. [1] According to estimates in 2002, there were 1,151,298 new cases, 410,712 deaths secondary to breast cancer, and more than 44 million women living with breast cancer worldwide. [2]

Nigella sativa (NS) or Black seed is considered as the curative black cumin in the Holy Bible and was described as Melanthion by Hippocrates and Dioscorides and as *Gith* by Pliny. [3] The seeds of NS (family: Ranunculaceae), has been used in folk medicine in the Arabian Gulf region, Far East Asia, and Europe. The pharmacological investigations of the seed extracts revealed a broad spectrum of activities including immuno-potentiation [4] and antihistaminic, [5] together with immunomodulation, [6] and anti-tumour activity. [7] Many of these activities have been attributed to the quinone constituents of the NS seed. [8,9] Furthermore, Nigella sativa preparations may have a chemopreventive effect and may reduce the toxicity of standard antineoplastic agents. [10]

Ali and Blunden [11] proved that *Nigella sativa* contain thymoquinone, which have cytoprotective and antioxidant actions. Furthermore, Salomi et al. [12] stated that the crude methanol extract of Nigella sativa, exhibited a strong new triterpene saponin (α-hederin), showed significant dose-dependent tumor inhibition.

Bio Bran, MGN-3, a non-toxic glyconutritional food supplement, was prepared by breaking down rice bran with enzymes isolated from the Shitake mushroom (Lentius edodes mycelia). Bio Bran act as a biological response modifier, increase natural killer cell activity in immunocompromised patients and may enhance the depleted immune systems. [15] Bio Bran has cytotoxic effect on the breast cancer cell line (MCF-7) and it increase the percentage of apoptotic MCF-7 cells and accelerates phagocytosis-induced apoptosis of cancer cells, proved by Ghoneum and Gollapudi. [16]

The vitamin E family is a group of compounds divided into two subgroups; tocopherols and tocotrienols. Tocotrienols display more potent apoptotic activity in mammary cancer cells. Although the mechanisms mediating tocotrienol-induced apoptosis is not fully understood, apoptosis is carried out by activation of initiator caspases (caspase-8 or -9) that subsequently activate effector caspases (caspase-3, -6, or -7), [17] It also inhibit the growth of preneoplastic and neoplastic mammary epithelial cells. [18]

The present study aimed to evaluate the chemo-preventive efficacy and reduce the risk of breast cancer recurrence and metastases by using three commonly dietary supplements, namely as Nigellar (Nigella sativa), and Lentin plus 1000 (Bio Bran), as well as vitamin E, on patients who received chemotherapy (CMF) followed by radiotherapy and/or tamoxifen.

METHOD

This work was performed at the Surgical Department, National Cancer Institute, Cairo University, Egypt. Hundred and sixty-eight patients with confirmed diagnosis of breast cancer, and twenty healthy control patients, with non-malignant pathology, were also included in this study. Informed written consent was obtained from all patients, and the local ethical committee of the National Cancer Institute approved the protocol. Patients were followed up from January 2003 till December 2007. Patients were checked regularly at a 3-6 month interval after the operation in the outpatient department.

Surgical Procedure

For tumors less than 3cm, breast conserving surgery was the procedure of choice, unless there was a contraindication (major intra-duct component, multi-centric tumors, or malignant-looking widespread micro-calcifications). For all other patients, modified radical mastectomy was the procedure of choice.

Drugs

For CMF; Cyclophosphamide (Cytoxan) was purchased from Bristol-Myers Squibb Company, Methotrexate and 5-fluorouracil were purchased from Sigma Hospira, Inc, US. Tamoxifen (Nolvadex) was purchased from ZENC Company. Nigellar capsules, vitamin E capsule were purchased from Pharco Company, Egypt. Lentin Plus 1000 (Bio Bran) (MGN-3), 1.4 gm sachets, was purchased from Daiwa Pharmaceutical Co, Japan.

Parameters under Investigations

The following parameters were performed in the sera for each patient. Analyses were performed using commercially kit with the ELISA technique. CEA and CA15-3 were performed using kit purchased from Abbott Japan Company. Interleukin-2, interleukin-6, and interleukin-18 were performed using kit purchased from Biosource International, US. The kit for interferon-γ was purchased from Diaclone Company, France. The kits for IgG1 and IgG2a were purchased from Southern Biotechnology, Birmingham, AL, US. Vascular endothelial growth factor (VEGF) was performed for estimating the angiogenesis assay using kit purchased from Biosource International. sFas and sFasL were performed using kit purchased from RandD Systems, Minneapolis, MN, US, for apoptosis. All of them were measured postoperatively and after six month post chemotherapy + hormonal therapy and supplements therapy. The patient's characteristics are given in Table 1.

Study Design

Group I composed of disease free, healthy women as control group (n=20). Group II composed of 168 breast cancer patients after surgery, and before start all types of treatment, blood were collected from all patient (n=168).

Table 1. Characteristic of breast cancer patients in the study group

Characteristics		*N*	%
Age	Median: 49 years		
	Range: 35-63 years		
Tumour size (T)	T1	4	2.46
	T2	134	79.69
	T3	16	9.52
	T4	14	8.33
Nodal status (N)	N0	26	15.48
	N1	128	76.19
	N2	13	7.71
	N3	1	0.62
	N4	0	0
Metastasis (M)	M0	168	100
	M1	0	0
Menopausal status	Pre-menopausal	0	0
	Post-menopausal	168	100
Histology	Ductal invasive	166	98.8
	Lobular invasive	2	1.2
Surgery	Conservative	2	1.2
	Radical Mastectomy	2	1.2
	Simple	11	6.55
	Modified radical	150	89.87
	No surgery	2	1.2
Family history of cancer	Yes	9	5.36
	No	159	94.64
Diet	Mixed	164	97.62
	Vegetarian	4	2.38
Treatment	Chemo+Radio therapy +tamoxifen	42	25
	Chemo+Radio therapy +tamoxifen+ Nigella Sativa	42	25
	Chemo+Radio therapy tamoxifen+ Bio Bran	42	25
	Chemo+Radio therapy +tamoxifen Vitamin E	42	25

N=number of patients, %=percentage of patients.

All patients were treated with CMF (cyclophosphamide 600mg/m^2, methotrexate 40mg/m^2, and 5-fluorouracil 600mg/m^2), every 3 weeks for 6 cycles, followed by radiotherapy, and then followed by tamoxifen 20 mg once daily (Nolvadex 10mg/tab, ZENC Company) for at least one year (n=42).

Then the patients were divided into four equal subgroups (N=42 each) as follows; Group A received no further treatment, Group B was treated with Nigellar capsule, 1 gm/8 hours for six months, Group C was treated with Lentin Plus 1000 (Bio Bran), 1.4 g sachets twice/day for six months, and Group D was treated with 100 mg vitamin E capsule (Pharco Company, Egypt) twice/day for six months. All supplements were taken one hour before meal.

Blood Collection

Five milliliters of venous blood were collected in a sera separator tube (Vacutainer, Becton Dickinson) and immediately centrifuged after clotting at 3000 rpm for 10 min. and stored at -80°C until used.

Statistical Analyses

Statistical Package for Social Sciences (SPSS), version 12, was used. Quantitative variables were summarized using mean and SD, median, minimum, and maximum values. Qualitative data were summarized using frequencies and percentage.

Comparison between different measurement values before and after treatment within each group were analyzed using paired-t test, combined with ANOVA. Comparison of differences between the means of the treatments were at the significance level of $p<0.05$, and highly significant when $p<0.01$. [19]

Result

In the present study, no serious side effect was observed during Nigellar administration together with chemotherapy and tamoxifen treatment compared to the other treatment groups. The commonest side-effect was increased body temperature in 4 patients reaching 38.8°C, and others in the range of 38.2-38.6°C responded to called foments and simple antipyretics. Vitamin E group showed no side effect.

Table 2 shows the serum value of tumor marker, cytokines, apoptotic markers and angiogenic factor levels. In group I, control subjects, the tumor marker levels were significantly lower ($p<0.05$) than the other groups and none of the control subjects had tumor marker levels higher than the cut off value. In group II, patients before start of treatment, the mean CEA and CA15-3 levels were found to be higher than other groups under investigation and healthy control ($p<0.05$).

CEA and CA15-3 levels of group A patients treated with chemotherapy followed by radiotherapy and tamoxifen for more than 1 year were found to be significantly lowered but its level were still higher than other groups ($p<0.05$).

Table 2. Serum levels of tumor marker apoptosis biomarkers and angiogenic factor in breast cancer patients

Groups	Tumor Marker		Apoptotic Marker		Angiogenic Factor
	CEA NV=<3 (ng/ml)	CA15-3 NV=5 (U/ml)	sFas NV=16 (ng/ml)	sFasL NV=19 (ng/ml)	VEGF NV=25 (pg/ml)
Group I (20)					
Control, normal,	2.07±0.18	5.59±1.36	38.3± 12	20.7±151	26.8±4.4
Group II (168)					
Pre-treatment	18.45± 4.13*	50.33±12.75	991.8±299*	382.3±166 *	78·9±118·0*
Group A (42)					
Treatment with R.th+ch.th+ tamoxifen	7.8±3.2*	36.42±9.64*	429.9±404*	202.3±142*	50·1±71·6*
Group B (42)					
Six month after treatment with Nigella Sativa along with ch.th + tamoxifen	3.76±1.52*	10.22±4.67*	69.3±89.7*	52.9±166*	24·1±72·6*
Group C (42)					
Six months after treatment with Bio Bran, along with ch.th + tamoxifen	4.8±1.5*	26.09±7.23*	85.4±97.3*	65.4±173	36.6±78·9*
Group D (42)					
Six months Vitamin E along with ch.th + tamoxifen	5.12±1.32*	18.22±4.67*	92.7±94.5*	67.5±166*	40·1±41.9*

Values are expressed as mean±SD. Number are expressed as mean±SD. Number of subjects are indicated in-groups and the Statistical significance expressed as *p<0.05.

Patients treated with Nigella sativa (group B), Bio Bran (group C), as well as Vitamin E (group D), showed a significant reduction (p<0.05) in CEA and CA15-3 levels.

The cytokines values were evaluated by estimating the IL-2 levels in pretreatment patients showing significant lower levels than healthy control group. However, in group A, IL-2 was elevated, but not to the control level. Addition of Nigella sativa, in group B, significantly elevated IL-2 (p<0.05) to near the normal value. In group C and D, the IL-2 levels were also improved, but less than patient received Nigella sativa (group B). IL-18 levels in pretreatment patients recorded significant higher (p<0.05) than control healthy group, and the mean value of IL-18 levels (n=168) was 441.6±193, which is much higher than the IL-18 levels in healthy control subject. In group A, which patients received chemotherapy followed by radiotherapy and tamoxifen, the IL-18 level was comparable to the control healthy group. In group B, further decrease of the IL-18 levels were induced by Nigella sativa. -6 level in group B was significantly reduced, compared with control. Its level was also improved in all treated groups, especially in group B treated with Nigella sativa, and reached near the normal level. The same pattern of effect was observed with IFN-γ, IgG1 and IgG2a levels. Patients treated with Nigella sativa showed the nearest value of them, compared to the control group.

Angiogenesis Assay

Serum VEGF levels increased significantly in untreated breast cancer patients. Serum VEGF levels decreased significantly after treatment with *Nigella sativa,* Bio Bran, and vitamin E, in comparison to tamoxifen treated group (group A) (p=0.051).

Apoptotic Assay

Serum levels of sFas and sFasL were low in healthy control subjects (group I), while they were significantly increased in pre-treatment breast cancer patients. In patients who were treated with standard dose of chemotherapy and hormonal therapy followed by radiation therapy alone (group A), both of them were lowered. The patients who were treated with standard dose of chemotherapy and hormonal therapy followed by radiation therapy along with *Nigella sativa* (group B), Bio Bran (group C), and vitamin E (group D), both of them were decreased much more than group A (Table 2). The concentration of IFN-γ in group of patients received chemotherapy, hormonal, and radiation therapy, and Nigella sativa (group B) had remarkably increased, compared with the breast cancer patients underwent standard dose of chemotherapy and hormonal therapy followed by radiation therapy alone (group A) (p=0.009) (Table 3). Figure 1 shows that breast cancer patients treated with Nigella sativa had better overall survival when compared to other treated groups under investigation.

DISCUSSION

Dietary and environmental factors are thought to play a role in the progression of a tumor from its latent into its clinical phase, and if certain dietary factors and supplements delay tumor progression in the latent phase, it is not unlikely that they may also be effective in delaying tumor progression in an established tumor. [20] A prospective study using CA15-3 as a discriminator of two populations after the patient has received the Nigellar, Lentin Plus 1000, and vitamin E capsule, after the patients received the courses of chemotherapy, in added to tamoxifen, should be performed to define a population that shows therapeutic response, and another population which has not responded to the chemotherapy or other type of treatment.

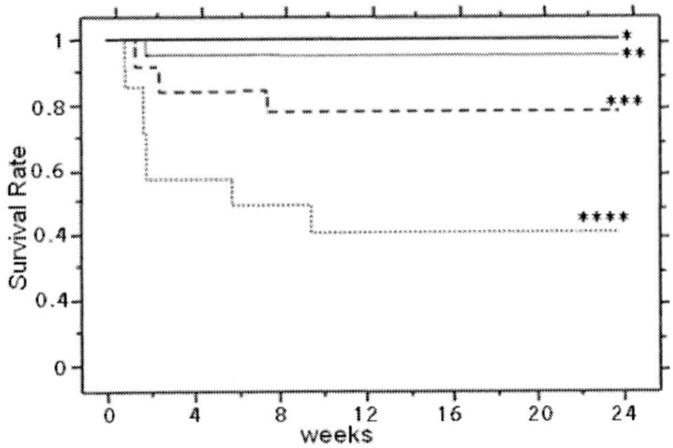

*Nigella SATIVA. **Lentin plus1000.
***Vitamin A. ****CH.th+Tamoxifen.

Figure 1. Survival rate in patients diagnosed with breast cancer, determined by the Kaplan-Meyer method. Patients were divided into five groups based on the type of treatment. The survival rate was compared between different types of treatment in breast cancer patient.

Table 3. Serum levels of cytokines and antibody marker in breast cancer patients

Groups	Cytokines				Sub-class of Immunoglubulin	
	IL-18 NV=255.6 (pg/ml)	IL-2 NV=7 (pg/ml)	IL-6 NV=<9.9 (pg/ml)	IFN-γ NV=24.58 (pg/ml)	IgG1 NV=17.31 (IU/ml)	IgG2 NV=4.5 (g/L)
Group I (20)						
Control, normal	256±106	7.50±1.37	9.70±1.75	24±0.01	17.26±0.01	4.32±0.03
Group II (168)						
Pre-treatment	441.6±193*	1.87±1.23*	2.5±19.4*	7.7±0.6*	1.52±0.25*	1.98±0.29*
Group A (42)						
Treatment with ch.th + tamoxifen	352.1± 139.8*	2.98±0.49*	4.74±1.29*	10±0.1*	5.05±0.04*	2.77±0.29*
Group B (42)						
Six month after treatment with Nigella Sativa along with ch.th + tamoxifen	265.5±80.7*	4.13±1.12*	8.19±1.46*	23.3±0.6*	13.23±0.0 1*	3.96±0.61*
Group C (42)						
Six months after treatment with Bio Bran, along with ch.th + tamoxifen	282.1±73.7*	3.35±0.96*	7.86±1.7*	19±0.31*	7.2±0.33*	2.63±0.28*
Group D (42)						
Six months Vitamin E along with ch.th + tamoxifen	291.9±71.0*	3.14±0.81*	5.94±1.9*	13±0.27*	6.1±0.27*	2.22±0.46*

Values are expressed as mean±SD. Number of subjects are indicated in-groups. Statistical significance expressed as *$p<0.05$.

The present data agree with several studies have shown that, in approximately 90% of the metastatic patients, serum levels of CEA and CA 15-3 correlate directly with the response of the disease to the treatment. [15,16] Hence, elevations of tumor marker levels during the treatment phase could be a promising indicative of disease progression or regression. [17]

In the present study, the administration of Nigella sativa, Bio Bran, as well as vitamin E, in breast cancer patients, had indicated a tumor suppressive effect, because it caused marked reduction completely compared with serum levels of CA15-3 and CEA. This was due to the specificity of Nigella sativa that could possibly be explained by the autoxidation of its flavonoids. It has been shown, that the rate of autoxidation of flavonoids decreases with increasing number of hydroxyl group at a neutral pH (ability to chelate oxygen radicals), by Miura et al. [21] Also, the present results agree with Siassi and Ghadirian [22] who reported that riboflavin influences epithelial integrity, tissue flavin concentrations, rate of prostaglandin biosynthesis, and glutathione metabolism, in addition to the study performed by Ziegler et al. [23] which proved that the riboflavin administration was found to decrease the risk of developing cancer by increasing the levels of total flavins, which has the capacity to capture reactive metabolites, and thereby decrease the carcinogen binding to cellular macromolecules, like DNA.

The tumor marker levels were significantly lower ($p<0.05$) than the other groups and none of the control subjects had tumor marker levels higher than the cut off value. Similar observation was reported by Robertson et al. [24] who demonstrate high tumor marker levels in patients with breast carcinoma than in control subjects, pregnant women, and in patients with benign disease. Earlier studies have reported elevated tumor marker levels of CEA and CA15-3 in preoperative and untreated breast cancer patients when compared to treated patients groups. [17] CEA and CA15-3 tumor marker levels above the cut off value are considered to yield poor prognosis to treatment and the risk of developing cancer recurrence and metastasis is increased in these patients. [24]

The supplements were well tolerated and no side effects were reported by any of the participant. The study suggests that supplementation of *Nigella sativa*, Bio Bran, as well as vitamin E, along with tamoxifen, to breast cancer patients, reduces the serum tumor marker levels of CEA and CA15-3, and VEGF, while there are high reductions in CEA and CA15-3, and VEGF levels in the group received *Nigella sativa* as supplement, thereby offering better cancer prognosis by reducing the risk of developing cancer recurrence and metastasis, because the seeds of *Nigella sativa* act as antiangiogenesis according to the study performed by Medenica et al. [25], which proved that the *Nigella sativa* seeds extract has antiangiogenic activity in cancer cell line.

Moreover, *Nigella sativa* seeds extract has highly antitumor principles. [12] Consequently, tamoxifen inhibits the expression of estrogen-regulated genes, including growth factors and angiogenic factors secreted by the tumor, that may stimulate growth by autocrine or paracrine mechanisms. [26]

The present data showed lower concentrations of VEGF and higher concentrations of IL-18 than those in the study group before treatment, as well as the patients received chemotherapy and tamoxifen alone. This agrees with study performed by Jablonska et al. [27] which demonstrated markedly elevated serum levels of VEGF as well as IL-18 according to the progression of the disease.

Fas and FasL also exist in a soluble form released from cell surfaces after cleavage by metalloproteinases. [28] sFas is formed due to cleavage of the external part of extracellular

Fas and acts as a FasL inhibitor to bind Fas and prevent Fas-mediated apoptosis. Tanaka et al. [28] claimed that immune cells, essentially lymphocytes, NK and T cells, were the most important sources of sFas, in response to a developing tumor. In cancer, malignant cells inhibit the expression of membrane-bound Fas and express FasL, which triggers tumor-infiltrating lymphocyte apoptotic cell death. [29] In contrast to their membrane-bound forms, soluble sFas and sFasL exhibit different patterns. The levels of sFas and sFasL have been measured independently in separate studies in different populations of normal subjects, [30] subjects with lung cancer, [31] and subjects with prostate cancer. [32]

The present data revealed that the serum sFas level was elevated in untreated breast cancer patients, then it was lesser decreased in the same patients after standard treatment, compared to those patients received the supplements. These results were supported by Naumnik et al. [33] who suggest that the cancerous cells are the main source of sFas in the serum of patients with lung cancer. The enhanced serum concentration of an apoptosis inhibitor (sFas), in patients with lung cancer, reflects an intense inhibition of cancerous cells apoptosis, which promotes the development of the tumor. Also, they revealed that the sFas concentration was higher after chemotherapy than before treatment, whereas patients with distant NSCLC metastases had higher sFas concentration than patients without metastases proved.

The elevated levels of sFas associated with a poor prognosis, but decreased serum sFas levels associated with good prognosis. Apoptosis mediated by the Fas-FasL system is suggested to be involved in the progression of malignant tumors and the response to therapy. It is reasonable to assume that the relationship between a high level of serum sFas and a worse prognosis in patients with ovarian carcinoma, may be dependent on a mechanism for the escape of tumor cells from the immune surveillance of NK cells and CTLs. [32]

The experiments performed by Gopalakhrisnamoorthy and Bhaskaran [34] showed that the IgG2a is indeed correlated with tumor eradication induced by the IL-12 vaccine, whereas IgGl induced by vaccination with IL-4-producing cells is not. The lack of correlation between the IgGl level and the antitumor effect remains to be studied. The present study found that there was a significant decrease in serum IgG1 and IgG2a levels after treatment with Bio Bran or *Nigella sativa* when compared with the standard treated group with tamoxifen and chemotherapy alone. From these results, we can suggest that *Nigella sativa* has an immunoregulatory effect more than Bio Bran, and this effect may be favorable for improving the performance status of cancer patients. It is well known that IgG1 are antibodies produced under activation of Th2. [35]

Tamoxifen inhibits allergic responses when given preventively and also other therapeutic types, and improves allergen-induced dermatitis. These data agree with the study performed by Babina et al. [36] which found that the tamoxifen inhibits allergic responses when given preventively and also therapeutically, and improves allergen-induced dermatitis. Because of its effectiveness, tamoxifen could bear significant therapeutic potential for the treatment of allergies leading to a reduction of allergen-specific Ig levels (IgE, IgG1 and IgG2a).

The present data revealed that, in the pretreatment, serum IL-2 levels have been shown to be lesser than serum IL-2 levels after treatment, in patients with breast cancer. This value in groups received Nigella sativa as supplemented increased and its value was near the value of control healthy. This was due to the IL-2 cytokine released from T helper lymphocytes. It promotes the generation, proliferation, and differentiation, of T lymphocytes, enhances the activity of natural killer cells, induces the generation of lymphokine-activated killer cells, and promotes the production of antibodies by B lymphocytes. Through these mechanisms, it plays an important role in antitumor immune responses. [37] Fu et al. [38] demonstrated that the level of some major Th1 cytokines in peripheral blood was significantly increased after therapy and could activate anti-tumor immune effect cells, leading to inducing the production of Th1 cytokines causing enhancement of the anti-tumor immune function of the body, and killing tumor cells.

The present results demonstrate that IL-2 and IL-6 were significantly elevated after receiving tamoxifen while IL-18 was decreased. In addition, the present results demonstrated that cytokines of Th1 (IL-2, IL-6, IFN-γ, and IL-18) were activated by both *Nigella sativa* and Bio Bran, and they induced significant changes in the levels of IL-18, IL-2, IL-6, and IFN-γ ($p<0.05$), when compared to healthy control persons. These results supported the findings of study performed by Haq et al. [39] which found that *Nigella sativa* may offer immunoregulatory effect and its purified proteins were found in mixed lymphocyte cultures to cause changes in the levels of IL-1beta, IL-8, and TNF-alpha. Also, the present data agreed with the several studies proving that the bronchial asthma, inflammation, and cancer development, are associated with elevated Th2 pattern of cytokine production. [40-42]

Conclusion

This study suggests the effectiveness of supplementing Nigellar, Lentin plus 1000, and vitamin E, on breast cancer patients under treatment with CMF every 3 weeks for 6 cycles and tamoxifen, which can reduce the serum tumor marker levels and hence reduce the risk of cancer recurrence and metastases. *Nigella sativa* induced down regulation of IgG isotypes and cytokines response in breast cancer patients than the other supplemented used in the present study. So, *Nigella sativa,* rather than Bio Bran, may be used as an effective immunoregulatory treatment for cancer patients. Moreover, the serum sFas level was elevated in patients with advanced cases treated with tamoxifen and cytotoxic drug, but in group treated with NS, it was diminished than group treated with Bio Bran or vitamin E. Following the strategy of the adjuvant therapy may lead to good prognosis.

Acknowledgment

Thanks are due to Dr Tark Ab-zaid and his El-Capten Pharm Company Cairo –Egypt for their supporting me all scientific services that we needed to complete that research

References

[1] Parkin, D. M., Pisani, P., Ferlay, J. Estimates of the worldwide incidence of 5 major cancers. *Int. J. Cancer* 1999; 80: 827-41.

[2] Veronesi, U., Boyle, P., Goldhirsch, P., Orecchia, R., Viale, G. Breast cancer. *Lancet* 2005; 365: 1727-41.

[3] Worthen, D. R., Ghosheh, O. A., Crooks, P. A. The *in vitro* anti-tumor activity of some crude and purified components of blackseed, *Nigella sativa* L. *Anticancer Res.* 1998; 18: 1527-32.

[4] Medinica, R., Mukerjee, S., Huschart, T., Corbitt, W. Immunomodulatory and anticancer activity of *Nigella sativa* plant extract in humans. *Proceedings of the American Association for Cancer Research Annual Meeting*; 1994: A2865.

[5] Mahfouz, M., Abdel-maguid, R., El-Dakhakhny, M. The effect of nigellon therapy on the histaminopexic power of blood sera asthmatic patients. *Arzneimittelforschung* 1965; 15: 1230-1.

[6] Al-Hader, A., Aqel, M., Hassan, Z. Hypoglycaemic effects of the volatile oil of *Nigella sativa*. *Int. J. Pharmacognosy* 1993; 31: 96-100.

[7] El Tahir, K. E., Ashour, M. M., Al-Harbi, M. M. The cardiovascular actions of the volatile oil of the black seed (*Nigella sativa*) in rats: elucidation of the mechanism of action. *Gen. Pharmacol.* 1993; 24: 1123-31.

[8] Houghton, P. J., Zarka, R., de las Heras, B., Hoult, J. R. Fixed oil of *Nigella sativa* and derived thymoquinone inhibit eicosanoid generation in leukocytes and membrane lipid peroxidation. *Planta Med.* 1995; 61: 33-6.

[9] El-Alfy, T., El-Fatatry, H., Toama, M. Isolation and structure assignment of an anti-microbial principle from the volatile oil *Nigella sativa* L. *Pharmazia* 1975; 30: 109-11.

[10] El-Kadi, A., Kandil, O. The black seed (Nigella sativa) and immunity: its effect on human T cell subset. *Fed. Proc.* 1987; 46: 1222.

[11] Ali, B. H., Blunden, G. Pharmacological and toxicological properties of *Nigella sativa*. *Phytother. Res.* 2003; 17: 299-305.

[12] Salomi, N. J., Nair, S. C., Jayawardhanan, K. K., Varghese, C. D., Panikkar, K. R. Antitumour principles from *Nigella sativa* seeds. *Cancer Lett.* 1992; 63: 41-6.

[13] Salomi, N. J., Nair, S. C., Panikkar, K. R. Inhibitory effects of *Nigella sativa* and saffron (*Crocus sativus*) on chemical carcinogenesis in mice. *Nutr. Cancer* 1991; 16: 67-72.

[14] Swamy, S. M. K., Tan, B. K. H. Extraction, isolation and characterization of anti-tumor principle, alpha-Hederin, from the seeds of *Nigella sativa*. *Planta Medica* 2001; 67: 29-32.

[15] Ghoneum, M. Enhancement of human natural killer cell activity by modified arabinoxylan from rice bran (MGN-3). *Int. J. Immunotherapy 1998*; 14: 89-100.

[16] Ghoneum, M., Gollapudi, S. Induction of apoptosis in breast cancer cells by Saccharomyces cerevisiae, the baker's yeast, in vitro 2004; 24 (3a): 1455-63.

[17] Shah, S., Gapor, A., Sylvester, P. W.. Role of Caspase-8 Activation in Mediating Vitamin E-Induced Apoptosis in Murine Mammary Cancer Cells. *Nutrition and Cancer* 2003; 45 (2): 236-46.

[18] Shah, S., Sylvester, P. W. Tocotrienol-induced caspase-8 activation is unrelated to death receptor apoptotic signaling in neoplastic mammary epithelial cells. *Exp. Biol. Med.* 2004; 229: 745-55.

[19] Saunders, D. B., Trapp, G. R. *Basic and clinical biostatistics*, 3rd edition. Connecticut: Appleton and Lang; 2001.

[20] Hoenjet, K. M., Dagnelie, P. C., Delaere, K. P. J., Wijckmans, N. E. G., Zambon, J. V., Oosterhof, G. O. N. Effect of Coenzyme Q_{10}, Riboflavin and Niacin on Serum CEA and

CA 15-3 Levels in Breast Cancer Patients Undergoing Tamoxifen Therapy. *Eur. Urol.* 2005; 47: 433-40.

[21] Miura, Y. H., Tomita, I., Watanabe, T., Hirayama, T., Fukui, S. Active oxygens generation by flavonoids. *Biol. Pharm. Bull.* 1998; 21: 93-6.

[22] Siassi, F., Ghadirian, P. Riboflavin deficiency and esophageal cancer: a case-control-household study in the Caspian Littoral of Iran. *Cancer Detection and Prevention*, 2005, 29 (5): 464-9.

[23] Ziegler, R. G., Lim, U. One-Carbon Metabolism, Colorectal Carcinogenesis, Chemoprevention with Caution. *J. Natl. Cancer Inst.* 2007; 99 (16): 1214-5.

[24] Robertson, J. F. R., Jaeger, W., Syzmendera, J. J., Selby, C., Coleman, R., Howell, A., Winstanley, J., Jonssen, P. E., Bombardieri, E., Sainsbury, J. R. C., Gronberg, H., Kumpulainen, E., Blamey, R. W. Cancer detection methods and reagents. *Eur. J. Cancer* 2003; 35: 47-53.

[25] Medenica, R., Janssens, J., Tarasenko, A., Lazovic, W., Carbitt, D., et al. Antiagiogenic activity of *Nigella sativa* plant extract in cancer therapy. *Proc. Annu. Meet. Am. Assoc. Canc. Res.* 1977; 38: A1377.

[26] Stearns, V., Yamauchi, H., Hayes, D. F. Circulating tumor markers in breast cancer: accepted utilities and novel prospects. *Breast Cancer Res. Tr.* 1998; 52: 239-59.

[27] Jablonska, E., Puzewska, W., Grabowska, Z., Jablonskiand, J., Talarek, L. VEGF, IL-18 and NO production by neutrophils and their serum levels in patients with oral cavity cancer. *Cytokine* 2005; 30 (7): 93-9.

[28] Tanaka, M., Suda, T., Takahashi, T., Nagata, S. Expression of the functional soluble form of human ligand in activated lymphocytes. *EMBO J.* 1995; 14: 1129-35.

[29] O'Brien, D. I., Nally, K., Kelly, R. G., O'Connor, T. M., Shanahan, F., O'Connell, J. Targeting the Fas/Fas ligand pathway in cancer. *Expert Opin. Ther. Targets* 2005; 9: 1031-44.

[30] Ichikura, T., Majima, T., Uchida, T., Okura, E., Ogawa, T., Mochizuki, H. Plasma soluble Fas ligand concentration: decrease in elderly men and increase in patients with gastric carcinoma. *Oncol. Rep.* 2001; 8: 311-314.

[31] Tanaka, T., Yoshimi, M., Maeyama, T., Hagimoto, N., Kuwano, K., Hara, N. Resistance to Fas-mediated apoptosis in human lung fibroblast. *Eur. Respir. J.* 2002; 20: 359-368.

[32] Abrahams, V. M., Kamsteeg, M., Mor, G. The Fas/Fas ligand system and cancer: immune privilege and apoptosis. *Mol. Biotechnol.* 2003; 25: 19-30.

[33] Naumnik, W., Iżycki, T., Ossolińska, M., Chyczewska, E. Serum levels of sFas and sFasL during hemotherapy of lung cancer. *Experimental Oncology* 2007; 29 (2): 132-6.

[34] Gopalakhrisnamoorthy, M., Bhaskaran, A. Woman With Edema, Vomiting, and Diarrhea. *Consultant* 2008; 48 (2): 1-2.

[35] Robinson, D. S., Hamid, Q., Ying, S., Tsicopoulos, A., Barkans, J., Bentley, A., Corrigan, C., Durham, S., Kay, B. Predominant Th2-like bronchoalveolar Tlymphocyte population in atopic asthma. *N Engl. J. Med.* 1992; 326: 298-304.

[36] Babina, M., Kirn, F., Hoser, D., Ernst, D., Rohde, W., Zuberbier, T., Worm, M. Tamoxifen counteracts the allergic immune response and improves allergen-induced dermatitis in mice. *Clin. Exp. Allergy* 2010; 40 (8): 1256-65.

[37] Mosmann, T. R., Cherwinski, H., Bond, M. W., Giedlin, M. A., Coffman, R. L. Two types of murine helper T cell clone. I. Definition according to profiles of lymphokine activities and secreted proteins. *J. Immunol.* 1986; 136 (7): 2348-57.

[38] Fu, Q. G., Meng, F. D., Shen, X. D., Guo, R. X. Efficacy of intraperitoneal thermochemotherapy and immunotherapy in intraperitoneal recurrence after gastrointestinal cancer resection. *World J. Gastroenterol.* 2002; 8 (6): 1019-22.

[39] Haq, A., Lobo, P. I., Al-Tufail, M., Rama, N. R., Al-Sedairy, S. T. Immunomodulatory effect of *Nigella sativa* proteins fractionated by ion exchange chromatography. *Intern. J. Immunophar.* 1999; 21: 283-95.

[40] Gong, F. L. *Medical Immunology*. 1st ed. Beijing, China: Science Publication; 2003: 238-52.

[41] Kalus, U., Pruss, A., Bystron, J., Jurecka, M., Smekalova, A., Lichius, J. J., Kiesewetter, H. Effect of *Nigella sativa* (black seed) on subjective feeling in patients with allergic diseases. *Phytother. Res.* 2003; 17 (10): 1209-14.

[42] Ormstad, H., Groeng, E. C., Duffort, O., Lovik, M. The effect of endotoxin on the production of IgE, IgG, and IgG2a antibodies against the cat allergen field in mice. *Toxicology* 2003; 30: 309-18.

In: Asia Pacific Life Sciences Compendium
Editor: Bandit Chumworathayi
ISBN: 978-1-62257-338-7

Chapter 15

EFFECT OF DIETARY LIPID LEVELS ON GROWTH PERFORMANCE, FEED UTILIZATION AND BODY COMPOSITION OF JUVENILE SILVER POMFRET, *PAMPUS ARGENTEUS* (EUPHRASEN, 1788)

M. A. Hossain,* S. M. Almatar and C. M. James
Mariculture and Fisheries Department, Kuwait Institute for Scientific Research, Safat, Kuwait

ABSTRACT

A study was conducted to determine the efficiency of lipid utilization by juvenile silver pomfret (Pampus argenteus). Duplicate groups of fish (34.3±1.2g) were fed three isonitrogenous (49% protein) diets containing 12, 16, and 20% lipid for 4 months. Significantly ($P<0.05$) higher weight gain and specific growth rates were observed in fish fed diet containing 16% lipid with a protein to energy (P:E) ratio of 22.6. A further increase in lipid level did not produce any improvement in the growth and feed utilization. There were no significant differences between the feed conversion ratio, protein efficiency ratio, apparent net protein utilization and energy retention in fish fed 16% and 20% lipid diets. Whole body lipid and fatty acid profiles were influenced by dietary lipid levels. Higher eicosapentaenoic acid (EPA) and docosahexaenoic acid (DHA) were accumulated in fish body fed 16 and 20% lipid diets. This study demonstrated that there is no beneficial effects in terms of growth and nutrient retention beyond 16% dietary lipid and a dietary lipid level of 16% in a 49% protein diet corresponding to a P:E ratio of 22.6 is optimum for better growth, feed utilization and whole body fatty acid profiles in silver pomfret.

Keywords: lipid, growth, feed, body composition, silver pomfret

---

* Corresponding author: M. A. Hossain. PO Box # 1638, 22017 Salmiya, Kuwait, Tel: +965 25711296 Ext 6327, Fax: +965 25711293 E-mail: drmdarshadhossain@yahoo.co.in.

BACKGROUND

Fish are known to utilize protein preferably to lipid or carbohydrate as an energy source. Therefore, it is important from a nutritional, environmental and economic point of view to improve protein utilization for tissue synthesis rather than for energy purposes [1]. The optimization of dietary protein to energy ratio (P:E) has proven to have an important role on protein and energy utilization [2]. Dietary lipids play an important role in fish nutrition as a source of energy and essential fatty acid to maintain biological structure and normal function of cell membranes [3].

It is well established that for the majority of the cultured fish species that the protein efficiency utilization can be improved through the utilization of lipids and carbohydrates in the diet [4]. It is important to provide adequate levels and ratios of protein, lipid and carbohydrate in diets in order to reduce catabolism of protein for energy [5,6]. Lipids are well utilized by most fishes and can improve diet utilization within certain limits [7,8], partly spare protein [9], and reduce organic matter and nitrogen loss to the aquatic environment [10]. Any imbalance with respect to non-protein energy source and/or their level of inclusion may have a direct effect on growth, conversion efficiencies, nutrient retention and body composition. If the dietary P:E ratio is unbalanced so that non-protein energy is inadequate, dietary protein may be catabolised and used as an energy source to satisfy maintenance before growth [11,12].

It is thus, essential to determine the optimum dietary lipid level as energy source that produces best growth, feed conversion and improve nutrient retention and body composition. Although dietary lipid can be used as a non-protein energy source, excessive dietary lipid levels can have a negative impact on body composition [13,14]. Excessive lipid in diets can lead to decrease in feed consumption, which in turn will reduce intake of other nutrients including protein [5]. Therefore, evaluating the effects of increasing non-protein energy through the supplementation of lipid is important.

The silver pomfret, *Pampus argenteus* (Euphrasen) is a valuable food fish of worldwide demand with a wide geographical distribution from East China sea to the Southeast Asia, Indian Ocean and the Arabian Gulf [15]. Recently Mariculture and Fisheries Department (MFD) of Kuwait Institute for Scientific Research (KISR), Kuwait has been successful in the spawning of captive silver pomfret [16] and presently research is being carried out to develop a commercial culture technology for this fish. However, the nutritional requirements of silver pomfret have not been studied in detail which is needed to develop a cost-effective commercial feed.

Some preliminary studies on the growth performance and feeding behavior of this fish under tank culture conditions using several commercially available feeds and feeds mixed with shrimp meat and feed additives have been tried [17,18,19]. However, no conclusive results were obtained on the efficiency of these feed components on the growth of this fish. It is important to define an optimal lipid level in diets for juveniles of silver pomfret. Dietary lipid levels also appear to have an important impact on carcass lipid and muscle quality. Accordingly, it is also important to evaluate the effect of dietary lipid levels on the muscle fatty acid profile of pomfrets.

In a recent study we estimated the optimum dietary protein requirement for silver pomfret to be as 49% [20] but a 12% dietary lipid level resulted in lower whole body lipid content

(3.9%) compared to that of the wild pomfret (8.0%) and we consider that such dietary lipid level may not be optimum for growth and feed utilization of silver pomfret.

The main objective of the present study was to determine the effects of increasing dietary lipid levels on growth, feed efficiency, whole body proximate composition, nutrient and energy retention, and whole body fatty acid composition of juvenile silver pomfret.

METHOD

Experimental System

The experiment was carried out using 1-m^3 capacity round fiber glass tanks each containing about 800 L of water. The tanks were painted black internally and the bottom had a 2° slope towards a 5-cm diameter central drainage stand pipe which provided water drainage from the bottom. The tanks were covered with nets to prevent the fish from jumping out. Filtered and UV treated ground water (with salinity close to that of sea water) and ambient sea water were mixed and flowed through the tanks of an open flow-through system at the rate of 10 L min^{-1}.

The water temperature in experimental tanks was maintained at $28.0^{0}C$ using in-line heater "Aqualogic" (model HTI-8-220, 8 KW from Aquatic Ecosystems, US) in the main inflow line. Continuous aeration was provided by two air stones using 1.5-inch PVC air-lift pipes inside the tanks to avoid air bubbles in the culture system. Fluorescent lights were used to provide a light intensity of 400-480 lux above the water surface and natural photoperiod of 12h light and 12h dark was maintained throughout the study period. The feeding trial was conducted for 4 months during November 2009 to February 2010.

Source of Fry and Acclimation

Hatchery reared silver pomfret juveniles originated from the eggs collected from wild spawners were obtained from the hatchery at MFD, KISR. For handling the fish while stocking and sampling 3 ppm of the tranquilizer "quinaldine" (Argent Chemical Laboratory, Redmond, US) was used. Prior to the initiation of the experiment, fish were randomly stocked in the respective experimental tanks and acclimated to the experimental condition for one week.

Feed Formulation and Diet Preparation

Three semi-purified experimental diets were formulated to contain 12, 16, and 20% lipid level with corresponding P:E ratio of 23.66, 22.62 and 21.60 for Diets 1, 2 and 3 respectively. The diets were formulated to contain in the order of 49% protein [20]. The experimental diets were prepared using a mixture of three commercial diets (Gemma diamond- 0.8mm, Gemma Micro- 300 and Biomar- 3mm). Prior to formulation of experimental diets, the proximate composition of these commercial diets were analyzed (Table 1).

Table 1. Proximate composition of feeds (% dry matter basis) used in experimental diet formulation

Feed/ingredients	Dry matter	Crude protein	Lipid	Ash	Crude fibre	NFE[a]
GEMMA diamond[b]	91.05	58.00	14.50	9.85	1.10	16.55
GEMMA Micro-300[c]	91.10	55.00	15.00	13.5	2.05	14.45
BIOMAR[d]	90.67	50.97	12.38	7.23	4.45	24.97

[a]Nitrogen free extract calculated as 100-%(crude protein+lipid+ash+crude fibre).
[b]Skretting, Cheshire, England (particle size-0.8mm).
[c]Skretting, Cheshire, England (particle size-300 micron).
[d]Biomar, France (particle size-3 mm).

Table 2. Formulation (%) of the experimental diets

Ingredients	Diet-1 (12% lipid)	Diet-2 (16% lipid)	Diet-3 (20% lipid)
GEMMA (Diamond)	40.00	40.00	40.00
GEMMA Micro-300	30.00	30.00	30.00
Biomar feed (3mm)	18.00	18.00	18.00
Fish oil from manhaden[a]	-	4.00	8.00
Vitamin premix[b]	1.00	1.00	1.00
Mineral premix[c]	1.00	1.00	1.00
Nucleosac[d]	0.50	0.50	0.50
Probiofeed[e]	0.50	0.50	0.50
Agar (binder)	1.00	1.00	1.00
Wheat flour	5.00	2.00	-
A-cellulose	3.00	2.00	-
Total	100.00	100.00	100.00

[a]From Sigma-Aldrich Chemie GmbH, Deisenhofen, Germany.

[b]Vitamin premix contained the following mixed with cellulose (g/ kg mix): thiamine, 2.5 g; riboflavin, 2.5 g; pyridoxine, 2.0 g; pantothenic acid, 5.0 g; inositol, 100 g; biotin, 0.30 g; folic acid, 0.75 g; para- amino benzoic acid, 2.5 g; choline, 200 g; niacin, 10.0 g; cyanocobalamine, 0.005 g; retinol palmitate (Vit-A), 100,000 IU; α-tocopherol acetate (Vit-E), 20.1 g; ascorbic acid, 50.0 g; menadione, 2.0 g, cholecalciferol (Vit-D), 500,000 IU.

[c]Mineral premix contained the following ingredients (g/ kg): calcium orthophosphate (2H20), 727.78; magnesium sulfate (7 H20), 127.50; sodium chloride, 60.00; potassium chloride, 50.00; ferrous sulfate (7 H20), 25.00; zinc sulfate (7 H20), 5.50; manganese sulfate (4 H20), 2.5375; copper sulfate (5 H20), 0.7875;cobalt sulfate (7 H20), 0.4775; calcium iodide, 0.2950; chromium chloride, 0.1275.

[d]Nucleosac-Antistress-growth promoter (Zymonutrient Pvt. Ltd, Kaernatak, India).

[e]Probiofeed- a probotics from Zymonutrient Pvt. Ltd, Karnataka, India.

Fish oil from menhaden (Sigma-Aldrich Co., Germany) was used as supplemental lipid. Diets were also supplemented with vitamin premix (1%), mineral premix (1%), binder (1%), nucleosac (0.5%) and probiofeed (0.5%).

In Diets 1 and 2, wheat flour and α-cellulose were used as filler or bulking material to keep the dietary lipid content at a desired level. In order to prepare diets, all the dietary ingredients were finely ground and sieved to pass through 0.5 mm mesh. All dietary ingredients were weighed as per formulae (Table 2) and mixed thoroughly with an electrical mixer machine (Hobart mixer, Model A200; Hobart Corporation, Ohio, US).

During mixing oil was poured gradually and mixed thoroughly to ensure homogeneity. The mixed feed was stored in air-tight container at -20°C until used for feeding. Before daily feeding each feed mixture was moist pelleted by adding about 35-40% fresh water and again made into paste/dough by hand before presentation to fish.

Addition of agar as binder resulted in a dough with suitable soft and sticky consistency which helped to adhere the feed at the bottom of the bowl (feeding tray) for more than two hour. This reduced the rate of disintegration of the feed. The diets were subjected to proximate and fatty acid analysis and the results are shown in Table 3.

Table 3. Analyzed proximate and fatty acid (% of total fatty acids) composition of the experimental diets (% dry matter basis)

Proximate composition	Diet-1 (12% lipid)	Diet-2 (16% lipid)	Diet-3 (20% lipid)
Dry matter	92.08	92.44	92.83
Crude protein	49.22	49.41	49.24
Lipid	12.37	16.51	20.32
Ash	9.37	9.21	9.07
Crude fibre	4.08	3.62	3.01
NFE[a]	24.96	21.25	18.36
Gross energy (kJ/g)[b]	20.80	21.84	22.80
P:E ratio[c]	23.66	22.62	21.60
Fatty acids[d]			
C14	5.92±0.12	7.21±0.14	7.35±0.14
C15	0.44±0.03	0.58±0.03	0.61±0.03
C16	31.17±0.51	30.32±0.48	28.00±0.44
C17	0.60±0.06	0.42±0.06	0.44±0.05
C18	6.19±0.12	5.98±0.15	6.07±0.12
C20	0.45±0.02	0.29±0.02	0.34±0.00
C16:1	3.43±0.09	5.09±0.10	5.96±0.05
C17:1	0.29±0.00	0.44±0.00	0.57±0.02

Table 3. (Continued)

Proximate composition	Diet-1 (12% lipid)	Diet-2 (16% lipid)	Diet-3 (20% lipid)
C18:1n-9	15.13±0.14	15.27±0.11	15.58±0.09
C20:1	4.62±0.11	3.79±0.08	3.83±0.07
C22: 1n-9	3.21±0.7	2.54±0.06	2.70±0.09
C24:1	Nd	Nd	Nd
C18:2n-6	16.37±0.14	12.60±0.14	11.57±0.10
C18:n-3	1.51±0.09	1.62±0.07	1.76±0.04
C18:3n-6	Nd	Nd	Nd
C20:3 n-3	0.29±0.03	0.45±0.04	0.48±0.01
C20:3n-6	0.11±0.002	0.19±0.00	0.31±0.01
C20:4 n-6	Nd	Nd	Nd
C20:5 n-3, EPA	6.25±0.09	7.76±0.10	8.41±0.09
C22:6 n-3, DHA	4.02±0.05	5.55±0.04	6.02±0.04
∑SFA[e]	44.47	44.70	42.81
∑MUFA[e]	26.68	27.13	28.64
∑PUFA[e]	28.55	28.17	28.55
∑n-3	12.07	15.38	16.67
∑n-6	16.48	12.79	11.88
n-3/n-6 ratio	0.73	1.20	1.40

[a]Nitrogen free extract (total carbohydrate) calculated as 100-%(crude protein+lipid+crude fibre+ash).
[b]Estimated according to NRC [12] using the values of 23.6, 39.5, and 17.2kJ/g for crude protein, lipid and total carbohydrate respectively.
[c]Protein to energy ratio in mg protein kJ/g of gross energy.
[d]Values are mean ±SD (n=3).
[e]SFA, saturated fatty acid; MUFA, mono-saturated fatty acids; PUFA, polyunsaturated fatty acids Nd, not detected.

Experimental Procedure, Feeding and Sampling of Fish

Juvenile silver pomfret (initial mean body weight 34.3±1.2g) were randomly distributed at a density of 13 fish per tank with two replicates. It may be mentioned that due to lack of physical facilities available, it was not possible to use more than two replicates.

Since the fish are reluctant to take pelleted feeds, they were offered the diets in paste/dough form. Fish were offered each diet to apparent satiation thrice daily, 7 days a week.

The feed was offered at 9.00 and 14.00 h and 18.00 h in white plastic bowls placed at the bottom of the tanks. As mentioned earlier, the feeds supplied in paste/dough form had a soft but sticky consistency which helped to adhere the feed at the bottom of the feeding bowl resulting in reduced disintegration of feeds.

However, complete prevention of disintegration and nutrient leaching could not be ascertained in such form of feeding used. Two hours after feeding the bowls left with any uneaten feed were lifted very carefully, filtered on fine-meshed (50 μ) plankton net and dried in an oven at 60°C to calculate the amount of feed consumed. Mortality of fish if any was recorded.

Besides weighing at the beginning and at the end of the feeding trial, the fish were weighed once in every month to observe the growth rate. All the fish in the experimental tanks were weighed individually during each sampling using a digital sensitive balance (Model SBC-61, Scaltec Instruments, Gottingen, Germany with 0.01 g sensitivity). At the beginning of experiment ten fish from the stock were collected and frozen at -80°C, freeze dried and finely ground for subsequent proximate and fatty acid analysis. At the end of the experiment, three fish from each replicate tank were sampled in the same manner for whole-body proximate and fatty acid analysis.

Water Quality Parameters

The water quality parameters such as water temperature, dissolved oxygen, pH were monitored daily through the experimental period. The total ammonia and salinity were measured weekly. All the parameters are found to be suitable for fish growth and survival. The ranges were: temperature 28.0-28.5°C, pH 7.1-7.8, dissolved oxygen 5.82-6.67 mg/L, salinity 38-40 g/L and total ammonia 0.04-0.07 mg/L.

Analytical Methods and Statistical Analysis

The proximate composition of fish samples, commercial and experimental diets was analyzed in triplicate according to the following procedure [21]: moisture content by drying in an oven at 105°C for 24h; crude protein content (Nx6.25) by the Kjeldahl method using an Auto Kjeldahl System (Kjeltec™ 2300 Foss Tecator AB, Hoganas, Sweden), lipid by ether extraction (Soxtec System HT6, Tecator AB, Hoganas, Sweden), ash by incineration in a muffle furnace at 600°C for 6h. Nitrogen free-extracts were calculated from the difference after determining crude fiber by digestion with 1.25% H_2S0_4 and 1.25% NaOH solutions (Fibertec System M, Tecator AB, Hoganas, Sweden). Lipid content of samples for fatty acid analysis was extracted by the Bligh and Dyer method [22].

Fatty acid composition was determined by preparing methyl esters and analyzing them by gas chromatography [23]. An HP 6890 gas chromatograph (Hewlett-Packard, Palo Alto, CA, US) equipped with an Chromapack column (CP-Sil 88 50 meter, ID 0.25mm, Varian Inc, Palo Alto, CA, US) was used for the analysis. The oven temperature program was as follows: initial temperature 50°C and increased 10°C per minute to final temperature 235°C, hold for 7 minutes. Fatty acid methyl ester standard mixture comprising 25 different fatty acids (ranging from C8 to C22:6) were obtained from Altech Associates, Deerfield, US. Fatty acids were identified by comparison of retention times with a mixture of standards containing all the fatty acids identified in this study. Each fatty acid was quantified by calculating its peak area relative to the total peak area. These values are referred to as fatty acid content (%) throughout the paper. An estimated amount of each fatty acid was calculated from the lipid content and fatty acid content, which approximately corresponds to g fatty acid per 100g lipid.

Specific growth rate (SGR), feed conversion ratio (FCR), protein efficiency ratio (PER) and apparent net protein utilization (ANPU) were calculated according to standard methods [24]. Data were subjected to one way ANOVA using MSTATC package and if significant differences were found, Tukey's test was used to rank the group. Statistical analyses were carried out using a package Super ANOVA, version 1.11, Abacus Concepts, Berkley, CA, US.

RESULT

Analyzed dietary crude proteins levels were similar and the lipid contents were 12.37, 16.52 and 20.32% with corresponding P:E ratio (mg protein/kJ energy) of 23.66, 22.62, and 21.60 for diets 1, 2, and 3 respectively. The individual fatty acid profile in different experimental diets varied depending on the dietary lipid contents (Table 3). Growth performances and feed utilization of juvenile silver pomfret fed experimental diets are shown in Table 4. The mean final weight of fish was significantly ($P<0.05$) higher with diet 2 (16% lipid) than with diet 1 (12% lipid) and diet 3 (20% lipid). The daily weight gain (DWG, g/fish/day) and SGR (%/day) was also higher in diet 2 than those of diets 1 and 3. However, there was no significant ($P>0.05$) differences in daily feed intake (DFI) among experimental groups.

There was no significant difference in FCR and PER values between diets 2 and 3. Apparent net protein utilization (ANPU) and energy retention (ER) in silver pomfret showed similar trend. There was no significant ($P>0.05$) differences in ANPU and ER values between diets 2 and 3 respectively, but these values were significantly higher than those in diet 1. The lipid retention (LR) values among different diets differed significantly ($P<0.05$) and diet 2 with 16% lipid resulted in the best LR values (83.95%). However, there was no significant ($P>0.05$) difference in survival (%) of fish among different dietary groups which ranged between 96 and 100%.

Fish fed diet 1 with 12% lipid had significantly ($P<0.05$) higher whole body moisture and lower lipid than those fed diet 2 with 16% lipid and diet 3 with 20% lipid (Table 5). However, there was no significant ($P>0.05$) differences in whole body moisture and protein contents of fish fed diets 2 and 3, respectively. Increase of dietary levels of lipids significantly affected the whole body lipid levels and fish fed diets 2 and 3 had significantly ($P<0.05$) higher levels of whole body lipid than those fed diet 1.

However, there was no significant difference in whole body ash contents in different dietary groups which ranged between 2.80 and 2.97%.

Except C18, there was no significant difference ($P>0.05$) among the whole body saturated fatty acids (SFAs) of different dietary groups and palmitic acid (C16:0) was the dominant SFA accounting about 26-27% of total fatty acids (Table 5).

Table 4. Growth performance and feed utilization of juvenile silver pomfret fed diets containing varying lipid levels and P:E ratios[1]

Parameters	Diet-1 (12% lipid)	Diet-2 (16% lipid)	Diet-3 (20% lipid))
Mean initial weight (g)	33.1±1.3^a	35.0±0.8^a	34.9±0.6^a
Mean final weight (g)	67.1±2.69^a	85.0±1.13^b	75.0±1.70^a
Mean weight gain (g)	34.0±1.41^a	50.0±1.91^b	40.1±1.13^a
DWG(g/fish/day)[2]	0.28±0.01^a	0.42±0.02^b	0.34±0.01^c
SGR (%/day)[3]	0.59±0.01^a	0.74±0.03^b	0.64±0.01^a
DFI (%)[4]	1.43±0.03^a	1.41±0.02^a	1.31±0.04^a
FCR[5]	2.33±0.06^b	1.88±0.04^a	2.00±0.06^a
PER[6]	0.87±0.03^a	1.08±0.02^b	1.02±0.03^b
ANPU(%)[7]	13.19±0.84^a	19.94±0.49^b	18.85±0.34^b
ER (%)[8]	11.61±0.46^a	21.82±1.12^b	20.42±0.35^b
LR(%)[9]	33.22±0.54^c	83.95±5.03^a	61.49±0.81^b
Survival (%)	96±5.6^a	100±0.0^a	100±0.0^a

[1]Values (mean±SD) in a row with different superscripts are significantly different determined by Tukey's test ($P<0.05$).

[2]DWG, daily weight gain (g/fish/day).

[3]Specific growth rate = 100 x (In [final body weight] – In [initial body weight] / experimental period (days)).

[4]DFI, daily feed intake (%) = Feed intake (dry matter) x 100 / [initial fish weight + final fish weight) x days fed/2].

[5]FCR, feed conversion ratio = Dry feed fed / live weight gain.

[6]PER, protein efficiency ratio = Live weight gain / crude protein fed.

[7]ANPU, apparent net protein utilization = (final fish body protein – initial fish body protein) / (total protein fed) x 100.

[8]ER, energy retention (%) = 100 x (energy gain, KJ/energy intake, K J).

[9]LR, lipid retention (%) = 100 x (lipid gain/lipid intake).

Among monosaturated fatty acids (MUFAs) only C16:1 and C22:1 content in whole body of fish fed diets 2 and 3 were significantly ($P<0.05$) higher than those fed diet 1. However, C18:1n-9 was the most dominant MUFA which accounted for about 22% of total fatty acids. Fish fed diets 2 and 3 had significantly ($P<0.05$) higher EPA and DHA content than those fed diet 1. The ∑SFA, ∑MUFA, and ∑PUFAs accounted 40-44%, 38-40%, and 18-21%, respectively, of the total fatty acids.

However, there was no significant ($P>0.05$) differences among the whole body ∑SFA, ∑MUFA and ∑n-6 fatty acid contents of different dietary groups. On the other hand, the whole body ∑PUFA, ∑n-3 fatty acids and n-3/n-6 ratios in fish fed diets 2 and 3 were significantly higher than those fed diet 1.

Table 5. Whole body proximate and fatty acid composition of silver pomfret fed diets containing varying dietary lipid levels and P:E ratios[1]

Components	Initial (all fish)	Diet-1 (12% lipid)	Diet-2 (16% lipid)	Diet-3 (20% lipid)
Proximate composition (%fresh matter basis):				
Moisture	80.52	78.98±0.13^{b}	72.48±0.76^{a}	73.530.36^{a}
Crude protein	13.41	14.28±30^{a}	16.27±0.22^{b}	16.00±0.02^{b}
Lipid	3.28	3.72±0.09^{a}	7.99±0.53^{b}	7.32± 0.08^{b}
Ash	2.31	2.80±0.09^{a}	2.91±0.15^{a}	2.97±0.17^{a}
Fatty acid composition (% of total fatty acids):				
C14	6.50	6.68±0.30^{a}	6.22±0.22^{a}	7.16±0.48^{a}
C15	0.69	0.60±0.04^{a}	0.56±0.01^{a}	0.58±0.12^{a}
C16	29.13	26.95±0.61^{a}	26.90±0.43^{a}	25.93±0.34^{a}
C17	0.35	0.81±0.01^{a}	0.77±0.2^{a}	0.78±0.18^{a}
C18	6.59	6.44±0.55^{b}	5.75±0.09^{ab}	5.13±0.37^{a}
C20	0.28	0.29±0.09^{a}	0.31±0.04^{a}	0.35±0.01^{a}
C16:1	5.99	4.75±0.16^{a}	5.85±0.12^{b}	6.82±0.02^{c}
C17:1	0.29	0.31±0.02^{a}	0.33±0.05^{a}	0.27±0.04^{a}
C18:1n-9	22.61	22.61±0.98^{a}	21.90±0.61^{a}	22.07±0.69^{a}
C20:1	6.36	6.14±0.28^{a}	5.96±0.10^{a}	5.90±0.28^{a}
C22: 1n-9	6.07	5.46±0.49^{b}	3.76±0.06^{a}	3.78±0.45^{a}
C24:1	0.44	0.75±0.06^{b}	0.68±0.05^{ab}	0.51±0.06^{a}
C18:2n-6	7.65	11.09±1.04^{a}	10.44±0.13^{a}	10.43±0.32^{a}
C18:n-3	0.48	0.64±0.03^{a}	0.70±0.03^{a}	0.73±0.08^{a}
C18:3n-6	Nd	Nd	Nd	Nd
C20:3 n-3	0.26	0.87±0.02^{a}	0.89±0.05^{a}	0.73±0.07^{a}
C20:3n-6	-	Nd	Nd	Nd
C20:4 n-6	-	Nd	Nd	Nd
C20:5 n-3, EPA	3.24	2.24±0.28^{a}	3.80±0.04^{b}	3.79±0.02^{b}
C22:6 n-3, DHA	2.97	3.4±0.09^{a}	5.18±0.11^{b}	5.03±0.21^{b}
∑SFA[2]	43.54	41.76±0.98^{a}	40.51±0.81^{a}	39.93±0.83^{a}
∑MUFA[2]	41.76	39.91±0.47^{a}	38.46±0.55^{a}	39.52±0.69^{a}
∑PUFA[2]	14.60	18.24±0.63^{a}	21.00±0.26^{b}	20.70±0.06^{b}
∑n-3	6.95	7.15±0.42^{a}	10.56±0.13^{b}	10.27±0.26^{b}
∑n-6	7.65	11.09±1.04^{a}	10.44±0.13^{a}	10.43±0.32^{a}
n-3/n-6 ratio	0.90	0.65±0.10^{a}	1.01±0.00^{b}	0.99±0.05^{b}

[1]Values (mean±SD) in row with different superscripts are significantly different determined by Tukey-Kramer's test ($P<0.05$).

[2]SFA, saturated fatty acid; MUFA, mono-saturated fatty acids; PUFA, polyunsaturated fatty acids Nd, not detected.

DISCUSSION

In the present study, increase of dietary lipid levels from 12 to 20% significantly improved the growth and feed utilization in silver pomfret juveniles. Results of number of studies with fish have shown improved growth and feed utilization efficiency with increasing dietary lipid [25,26,27]. In contrast, no differences in growth rate and feed efficiency was observed in European sea bass juveniles fed diets containing lipid (cod-liver oil) at levels ranging from 12 to 24% [1]. Similarly, no significant effect was found on weight gain, feed and protein efficiency ratios of channel catfish with supplementation of 0, 3, 6, and 9% lipid (menhaden oil) to a commercial floating catfish diet containing 5.61% lipid [28].

The best growth performance and feed utilization was obtained with diet 2 having 16% lipid and the decline in growth and feed utilization with increasing dietary lipid above this level was observed in the present study. Similar results have been reported in turbot [29], salmon [30], rainbow trout [31] and carp [32]. However, as mentioned earlier some reports showed no effect of increasing dietary lipid on body weight gain in juvenile turbot [33], Atlantic cod [34] and surubim [35].

It is assumed that fish as well as homoeothermic animals adjust feed intake in order to satisfy their digestible energy requirements [4]. The gross energy level in diets for the present study varied from 20.8 to 22.8 kJ/g and the feed intake decreased with increase of dietary lipid level although the differences were not statistically significant. A negative correlation between dietary energy level and feed intake was reported for rainbow trout [36], red sea bream [37], common dentex [38], Atlantic cod [39] and tiger puffer [27]. In European sea bass, a significant reduction of feed intake occurred only when dietary lipid levels were increased from 12 to 30% but not from 12 to 24% [8]. Some authors believed that lack of protein sparing effect in sea bass by dietary lipid may be related to the high protein level of diet [1] while according to others [40], the beneficial effect of an increase of lipid level from 10 to 18% in sea bass diets were significant only when with a low protein diet, but not with a high protein diet. But in the present study, although the dietary protein content (490g/kg) was relatively high, when lipid level was below 20%, the protein utilization increased with increasing lipid level. This suggests, even in high protein level diets, the protein sparing effect by lipid is possible with a low upper limit. The significant decrease in lipid retention with increasing lipid level (above 16%) suggests an increased proportion of lipid used for energy. This agrees with observation in rainbow trout that the highest lipid diet did not promote the highest lipid retention [41]. Similarly decreasing lipid retention were also reported in European sea bass when dietary lipid increased from 12 to 30% [1]. A decrease in energy retention in striped bass, *Morone saxatilis* [42] and mutton snapper, *Lutjanus analis* [43] fed highest dietary lipid levels, suggesting that at these lipid levels the fish had reached a metabolic limit for efficient catabolism. Thus, in the present study no further protein sparing was observed for fish diet containing lipid levels higher than 16% suggesting that silver pomfrets had reached their limit of lipid utilization.

Since fish can meet a substantial part of their energy requirement from dietary protein, the P:E ratio is extremely influential on the efficiency of protein and energy utilization. In the present study, the best results in terms of growth and feed utilization were obtained with 16% lipid diet, corresponding to a P:E ratio of 22.66. This value is similar to the P:E ratio of 21-22 reported for European sea bass [1,40] but slightly lower than reported optimum value of 23.7

for sunshine bass [44]' 25.9 for Japanese sea bass [45], 30.62 for Asian bass [46] and 27.5–29.5 for mutton snapper [43]. However, the optimum P:E ratios in fish may vary depending on the species, dietary P:E levels and ingredients used in feed formulation [45].

The SGRs in the present study (0.59-0.74%) were lower than those observed in a previous study with the same species (1.5-2.1%) [20]. As specific growth rates decrease with increasing fish size, these differences in SGR are probably due to a higher initial weight of fish used in the present study (Table 4). However, the overall growth rate of silver pomfrets fed the experimental diets in the present study was slightly better than those observed in silver pomfret fingerlings (DWG of 0.34g/fish/day) fed a Turbot feed containing520g/kg protein [17]. The DWG of fish in diet 2 (16% lipid) in the present study was 0.42g/fish/d.

The FCRs in the present study (1.88-2.33) were very similar to those observed with the same species fed fish meal based experimental diets (1.70-2.50) in our previous study [20]. The best FCR (1.88) obtained with diet 2 in the present study is also similar to that of 1.87, and lower than 2.08 reported for pomfret juveniles fed gemma and salmon feeds respectively [19]. While increasing the lipid content up to certain level with diets having same protein content in the present study led to an improvement in growth performance and feed utilization suggesting that protein may be utilized for growth rather than energy. This conclusion is supported by the PER values observed with increased lipid levels in our study. A similar trend was also found with ANPU values. ANPU values (13.19-19.94%) in the present study are more or less similar to those observed (17.5-22.7%) in our previous study [20]. In the present study, ANPU or protein retention of pomfret fed diets containing more than 16% lipid tended to decrease which agreed with other studies [29,35] indicating that highest levels of dietary lipid does not always improve protein utilization.

The increase of dietary lipid levels is usually associated with an increase of whole-body lipid content [1]. In this study, the increase of dietary lipid level from 12 to 16% also increased the whole body lipid deposition, although a further increase of dietary lipid to 20% did not improve any further increase of lipid deposition. This agrees with the findings in rainbow trout that the highest lipid diet did not promote highest lipid retention [41]. The whole body crude lipid content in fish fed diet 2 in the present study was higher than those reported in silver pomfret [47,48,49]. Based on the moisture and lipid contents, the silver pomfret is a medium-fat fish, with a fat content of 5-10% by weight [50]. The negative correlation between moisture and lipid in the present study corroborates the findings of earlier studies [51,52].

In the present study, the whole body fatty acid reflected the content of dietary lipid which is similar to the previously reported studies [53,54]. The fatty acid profile of the pomfret was dominated by saturated fatty acids (SFA) which comprised about 40% of the total fatty acids. The SFA contents are similar to those of wild pomfrets but slightly lower than those of cultured pomfrets [48]. Marked changes in the whole body tissue fatty acid composition as a response of increasing dietary fish oils levels, particularly EPA and DHA rather than linolenic acid (C18:3n-3) was observed in the present study indicating that direct incorporation was likely. Similar results with increased levels of EPA and DHA were reported in catfish fed a practical diet supplemented with menhaden oil [55]. Compared to the initial whole body composition, the deposition of DHA was much higher than the EPA. Similar results of higher incorporation of DHA into body lipid have also been found in gilthead bream [56] and starry flounder [57]. The whole body n-3 PUFAs in fish fed diets 2 and 3 in the present study were very similar (10.56 and 10.27% respectively), which are lower than those of 13.46% and

17.10% reported for wild pomfrets [58,48]. On the other hand, the whole body n-6 PUFAs in fish fed diets 2 and 3 were higher than those of *Pampus punctatissimus* [49] and slightly lower than Malaysian silver pomfrets [58]. However, these differences could be due to the geographical differences, food, environmental factors, as well as the size and maturity of the fish [3]. In conclusion, the results of the present study suggest that a dietary lipid level of 16% in a 49% protein diet corresponding to a P:E ratio of 22.6 is optimum for better growth, feed utilization and whole body fatty acid profile in silver pomfret.

ACKNOWLEDGMENTS

This research was funded by the project "Culture of Zobaidy (*Pampus argenteus*) in Kuwait" Phase II (Project No. FM 046K) supported by Kuwait Foundation for the Advancement of Sciences (KFAS), Arab Fund for Economic and Social Development (AFESD) and KISR, Kuwait. Thanks are due to Ms Sharifa Barak Yaseen of MFD, KISR, for her assistance with the proximate analysis. The authors would like to thank Mr. Nisar Ahmed of Central Analytical Laboratory (CAL), KISR, for fatty acid analysis of feed and fish samples.

REFERENCES

[1] Peres, H. and Oliva-Teles, A. (1999a). Effect of dietary lipid level on growth performance and feed utilization by European sea bass juveniles (*Dicentrarchus labrax*). *Aquaculture,* 179, 325–334.

[2] Kaushik, S. J. (1994). *Nutritional strategies for the reduction of aquaculture wastes. Proceedings of the Third International Conference on Fisheries and Ocean Industrial Development for Productivity Enhancement of the Coastal Waters*. Pusan, Korea, pp115-132.

[3] Sargent, J., Bell, G., McEvoy, L., Tocher, D., and Estevez,. A. (1999). Recent developments in the essential fatty acid nutrition of fish. *Aquaculture,* 177, 191-199.

[4] Kaushik, S. J. and Medale, F. (1994). Energy requirements, utilization and dietary supply to salmonids. *Aquaculture*, 124, 81-97.

[5] Lee, S. M., Jeon, I. G. and Lee, J. Y. (2002). Effects of digestible protein and lipid levels in practical diets on growth, protein utilization and body composition of juvenile rock fish (*Sebastes schlegeli*). *Aquaculture*, 211, 227-239.

[6] Kim, K. W., Wang, X., Choi, S. M., Park, G. J., and Bai, S. C. (2004). Evaluation of optimum dietary protein-to-energy ratio in juvenile olive flounder *Paralichthys olivaceus* (Tmminck et Schlegel). *Aquacult. Res*., 35, 250 - 255.

[7] Johnsen, F., Hillstead, M., Austreng, E. (1993). High energy diets for Atlantic salmon. Effect on pollution. In: *Fish Nutrition in Practice* (Kaushik, S. J. and Luquet, P. eds), pp391-401. Proceedings of the International Symposium on Fish Nutrition and Feeding, Biarritz, France.

[8] Peres, H. and Oliva-Teles, A. (1999b). Influence of temperature on protein utilization in juvenile European sea bass (*Dicentrarchus labrax*). *Aquaculture*, 170, 337-348.

[9] Hillestad, M. and Johnsen, F. (1994). High-energy/low-energy diets for Atlantic salmon: effects on growth, nutrient retention and slaughter quality. *Aquaculture*, 124, 109–6.

[10] Peres, H. and Oliva-Teles, A. (2005). Protein and energy metabolism of European seabass (*Dicentrarchus labrax*) juveniles and estimation of maintenance requirements. *Fish Physiol. Biochem*., 31, 23 - 31.

[11] Cowey, C. B. and Sargent, J. R. (1979). Nutrition. In: *Fish Physiology: Bioenergetics and Growth*. Vol. VIII (Hoar, W. S., Randall, D. J. and Brett, J. R. eds.), pp 1- 69, Academic press., New York, London.

[12] NRC, (1993). *Nutritional Requirements of Fish*. National Academy Press, Washington DC, US. 114p.

[13] Nematipour, G. R., Brown, M. L. and Gatlin III, D. M (1992). Effects of dietary energy: protein ratio on growth characteristics and body composition of hybrid striped bass (*Morone chrysps* X *M. saxatilis*). *Aquaculture*, 107, 359-368.

[14] Tidwell, J. H., Webster, C. D. and Cole, S. D.(1996) Effects of dietary protein level on second year growth and water quality for largemouth bass (*Micropterus salmodes*) raised in ponds. *Aquaculture,* 145, 213-223.

[15] Haedrich, R. L. (1967). The Stromatoid fishes: Systematics and Classification. *Bull Museum Comp. Zool*., 135, 31–39.

[16] James, C. M. and Almatar, S. M. (2007) A breakthrough in the spawning of domesticated silver pomfret. *Aquacult. Asia Pac*., 3(1), 26-28.

[17] Cruz, E. M., Almatar, S. M., Abdul-Elah, K., and Al-Yaqout, A. (2000). Preliminary studies on the performance and feeding behavior of silver pomfret (*Pampus argenteus*) fingerlings fed with commercial feed and reared in fiberglass tanks. *Asian Fish. Sci*., 13, 191-199.

[18] Cruz, E. M., Almatar, S. M., Abdul-Elah, K., and Al-Yaqout, A. (2003) Indoor overwintering of silver pomfret (*Pampus argenteus* Euphrasen) fingerlings in fiberglass tanks. *Asian Fish. Sci*., 16, 33- 40.

[19] Almatar, S. M. and James, C. M. (2007). Performance of different types of commercial feeds on the growth of juvenile silver pomfret, *Pampus argenteus*, under tank culture conditions. *J. World Aquacult. Soc*., 38, 550-556.

[20] Hossain, M. A., Almatar, S. M. and James, C. M. (2010) Optimum dietary protein level for juvenile silver pomfret, *Pampus argenteus* (Euphrasen). *J. World Aquacult. Soc*., 41(5), 710-720

[21] AOAC. (2000). *Official Methods of Analysis*. Association of the Official Analytical Chemists. 17th edn. AOAC, Washington, DC.

[22] Bligh, E. G. and Dyer, W. J. (1959). A rapid method of total lipid extraction and purification. *Can. J. Biochem. Physiol*., 37, 911 – 917.

[23] AOCS. (1992). *Fatty acid composition by GLC. Marine oils*. American Oil Chemists' Society. AOCS Official Methods Ce 1b-89, AOCS, Champaign, IL.

[24] Castell, J. D. and Tiews, K. (1980). Report on the EIFAC, IUNS and ICES working group on the standardization of methodology in fish nutrition research. Hamburg, Federal Republic of Germany. 21-23 March. *EIFAC Technical Paper*, 36.

[25] Williams, C. D. and Robinson, E. H. (1988). Responses of red drum to various dietary levels of menhaden oil. *Aquaculture*, 70, 107–120.

[26] Santha, C. R. and Gatlin III, D. M. (1991). Growth response and fatty acid composition of channel catfish fry fed practical diets supplemented with menhaden fish oil. *Progr. Fish-Cult.*, 53, 135–40.

[27] Kikuchi, K., Furuta, T., Iwata, N., Onuki, K., and Noguchi, T. (2009). Effect of dietary lipid levels on the growth, feed utilization, body composition and blood characteristics of tiger puffer, *Takifugu rubripes*. *Aquaculture*, 298, 111–117.

[28] Yildirim-Aksoy, M,. Shelby, R., Lim, C. and Klesius, P. H. (2007). Growth performance and proximate and fatty acid composition of channel catfish, *Ictalurus punctatus*, fed for different duration with a commercial diet supplemented with various levels of menhaden fish oil. *J. World Aquacult. Soc.*, 38(4), 471-474.

[29] Regost, C., Arzel, J., Cardinal, M., Robin, J., Laroche, M., and Kaushik, S. J. (2001) Dietary level, hepatic lipogenensis and flesh quality in turbot (*Psetta maxima*). *Aquaculture*, 193, 291–309.

[30] Silverstein, J. T., Shearer, K. D., Dickhoff, W. W., and Plisetskaya, E. M. (1999). Regulation and nutrient intake and energy balance in salmon. *Aquaculture*, 177, 161 – 169.

[31] Weatherup, R. N., McCraken, K. J., Foy, R., Ride, D., McKendry, J., Maris, R. J., and Hoey, R. (1997). The effects of dietary fat content on performance and body composition of farmed rainbow trout *Oncorhynchus mykiss*. *Aquaculture*, 15, 173–104.

[32] Murai, T., Watanabe, T. and Nose, T. (1980). Effects of dietary protein and lipid levels on performance and carcass composition of fingerling carp. *Bull. Jpn. Soc. Sci. Fish.*, 51, 605–608.

[33] Danielssen, D. S. and Hjertnes, T. (1993) Effects of protein levels in diets for turbot (*Scophthalmus maximus*) to market size. In: *Fish Nutrition in Practice (*Kaushik, S. J. and Liquet, P. eds.) pp89-96. Proceedings of the International Symposium on Fish Nutrition and Feeding, Biarritz, France.

[34] Berge, G. M. and Storebakken, T. (1991). Effects of dietary fat on weight gain, digestibility and fillet composition of Atlantic halibut. *Aquaculture*, 99, 331–338.

[35] Martino, R. C., Cyrino, J. E. P., Portz, L., and Trugo, L. C. (2005). Performance, carcass composition and nutrient utilization of surubim, *Pseudoplatystoma coruscans* (Agassiz) fed diets with varying carbohydrate and lipid levels. *Aquacult. Nutr.*, 11, 131-137

[36] Boujard, T. and Medale, F. (1994). Regulations of voluntary feed intake in juvenile rainbow trout fed by hand or by self feeders with diets containing two different protein/energy ratios. *Aquat. Liv. Resour.*, 7, 21–215.

[37] Ogata, H. Y. and Shearer, K. D. (2000). Influence of dietary fat and adiposity on feed intake of juvenile red sea bream *Pagrus major*. *Aquaculture*, 189, 237–49.

[38] Skalli, A., Hidalgo, M. C., Abellan, E., Arizcum, M. and Cardenete, G. (2004). Effects of the dietary protein/lipid ratio on growth and nutrient utilization in common dentex (*Dentex dentex* L.) at different growth stages. *Aquaculture*, 235, 1-11.

[39] Grisdale-Helland, B., Shearer, Gatlin III, D. M. and Helland, D. M. (2008). Effects of dietary protein and lipid levels on growth, protein and lipid levels on growth, protein digestibility, feed utilization and body composition of Atlantic cod (*Gadus morhua*). *Aquaculture*, 283, 156–162.

[40] Dias, J., Alvarez, M. J., Diez, A., Arzel, J., Corraze, G., Bautista, J. M., and Kaushik, S. J. (1998) Regulation of hepatic lipogenensis by dietary protein/energy in juvenile European sea bass (*Dicentrarchus labrax*). *Aquaculture*, 161, 169–186.

[41] Cho, C. Y. and Watanabe, T. (1985) Dietary energy and lipid requirement of rainbow trout (*Salmo gairdneri*) at different water temperature. In: *Energy Metabolism of Farmed Animals* (Moe, P. V., Tyrrell, H. F. and Reynolds, P. J. eds.) pp206-209. Proceedings of the 10th Symposium EAAP, Airlie, Virginia.

[42] Berger, A. and Halver, J. E. (1987). Effect of dietary protein, lipid and carbohydrate content on the growth, feed efficiency and carcass composition of stripped bass *Morone saxatilis* fingerlings. *Aquacult. Fish. Managt.*, 18, 345–356.

[43] Watanabe, W. O., Ellis, S. C. and Chaves, J. (2001). Effects of dietary lipid and energy to protein ratio on growth and feed utilization of juvenile mutton snapper *Latjanus analis* feed iso-nitrogenous diets at two water temperatures. *J. World Aquacult. Soc.*, 32,30–40.

[44] Webster, C. D., Tiu, L. G., Tidwell, J. H., Wyk, P. V. and Howerton, R. D. (1995). Effects of dietary protein and lipid levels on growth and body composition of sunshine bass (*Morone chrysops* x *M. saxatilis*) reared in cages. *Aquaculture*, 131, 291–301.

[45] Ai, Q., Mai, K., Li, H., Zhang, C., Duan, Q., Tan, B., Xu, W., Ma, H., Zhang, W., and Lufufu, Z. (2004). Effects of dietary protein to energy ratios on growth and body composition of juvenile Japanese sea bass, *Lateolabrax japonicus*. *Aquaculture,* 230, 507–516.

[46] Catacutan, M. R. and Coloso, R. M. (1995). Effects of dietary protein to energy ratio on growth, survival, and body composition of juvenile Asian sea bass, *Lates calcarifer*. *Aquaculture*, 131, 125- 31.

[47] Chakraborty, S., Ghosh, S. and Bhattacharyya, D. K. (2005). Lipid profiles of pomfret fish (*Pampus argenteus*) organs. *J. Oleo Sci.*, 54(2), 85-88.

[48] Zhao, F., Zhuang, P., Zhang, L. and Shi, Z. (2010). Biochemical composition of juvenile cultured vs wild pomfret, *Pampus argenteus*: determining the diet for cultured fish. *Fish Physiology and Biochemistry* DOI 10.1007/s10695-010-9388-5.

[49] Zhao, F., Zhuang, P., Song, C., Shi, Z. and Zhang, L. (2010). Amino acid and fatty acid compositions and nutritional quality of muscle in the pomfret, *Pampus punctatissimus*. *Food Chem.*, 118, 224-227.

[50] Bennion, M. (1997). *Introductory Foods*. (7th Edition). Macmillan, New York, US.

[51] Martinez-Palacios, C. A., Harfush-Mandez, M., Chavez-Sanchez, C., and Ross, L. G. (1996).The optimum dietary protein level for the Mexican cichlid *Cichlasoma uropthalmus* (Gunther): a comparison of estimates derived from experiments using fixed rate feeding and satiation feeding. *Aquacult. Nutr.*, 2, 11-20.

[52] Ng, W. K., Soon, S. C. and Hashim, R. (2001). The dietary protein requirement of a bagrid catfish, *Mystus nemurus* (Cuvier and Valenciennes), determined using semi-purified diets of varying protein level. *Aquacult. Nutr.*, 17, 45-51.

[53] Gatlin III, D. M and Stickney, R. R. (1982). Fall-winter growth of young channel catfish in response to quantity and source of dietary lipid. *Am. Fish. Soc.*, 111, 90–93.

[54] Lee, S. M. and Kim, K. D. (2001). Effects of dietary carbohydrate to lipid ratios on growth and body composition of juvenile and grower rockfish, *Sebastes schlegeli*. *Aquacult. Res.*, 40, 1830-1837.

[55] Morris, C. A., Hynes, K. C., Keeton, J. T., and Gatlin III, D. M. (1995). Fish oil effects on fatty acid composition and flavor of channel catfish. *J. Food Sci.*, 60, 1225-1227.

[56] Kalogeropoulos, N., Alexis, M. N. and Henderson, J. J. (1992). Effect of dietary soybean and cod-liver oil levels on growth and body composition of gilthead bream (*Sparus aurata*). *Aquaculture*, 104, 293-308.

[57] Lee, S. M., Lee, J. H. and Kim, K. D. (2003). Effect of dietary essential fatty acids on growth, body composition and blood chemistry of juvenile starry flounder (*Platichthys stellatus*). *Aquaculture*, 225, 269-281.

[58] Osman, H., Suriah, A. R. and Law, E. C. (2001) Fatty acid composition and cholesterol content of selected marine fish in Malaysian waters. *Food Chem.*, 73(1), 55-60.

In: Asia Pacific Life Sciences Compendium
Editor: Bandit Chumworathayi ISBN: 978-1-62257-338-7

Chapter 16

PROTECTIVE EFFECT OF AQUEOUS EXTRACT OF *TRIGONELLA FOENUM* GRAECUM (FENUGREEK) IN CYPERMETHRIN INDUCED GENOTOXICITY: A STUDY BY ALKALINE COMET ASSAY AND *IN VIVO* MICRONUCLEUS ASSAY

***Navayath Sushma*[1*] *and Thiyagarajan Devasena*[2†]**
[1]Department of Biochemistry, Vels University, Chennai, India
[2]Centre for Nanoscience and Technology, Anna University, Chennai, Tamilnadu, India

ABSTRACT

Cypermethrin (CM) is a potent Type II pyrethroids pesticide extensively used in pest control. Studies have shown that cypermethrin induces genotoxicity in experimental animals. Antioxidants are known to protect the integrity of DNA from genotoxicants. In the present study, we have shown the potential protective effect of aqueous extract of fenugreek against cypermethrin induced DNA damage in femoral bone marrow lymphocytes, by *in vivo* micronucleus assay and alkaline comet assay. Male Wistar rats were treated with 1/10 LD_{50} (25 mg/kg body weight) of cypermethrin and 10% aqueous extract of fenugreek (GFaq) for 60 days. *In vivo* micronucleus assay showed that cypermethrin treatment caused a significant increase ($p<0.05$) in % MNPCE and a significant decrease ($p<0.05$) in % PCE and P/N ratio. Alkaline comet assay showed a significant increase ($p<0.05$) in tail length, % DNA in tail and Olive tail moment in cypermethrin treated rats, indicating the clastogenic effect of cypermethrin. Treatment with 10% GFaq showed significant reduction in DNA damage. The protective effect of GFaq could be attributed to the active flavonoids and polyphenols present in fenugreek.

Keywords: Comet assay; cypermethrin; fenugreek; flavonoids; genotoxicity; in vivo micronucleus assay; pyrethroids

* Navayath Sushma: Department of Biochemistry, Vels University, Chennai, 600117.
† Corresponding author: Thiyagarajan Devasena. Centre for Nanoscience and Technology, Anna University, Chennai, 600025, Tamilnadu, India. Tel: +91 9444809634, E-mail: tdevasenabio@sify.com.

Abbreviations

CM	Cypermethrin;
GFaq	Aqueous extract of germinated fenugreek seeds;
ROS	Reactive oxygen species;
PCE	Polychromatic erythrocytes;
NCE	Normochromatic erythrocytes;
MPCE	Micronucleated polychromatic erythrocytes;
P/N Ratio	Ratio of polychromatic and normochromatic erythrocytes.

Introduction

Pyrethroid pesticides have been used in agricultural and home formulations and account for $1/4^{th}$ of the world wide insecticide market. They are increasingly used over organochlorines, organophosphates and carbamates due to their greater stability, rapid metabolism and elimination from mammalian system, limited persistence in soil and greater potency [1]. Cypermethrin is a potent Type II pyrethroid that is extensively used in agricultural farming, livestock industry and to control house hold ectoparasites.

Studies in rats show that CM is rapidly metabolized and over 99% eliminated within hours, but accumulation of CM and its fatty acid conjugates in adipose tissue, brain and liver of rats was reported [2]. Cypermethrin has been classified by EPA as possible human carcinogen. A number of reports have shown that cypermethrin can induce genotoxicity [3,4,5,6]. The induction of oxidative stress and alteration of antioxidant system by pyrethroids in animals have been reported [7]. Pyrethroids have also been reported to cause free radical mediated damage of tissues, liver, brain, kidney and erythrocytes *in vitro* and *in vivo* [8,9]. DNA is a key cellular component susceptible to oxidative damage. The ROS attack on DNA leads to lesion which includes single and double strand breaks [10].

Under normal conditions, excessive formation of free radicals and concomitant damage at cellular and tissue concentrations is controlled by cellular defense systems. It is reported that dietary antioxidants can prevent the oxidative DNA damage [11]. Fenugreek (*Trigonella foenum graecum*), a popular spice and common medicinal herb, has been extensively used as food in Indian sub-continent and in Indian system of medicine for centuries. Fenugreek seeds are a rich source of flavonoids and polyphenols compounds such as quercetin, luteolin, kaempferol, tricin, gallic acid etc.] Various components of fenugreek are reported to have restorative and chemopreventive effects. Prior workers have also reported the antioxidant potential of fenugreek [12]. The chemopreventive effects of fenugreek is attributed to its antioxidant potential [13,14,15]. Earlier we have reported that aqueous extract of germinated fenugreek seeds prevents cypermethrin induced hepatic and renal toxicity [16]. The present study was designed to study the protective effect of aqueous extract of germinated fenugreek seed powder in preventing the cypermethrin induced genotoxicity.

METHODS

Chemicals

Technical grade cypermethrin [(RS)-3-phenoxybenzyl (IRS)-cis-trans-3-(2,2-dichlorovinyl)-2,2-dimethyl cyclopprop ane carboxylate] of greater than 95% purity was obtained as a gift from IIBAT, Padappai, Chennai. All fine chemicals were obtained from Sigma Chemical Co, US. All other reagents used were of analytical grade and was obtained either from Merck India Ltd. or from Hi Media India Ltd.

Germinated Fenugreek Seeds Powder (Aqueous Extract)

Fenugreek seeds were purchased from a local grocery shop, cleared of extraneous matter. Aqueous extract of fenugreek (GFaq) was prepared by the method of Dixit *et al.* [17]. Fenugreek seeds were soaked in water and germinated for 24 hours. They were kept at 4°C for two days, dried in shade and powdered. 10% GFaq was prepared by weighing 1g of powder in 10 ml of distilled water. The solution was stirred on a magnetic stirrer for one hour. It was then centrifuged and supernatant stored at -20°C until use.

Animals and Treatment

Male albino Wistar rats weighing 120-160 g and six weeks old were taken for the present study. They were maintained in polypropylene cages on a 12h light-12h dark cycle. All animals were kept under controlled conditions of temperature (30±2°C) and humidity. They were fed with pellet food and water *ad libitum*. The procedures employed in the study were accepted by the animal ethical committee, Committee for the Purpose of Control and Supervision on Experiments on Animals (CPCSEA). Animals were randomly divided into four groups containing six animals each. They were treated with 1/10 LD_{50} (LD_{50} of cypermethrin is 250 mg/kg body weight) of cypermethrin, 25 mg/kg body weight/per day [18] and 10% aqueous extract of fenugreek for 60 days by intragastric tubing as follows;

Group I (normal control) – They were fed with standard pellet food and water *ad libitum*.
Group II (CM treated) – They were treated with 25 mg/kg body weight of cypermethrin in corn oil.
Group III (CM + GFaq treated) – They were treated with 25 mg/kg body weight of CM in corn oil and 5 ml of 10% GFaq.
Group IV (GFaq treated) – They were treated with 5 ml of 10% GFaq.

Bone Marrow Micronucleus Assay

The bone marrow was prepared according to the method described by Schmid [19] and Salamone *et al.* [20]. After having sacrificed the animals by cervical decapitation, the femora were removed and freed from muscles. The bone marrow was flushed out of the diaphysis into centrifuge tubes using a cannulated syringe filled with fetal bovine serum (about 2 ml of serum/femur). After thorough mixing, the suspension was centrifuged at 1,500 rpm for 5 min and the sediment was re-suspended with 50 μl of fresh fetal bovine serum and dropped onto a clean slides. Smears were prepared, air-dried and stained with May- Grunwald and Giemsa solutions. Three slides per animal were prepared.

Scoring of Micronuclei

All the slides were coded. Viewing under 40X magnification checked the quality of the slide initially. At 100X oil immersion, 2000 PCEs with or without micronuclei and corresponding NCEs per animal were scored.

Alkaline Comet Assay

The femoral bone marrow, flushed into fetal bovine serum were counted using a hemocytometer and diluted with phosphate buffered saline (PBS to achieve a concentration of 0.2×10^6 cells/ml). The cells were checked for viability by trypan blue exclusion staining. Alkaline comet assay was conducted according the original method of Singh *et al.* [21]. Conventional slides (size 75mm×25 mm) with 19-mm frosted ends were used. Briefly, femoral bone marrow containing 0.2×10^6 lymphocytes were embedded in 90 µl of 0.7% low-melting agarose and sandwiched between a layer of 0.5% normal-melting agarose and a top layer of 0.5% low-melting agarose (90 µl) onto conventional microscope slides. To lyse cellular and nuclear membranes of the embedded cells and to permit DNA unfolding in alkaline conditions, the slides were immersed in ice-cold freshly prepared lysis solution (2.5 M NaCl, 100 mM EDTA, 10 mM Tris, pH 10 with 10% DMSO and 1% Triton X-100 being added just before use) for 1 h, at 4°C in the dark. The slides were then placed in electrophoresis alkaline buffer (300 mM NaOH, 1 mM Na_2EDTA) and the embedded cells were exposed to alkali for 20 min to allow DNA unwinding. Electrophoresis was performed in the same buffer (pH>13) for 20 min by applying an electric field of 25 V and adjusting the current to 300 mA. After electrophoresis, the slides were first washed gently with 0.4 M Tris-HCl buffer (pH 7.5) to neutralise the alkali, and the DNA was then stained by adding 75 µl of ethidium bromide (20 µg/ml). All the steps were conducted under yellow light to prevent the occurrence of additional DNA damage. The slides were kept in a humidified sealed box to prevent drying of the gel and analysed within 48 h.

Scoring of Slides

Slides were scored using an image-analysis system attached to a fluorescence microscope (Olympus CKX 41) equipped with appropriate filters (N2.1, excitation wavelength of 515-560 nm and emission wavelength of 590 nm). The microscope was connected to a computer through a charge-coupled device (CCD) camera to transport images to software for analysis. Slides were scored using CometScore software (TriTek Corp., US). The comet parameters recorded were Olive tail moment (OTM, arbitrary units), tail DNA (%) and tail length (migration of the DNA from the nucleus, µm). Images from 50 cells (25 from each replicate slide) were analyzed.

Statistical Analysis

The values were expressed as mean ± SE of six animals per group. The data were analyzed by use of one-way analysis of variance (ANOVA). The group means were compared by Duncan's Multiple Range Test (DMRT). The results were considered statistically significant if "p" value was 0.05 or less.

Results

The results of treatment with CM (25 mg /kg body weight) and 10% GFaq for 60 days on the DNA damage in lymphocytes from the femoral bone marrow of male Wistar rats are as follows;

In Vivo Micronucleus Assay

The results of micronucleus assay show that CM induces significant DNA damage in the lymphocytes. There was a significant increase ($p<0.05$) in the % MNPCE and a significant decrease ($p<0.05$) in % PCE and P/N ratio in CM treated rats when compared with the control. Simultaneous treatment with 10 % GFaq for 60 days showed significant decrease ($p<0.05$) in % MNPCE and significant increase ($p<0.05$) in % PCE and P/N ratio when compared to the CM treated rats (Table 1).

Alkaline Comet Assay

The results of alkaline comet assay on bone marrow lymphocyte of the experimental animals showed a significant increase ($p < 0.05$) in tail length, % DNA in tail and Olive tail moment in the CM treated rats when compare to the normal control (Table 2). The extent of DNA damage is proportional to the migration of DNA towards the anode or tail length [21] and the intensity of tail (% DNA in tail) is proportional to the frequency of DNA strand breaks [22]. Olive tail moment (OTM), the product of percentage DNA in tail and tail length, is the most statistically significant comet measurement [23]. Treatment with 10% GFaq showed significant decrease ($p<0.05$) in comet parameters when compared to the CM treated rats (Table 2).

Table 1. Effect of cypermethrin and GFaq on nuclear DNA of rat bone marrow lymphocytes with respect to PCE %, NCE %, MPCE % and P/N ratio in control and experimental groups

	PCE%	NCE%	MPCE%	P/N
NORMAL CONTROL	80.45 ± 1.2	18.818 ± 1.2	0.936 ± 0.02	4.2675 ± 0.3
CM	$67.21 \pm 1.1^{a*}$	$29.083 \pm 0.7^{a*}$	$3.6925 \pm 0.5^{a*}$	$2.3625 \pm 0.18^{a*}$
CM + GFaq	$77.035 \pm 2^{b*}$	$22.258 \pm 1.4^{b*}$	$1.1178 \pm 0.2^{b*}$	$3.4675 \pm 0.3^{b*}$
GFaq	86.918 ± 1.1	12.468 ± 1	0.6607 ± 0.03	7.0075 ± 0.65

PCE=Polychromatic erythrocytes; NCE=normochromatic erythrocytes; MPCE=Micronucleated polychromatic erythrocytes; P/N Ratio=Ratio of polychromatic and normochromatic erythrocytes. Values represent mean ± S.E. of six animals in each group, a as compared with Group I (Normal control); b as compared with Group II (CM Treated); ($^{*}p<0.05$).

Table 2. Effect of cypermethrin and GFaq on nuclear DNA of rat bone marrow lymphocytes with respect to Comet Length (μm), Tail DNA (%), Tail Length (μm) and Olive Tail Moment (Arbitrary units) in control and experimental groups

	Comet length (μm)	Tail length (μm)	% DNA in tail	Olive tail moment
NORMAL CONTROL	175.46 ± 13.22	2.88 ± 0.80	1.80 ± 0.51	1.09 ± 0.32
CM	$334.78 \pm 12.38^{a*}$	$199.70 \pm 9.66^{a*}$	$66.11 \pm 2.23^{a*}$	$83.49 \pm 3.90^{a*}$
CM + GFaq	$178.70 \pm 13.51^{b*}$	$43.60 \pm 4.86^{b*}$	$24.80 \pm 2.96^{b*}$	$18.23 \pm 2.54^{b*}$
GFaq	187.88 ± 17.41	2.80 ± 0.79	1.73 ± 0.56	1.16 ± 0.40

Values represent mean ± S.E. of six animals in each group, a as compared with Group I (Normal control); b as compared with Group II (CM Treated); ($^{*}p < 0.05$).

DISCUSSION

CM is a type II pyrethroid containing an alpha cyano group, which may release cyanohydrins that are unstable and further decompose to give cyanides and aldehydes which may produce free radicals [7]. Earlier studies have shown that CM can induce DNA damage in cells and tissues. CM increased the number of micronucleated cells in whole blood lymphocyte cultures [24]. DNA damage was reported in the lymphocytes of workers occupationally exposed to CM [25]. CM showed a dose dependent increase in DNA damage in male albino mice in all the vital organs [4]. CM was also shown to induce DNA strand breaks in CHO cells *in vitro* at a concentration of 1000M and 5000M [26]. The results of our study show that rats treated with 25 mg/kg body weight (1/10 LD_{50}) of CM for 60 days can induce significant DNA damage in the lymphocytes. Our results are consistent with the previous reports.

Following the exposure to CM, structural damage of DNA could result in the formation of acentrics (chromosome fragments). When the enucleation of erythrocytes occurs, the acentrics are not extruded from the cells; instead they appear as small bodies called micronuclei. Our results show an increased micronucleated PCE (MNPCE %), decrease in PCE% and decrease in P/N ratio in the lymphocytes in CM treated rats. An increased frequency of micronucleated cells (MNPCE) reflects the CM induced chromosome breakage or non-disjunction. These effects observed in micronucleus assay have been correlated with tumour initiation [27].

The frequency of PCEs observed in bone marrow provides an index of mitotic activity [28]. The PCE% declined significantly in CM treated rats suggesting the suppression of erythrocyte production. The decreased PCE% also indicates the extent of cell death. The micronuclei formation indicates that a considerable amount of genetic information is no longer available to the cell, which would result in cell death. Reduction in P/N ratio in cypermethrin treated rats is indicative of bone marrow cytotoxicity and changes in erythropoiesis. The exact mechanism of genotoxic effects of CM is not known, but it could be due to the oxidative DNA damage as various studies have pointed out that pesticides induce the formation of reactive oxygen species. Oxidative stress has also been reported to be involved in induction of micronuclei in mouse lung fibroblasts by pesticides dieldrin and phosphamidon [29]. Earlier we have reported the cypermethrin induced oxidative stress and inhibition of antioxidant enzymes in rat liver, kidney [16] and circulation [30]. Hence the genotoxicity of cypermethrin could be attributed to the oxidative stress and the resulting compromised antioxidant system.

Single-cell gel electrophoresis (the comet assay) is a sensitive and rapid method for the detection of germ cell genotoxicity induced by chemicals in individual cells [31]. Significant increase in tail length, % DNA in tail and olive tail moment (OTM) in the CM treated rats when compared to control rats shows that CM induces DNA strand breaks. Our results are in agreement with the observations of Patel *et al.* [4]. They have reported significant DNA damage in bone marrow, liver, kidney and brain in rats at doses of 12.5, 25, 50, 100 and 200mg/kg body weight.

The varying intensities of fluorescence in comet tail show the extent of DNA damage in these cells (Figure 1). Control after 60 days (A), Cypermethrin treated for 60 days (B), 10% GFaq treated for 60 days (C), Cypermethrin + 10% GFaq treated for 60 days (D).

The results of our study indicate that GFaq confers substantial protection to prevent the DNA damage in CM treated rats. In alkaline comet assay, we observed significant decrease in tail length, % DNA in tail and OTM in the treatment group (Figure 1), while in micronucleus assay, GFaq treatment reduced the % MNPCE. This is a clear indication that GFaq can protect the genetic integrity from oxidative damage due to CM.

The main constituents of fenugreek are fibre (galactomannan), steroidal saponins (diosgenin), flavonoids, polyphenols, tannins, lysine and L- tryptophan rich proteins which are thought to be responsible for its many medicinal effects.

Fenugreek has been reported to possess antioxidant potential comparable to Glutathione and α-tocopherol [12]. Chemopreventive effect of fenugreek was observed in 7,12-dimethylbenz (a) anthracene (DMBA) induced breast cancer, which is attributed to the flavonoids in the water extract [14]. Fenugreek seeds extract exhibited significant antineoplastic effect in T-lymphoblastic leukemia cell lines [15].

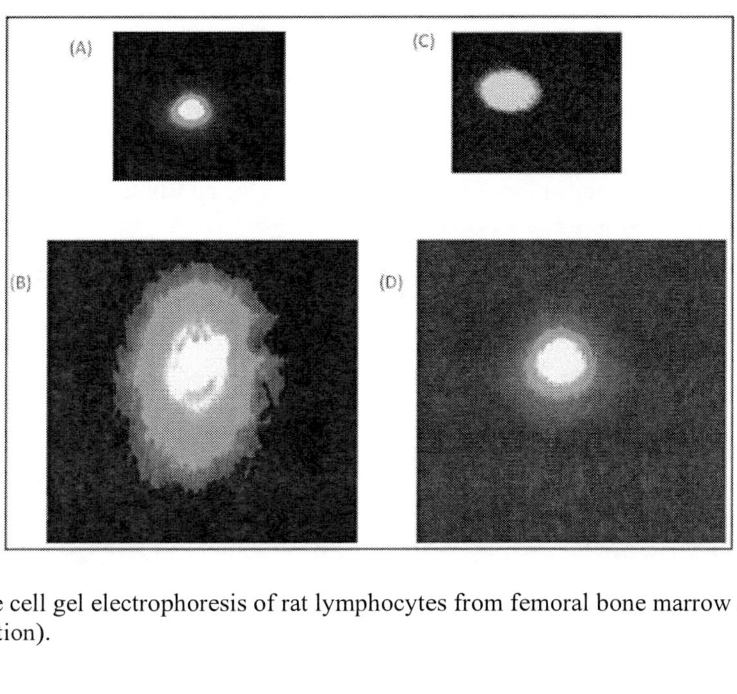

Figure 1. Single cell gel electrophoresis of rat lymphocytes from femoral bone marrow showing comets (40× magnification).

It has been reported that inclusion of fenugreek powder in the diet of DMH (1,2-Dimethyl hydrazine) treated rats reduced colon tumour incidence 16.6% that was attributed to the antioxidative effect of phytochemicals present in fenugreek [32,33].

Fenugreek itself was tested by the standard battery of tests and proved to be non genotoxic [34]. The protective effect of GFaq may be attributed to the phytochemicals including active flavonoids present in the extract which actively scavenge the free radicals and protects the DNA. An extract of fenugreek exhibited hydroxyl radicals (HO°) scavenging activity and antimutagenic activity by protecting plasmid pBR 322 from γ- radiation induced DNA strand breaks *in vitro* which is attributed to the polyphenolic components present in the extract [35].

Various reports have proved that flavonoids like quercetin, rutin, naringenin and kaempferol protects the DNA against oxidative damage [36]. 10% GFaq used in our investigation was reported to contain the flavonoids quercetin, naringenin, vitexin, orientin, tricin and gallic acid [30]. The abatement of DNA damage conferred by the aqueous extract of germinated fenugreek could be a direct antioxidant effect that scavenges the ROS or stimulation of DNA repair by the phytochemicals in the extract. Further research is needed to know the exact mechanism and the actual component of GFaq that exerts the protective effect. Because of its widespread availability and relative lack of toxicity, fenugreek is an ideal herb for protection against pesticide induced genotoxicity.

Conflict of Interests

None. This research received no specific grant from any funding agency in the public, commercial, or non-for-profit sectors.

References

[1] Elliott, M. Properties and applications of Pyrethroids. *Environmental Health Perspectives*, 1976 14, 3-13.

[2] Marei, A. E. M., Ruzo, L. O., Casida, J. E. Analysis and persistence of permethrin, cypermethrin, deltamethrin and fenvalerate in the fat and brain of treated rats. *Journal of Agricultural and Food chemistry*, 1982 30, 558-562.

[3] Amer, S. M., Ibrahim, A. A., Sherbeny, K. M. Induction of chromosomal aberrations and sister chromatid exchange in vivo and in vitro by the insecticide cypermethrin. *Journal of Applied Toxicology*, 1993 13, 341–345.

[4] Patel, S., Pandey, A. K., Bajpayee, M., Parmar, D., Dhawan, A. Cypermethrin-induced DNA damage in organs and tissues of the mouse: Evidence from the comet assay. *Mutation Research,* 2006 607, 176–183.

[5] Bhunya, S. P., Pati, P. C. Genotoxic effects of a synthetic pyrethroid insecticide, cypermethrin, in mice in vivo. *Toxicology Letters*, 1988 41, 223–230.

[6] Giri, S., Giri, A., Sharma, G. D., Prasad, S. B. Induction of sister chromatid exchanges by cypermethrin and carbosulfan in bone marrow cells of mice in vivo. *Mutagenesis,* 2003 18, 53–58.

[7] Kale, M., Rathore, N., John, S., Bhatnagar, D. Lipid peroxidative damage on pyrethroid exposure and alterations in antioxidant status in rat erythrocytes: a possible involvement of reactive oxygen species. *Toxicology Letters,* 1999 105, 197–205.

[8] Giray, B., Gurbay, A., Hincal, F. Cypermethrin-induced oxidative stress in rat brain and liver is prevented by Vitamin E or allopurinol. *Toxicology Letters,* 2001 118, 139–146.

[9] Rehman, H., Ali, M., Hatif, F., Kaur, M., Bahtia, K., Sheikh, R. The modulatory effect of deltamethrin on antioxidants in mice. *Clinica Chimica Acta*, 2006 369, 61-65.

[10] Demple, B., Levin, J. Repair systems for radical-damaged DNA. In: Sies H., editors. *Oxidative Stress: Oxidants and Antioxidants*. New York: Academic Press; 1991; 119-154.

[11] Horvathova, K., Novotny, L., Vachalkova, A. The free radical scavenging activity of four flavonoids determined by the comet assay. *Neoplasma*, 2004 50, 291– 295.

[12] Thirunavukkarasu, V., Anuradha, C. V., Viswanathan, P. Protective effect of fenugreek (Trigonella foenum graecum) seeds in experimental ethanol toxicity. *Phytotherapy Research*, 2003 17(7), 37-43.

[13] Devasena, T., Venugopal, P. M., Rajasekharan, K. N. Enhancement of circulatory antioxidants by fenugreek during 1,2-DMH- induced colon carcinogenesis. *Journal of Biochemistry Molecular biology and Biophysics*. 2006 4, 289- 292.

[14] Amin, A., Alkaabi, A., Al-Falasi, S., Daoud, S. A. Chemopreventive activities of Trigonella foenum graecum (Fenugreek) against breast cancer. *Cell Biology International,* 2005 29, 687–694.

[15] Shaban, A., Jahanmehr, S. A., Rezaeeian, M., Einolahi, N., Arjomand, A., Monsefesfehani, H. R., Kazemi, V., Akrami, M. Fenugreek (Trigonella foenum graecum L.) seed extract induces cell death, growth inhibition and morphological change indicative of apoptosis in acute lymphoblastic leukemia. *Research journal of biological sciences,* 2007 2(4), 438-443.

[16] Sushma, N., Devasena, T. Aqueous extract of *Trigonella foenum graecum* (fenugreek) prevents cypermethrin induced hepatotoxicity and nephrotoxicity. *Human and Experimental Toxicology*, 2010a 29(4), 311-320.

[17] Dixit, P., Ghaskadbi, S., Mohan, H., Devasagayam, T. Antioxidant properties of germinated feugreek seeds. *Phytotherapy Research*, 2005 19(11), 977-983.

[18] Cantalamessa, F. Acute toxicity of two pyrethroids, permethrin and cypermethrin in neonatal and adult rats. *Archives of Toxicology*, 1993 67, 510-513.

[19] Schmid, W. The micronucleus test for cytogenetic analysis. In: Hollaender A, editor. *Chemical mutagens principles and methods for their detection*. New York; Plenum Press; 1976; 31-53.

[20] Salamone, M., Heddle, J., Stuart, E., Katz, M. Towards an improved micronucleus test - studies on 3 model agents, mitomycin C, cyclophosphamide and dimethyl benzanthracene. *Mutation Research*, 1980 74, 347-356.

[21] Singh, N. P., Mc Coy, M. T., Tice, R. R., Schneider, E. L. A simple technique for quantitation of low levels of DNA damage in individual cells. *Experimental Cell Research*, 1988 175, 184–191.

[22] Olive, P. L., Banath, J. P., Durand, R. E. Heterogeneity in radiation-induced DNA damage and repair in tumor and normal cells measured using the Comet assay. *Radiation Research*, 1990 122, 86–94.

[23] Kumaravel, T. S., Jha, N. Reliable Comet Assay measurements for detecting DNA damage induced by ionising radiation and chemicals. *Mutation Research*, 2006 605, 7–16.

[24] Surralles, J., Xamena, N., Creus, A., Catalfin, J., Norppa, H., Marcos, R. Induction of micronuclei by five pyrethroid insecticides in whole-blood and isolated human lymphocyte cultures. *Mutation Research*, 1995 341, 169-184.

[25] Undeger, U., Basaran, N. Assessment of DNA damage in workers occupationally exposed to pesticide mixtures by the alkaline comet assay. *Archives of Toxicology*, 2002 76, 430–436.

[26] Patel, S., Pandey, A. K., Bajpayee, M., Parmar, D., Dhawan, A. In vitro induction of cytotoxicity and DNA strand breaks in CHO cells exposed to cypermethrin, pendimethalin and dichlorvos. *Toxicology in vitro*, 2007 21: 1409–1418.

[27] Manjula, S. D., Benjamin, S., Bairy, K. L. Modulatory Effect of Vitamin C on Genotoxic Effect of Endosulfan in Developing Albino Rats. *Indian Journal of Pharmacology,* 2006 5(2), 113-116.

[28] Suzuki, Y., Nagae, J., Li, H., Sabaka, K., Mozawa, A., Takahashi, H., Shimizu. The micronucleus test and erythropoiesis: effects of erythropoietin and a mutagen on the ratio of polychromatic to normochromatic erythrocytes (P:N ratio). *Mutagenesis*, 1989 4, 420–424.

[29] Cicchetti, R., Argentin, G. The role of oxidative stress in the in vitro induction of micronuclei by pesticides in mouse lung fibroblasts. *Mutagenesis*, 2003 18(2), 127-132.

[30] Sushma, N., Devasena, T. Aqueous extract of *Trigonella foenum graecum* ameliorates cypermethrin induced circulatory oxidative stress in male Wistar rats. *International Journal of Biological and Chemical Sciences*, 2010b 4(1), 25-33.

[31] Speit, G., Vasquez, M., Hartmann, A. The comet assay as an indicator test for germ cell genotoxicity. *Mutation Research*, 2009 681, 3–12.

[32] Devasena, T., Venugopal, M. P. Fenugreek seeds modulate 1,2-dimethylhydrazine induced hepatic oxidative stress during colon carcinogenesis. *Italian Journal of Biochemistry,* 2007 56(1), 28-34.

[33] Devasena T. Anitumorigenic action of fenugreek seeds. Asian Journal of Bioscience. 2009 4(1), 83-87.

[34] Flammang, A. M., Cifone, M. A., Erexson, G. L., Stankowski, Jr, L. F. Genotoxicity testing of fenugreek extract. *Food and Chemical Toxicology*, 2004 42, 1769–1775.

[35] Kaviarasan, S., Naik, G. H., GangaBhagirathi, R., Anuradha, C. V., Priyadarshini, K. I. In vitro studies on antiradical and antioxidant activities of fenugreek seeds. *Food Chemistry*, 2007 103, 31-37.

[36] Wilms, L. C., Hollman, P. C., Boots, A. W., Kleinjans, J. C. Protection by quercetin and quercetin-rich fruit juice against induction of oxidative DNA damage and formation of BPDE-DNA adducts in human lymphocytes. *Mutation Research*, 2005 582, 155–162.

In: Asia Pacific Life Sciences Compendium
Editor: Bandit Chumworathayi
ISBN: 978-1-62257-338-7

Chapter 17

AGE-RELATED CHANGES IN FEMORAL ARTERY: A HISTOLOGICAL STUDY

Shankar Dayal Gupta,[1] Sanjeev Kumar Gupta,[2*] Dinesh Kumar Pal,[3] Radha Sarawagi[4] and Prashant Gupta[5]

[1]Department of Anatomy, LN Medical College, Kolar Road, Bhopal (MP) India
[2]Department of Community Medicine,
Mahatma Gandhi Medical College and Research Institute,
Pillaiyarkuppam Pondicherry, India
[3]Department of Community Medicine, Gandhi Medical College,
Bhopal, Madhya Pradesh, India
[4]Department of Radiology, Mahatma Gandhi Medical College and Research Institute,
Pillaiyarkuppam Pondicherry, India
[5]Post Graduate Diploma in Clinical Research (PGDCR),
Indian Institute of Public Health, Delhi, India

ABSTRACT

A detailed microscopic study was carried out to see the changes in histological pattern of femoral artery in different age group ranging from new born to 70 years. 62 formalin-fixed specimens of either sex were studied. Thickness of tunica intima was minimal during infancy and it could not be measured, whereas in the age group 60 years and above, thickness of tunica intima ranged between 415 μs and 435 μs. While during infancy, mean thickness of tunica media was 115 μs [ranged 50-185 μs]. It reaches to the maximum mean thickness of 512 μs, in the age group 40-60 years. But in 60 years and above age group, mean thickness of tunica media decreased to 428 μs. All three layers of the arterial wall can be identified in infants. Tunica intima consists of internal elastic lamina and endothelium. Sub-endothelial layer is scarcely perceptible. Elastic fibers on tunica media are long and parallel, uniformly arranged and thicker than adult. In the earlier childhood, sub-endothelial connective tissue layer becomes more apparent. In

* Corresponding author: Sanjeev Kumar Gupta. Tel: +91-7639656802, Fax: +91-4132615457, E-mail: drsanjeev15@gmail.com.

adults, thickness of tunica intima increases while tunica media appears thin over regional intimal hypertrophy, but in the old age diffuse intimal hypertrophy is seen. Tunica intima become half, to two third of tunica media. Hyalination of collagen tissue of media occurs with loss of normal staining properties of elastic tissue. Elastic laminae become thinned and show fragmentation with occasional clumping. These changes are more prominent near inner media. Adventitia is thicker than media in infants and becomes thinner with age.

Keywords: artery, histology, tunica intima, tunica media, elastic laminae

INTRODUCTION

Aging is the process of growing older and includes changes in both biology and psychology. Biology refers to the way the body functions. Psychology describes how the mind functions. How people age has to do with genetics, environment, and lifestyle over a lifetime. The process of aging is complex, and may derive from a variety of different mechanisms and exist for a variety of different reasons. Aging is a universal biological phenomenon, at least amongst eukaryotic organisms. Yet the average lifespan within and between species can vary greatly. This suggests that both genetic and environmental factors contribute to aging. Theories that explain aging can generally be divided between the programmed and error theories of aging. Programmed theories imply that aging is regulated by biological clocks operating throughout the life span.

This regulation would depend on changes in gene expression that affect the systems responsible for maintenance, repair, and defense responses. Error theories blame environmental insults to living organisms that induce cumulative damage at various levels as the cause of aging (e.g., DNA damage, oxygen radicals, cross-linking). One potential cause of senescence is the accumulation of mutations in DNA, eventually leading to the progressive loss of key genes. Another is the shortening of telomeres in the process of DNA replication during cell division.

Senescence (derived from the Latin word *senex*, meaning *old man*, *old age*, or *advanced in age*) or biological aging is the change in the biology of an organism as it ages after its maturity. Such changes range from those affecting its cells and their function to that of the whole organism. There are a number of theories as to why senescence occurs, including ones that claim it is programmed by gene expression changes and that it is the accumulative damage of biological processes.

Theories of Aging

Evolutionary theories: A gene can be expressed at various life-stages. Therefore, natural selection can support lethal and harmful alleles, if their expression occurs after reproduction. Senescence may be the product of such selection. [1,2,3] In addition, aging is believed to have evolved because of the increasingly smaller probability of an organism still being alive at older age, due to predation and accidents, both of which may be random and age-invariant. It is thought that strategies that result in a higher reproductive rate at a young age, but shorter overall lifespan, result in a higher lifetime reproductive success and are therefore favored by natural selection. In essence, aging is, therefore, the result of investing resources in reproduction, rather than maintenance of the body (the "Disposable Soma" theory [4]); in light of the fact that accidents, predation, and disease will eventually kill the organism no matter how much energy is devoted to repair of the body. Various other theories of aging exist, and are not necessarily mutually exclusive.

Gene regulation: Despite the lack of a connection between circular DNA and aging in higher organisms, extra copies of Sir2 are capable of extending the lifespan of both worms and flies (though, in flies, this finding has not been replicated by other investigators, and the activator of Sir2, Resveratrol did not reproducibly increase lifespan in either species [5].

Gene expression is imperfectly controlled, and it is possible that random fluctuations in the expression levels of many genes contribute to the aging process as suggested by a study of such genes in yeast. [6]

Cellular Senescence

As noted above, senescence is not universal, and senescence is not observed in single-celled organisms that reproduce through the process of cellular mitosis. [7] Moreover, cellular senescence is not observed in many organisms, including perennial plants, sponges, corals, and lobsters. One of the earliest aging theories was the *Rate of Living Hypothesis* described by Raymond Pearl in 1928, [8] which states that fast basal metabolic rate corresponds to short maximum life span.

Metabolic rate is a poor predictor of lifespan for birds, bats and other species that, it is presumed, have reduced mortality from predation, and therefore have evolved long lifespan even in the presence of very high metabolic rates. [9]

More recently, it was shown that, when modern statistical methods for correcting for the effects of body size and phylogeny are employed, metabolic rate does not correlate with longevity in mammals or birds. [10]

Reliability theory: Reliability theory suggests that biological systems start their adult life with a high load of initial damage. Reliability theory is a general theory about systems failure. It allows researchers to predict the age-related failure kinetics for a system of given architecture (reliability structure) and given reliability of its components. Reliability theory predicts that even those systems that composed entirely of non-aging elements (with a constant failure rate) will nevertheless deteriorate (fail more often) with age, if these systems are redundant in irreplaceable elements. Aging, therefore, is a direct consequence of systems.

Miscellaneous: Recently, a kind of early senescence has been alleged to be a possible unintended outcome of early cloning experiments. The issue was raised in the case of Dolly the sheep, following her death from a contagious lung disease. The claim that Dolly's early death involved premature senescence has been vigorously contested, [11] and Dolly's creator, Dr Ian Wilmut, has expressed the view that her illness and death were probably unrelated to the fact that she was a clone.

None of the many theories of senescence can account for the aging process in its entirety. These theories include the evolutionary theory of aging and its EvoDevo corollary, the replicative senescence and free radicals theory, theories based on large-scale analyses (proteomics, etc.), and systemic theories. Cardiovascular senescence is first and foremost due to vascular senescence. [12]

Advancing age is unequivocally the dominant risk factor for atherosclerosis and its major clinical consequences of ischemic heart disease and cerebro-vascular disease. [13]

Among the cardiovascular diseases, degenerative diseases of blood vessels are responsible for about 20-25% cases. There is a considerable morphological difference between vessels fixed in a collapsed or physiological distended state. In collapsed vessels, the endothelium is thick; the sub-endothelium gap appears distended near nuclei, the internal elastic lamina and in elastic vessels the elastic fibers are undulating and loosely arranged and the gap between them varies in width. The fibers of distended elastic type vessels are straight and the gaps between them are uniformly wide.

The final differentiation of various components of vessels wall is not complete until adult life. The three main coats of the arteries are acquired during the 4th month of fetal life but intima of aorta were not completed for about 30 years. In an aorta of 4th month old human embryo, the intima consists only of endothelium and of rather thick internal elastic membrane while the media consists of several layers of circular smooth muscles cells between which are flat networks of elastic fibers embryonic connective tissue. At the end of embryonic life, the internal elastic membrane becomes thicker, and the network of elastic fibers in media becomes thick elastic membrane. The smooth muscles have increased slightly in number but are still inconspicuous. The adventitia, by this time, becomes smaller. After birth, in muscular arteries, in addition to the thickening of the wall as a whole, a connective tissue layer gradually develops between endothelium and internal elastic membrane. [14]

Other study also describes that, at birth, endothelium lies almost directly upon the internal elastic lamina, but even in childhood, these layers become separated by loose mesh of connective tissue. [15] Intimal thickening is the most important aging change for it appears to carry the seeds of arterial disease. The elements added to in time include fibrocytes, smooth muscle cells collagen fibers and ground substance rich in mucopolysaccharides.

In this way, the intima comes to approach in thickness of media and it is not uncommon to find that in the coronary arteries of men aged 60 years or so, intima equals or exceeds the media without any disease process being identifiable. In the middle and later age, degenerative changes develop in the elastic fibers of internal and external elastic laminae and in the media. [16] There is often some reparative process shown by the development of thinner fibers irregularly oriented around the original membrane. [17]

METHOD

The present work was carried out in the Department of Anatomy at Gandhi Medical College, Bhopal, Madhya Pradesh, India. 62 human femoral arteries sample from of all age groups, beginning infant to seventy years, were collected from unselected postmortems. 41 were from male and 31 belonged from female. These blood vessels were preserved in 10% formal saline immediately after its removal at room temperature before it was taken for histological examination. Dehydration was effected by immersing the tissue in ascending grades of alcohol beginning with 70% alcohol for 8 hrs and 90% alcohol for 16 hrs, followed by 3 changes of absolute alcohol, Absolute alcohol I for 2 hrs, absolute alcohol II for 3 hrs, and absolute alcohol III for 3 hrs. After the dehydration, tissue was cleared in cedar wood oil. Tissue was impregnated in paraffin wax of 60 °C melting point. Three such changes were given in wax at an internal of half hours each. Tissue blocks were prepared by filling with malted as a pair of L shaped pieces of melted mould placed on a metal or glass plate, which acts as a base of the mould.

After preparing the tissue blocks, serial section were cut of 5 micron thickness with the help of rotatory microtome. Sections were mounted on microscopic slides and the slides were kept at 37 °C in an incubator.

Staining

Various staining techniques were employed for the demonstration of general differentiation of the tissue and for specific tissue component of the organ. These are Haematoxylin and Eosin, Masson's Trichome, Orcein, PATH, Gomeri. Observations were recorded under light microscope. Thickness of intima and media were measured with standardized micrometer; mean values were calculated. The present study has been carried out on the Human blood vessels. 62 arteries sample were collected from cadavers of different age groups. They are classified as follows:-

Group I	0-1 year
Group II	1-5 years
Group III	5-15 years
Group IV	15-40 years
Group V	40-60 years
Group VI	above 60 years

We have measured the thickness of tunica intima and media of femoral artery. We have also measured the localized intimal hypertrophy of femoral artery.

Ethical Clearance

Ethical clearance for conducting the study was taken from the ethical committee of the institution, i.e., Gandhi Medical College, and associate Hospital Bhopal, with the assurance that confidentiality will be maintained and the information obtained for this study will not be used for any other purpose except for academic purpose.

Result

Group I (0-1 year): A prominent histological feature in this group was absence of sub endothelial connective tissue between internal elastic lamina and endothelium in tunica intima of femoral artery. Internal elastic lamina was continuous and uniformly thick. All the slides from this group did not show intimal thickening and there was no localized intimal hypertrophy. Tunica media shows very well developed circularly arranged smooth muscle cells with occasional or no elastic fibers. Collagen tissue was also very scanty.

The average thickness of tunica media was 115 micron. Tunica adventitia in this age group is even thicker than the media. It shows longitudinally arranged collagen and elastic tissue.

Group II (1-5 years): It was observed that thickness of tunica intima starts appearing at this age, i.e., sub-endothelial connective tissue appears between internal elastic lamina and endothelium. Average thickness of tunica intima was 50 micron. Tunica media shows circularly arranged smooth muscle cell layer. There is a consistent increase of collagen tissues and elastic fiber. Average thickness of tunica media was 165 micron.

Group III (5-15 years): Microscopic appearance of femoral artery was very much similar to the earlier group except: (i) Thickness of tunica intima was more than the previous group, average thickness being 22 Micron. (ii) Splitting and stretching of internal elastic lamina was noted at few places. (iii) Localized thickening of tunica intima was more prominent with mean value of 85 micron. (iv)Increase in the elastic fibers of tunica media was observed and mean thickness of tunica media was 300 micron.

Table 1. Distribution of blood vessels according to age

Group	Age	Number of artery	%
1	0-1	3	50
2	1-5	5	8
3	5-15	4	6
4	15-40	14	22
5	40-60	25	40
6	60+	11	19
Total		62	100

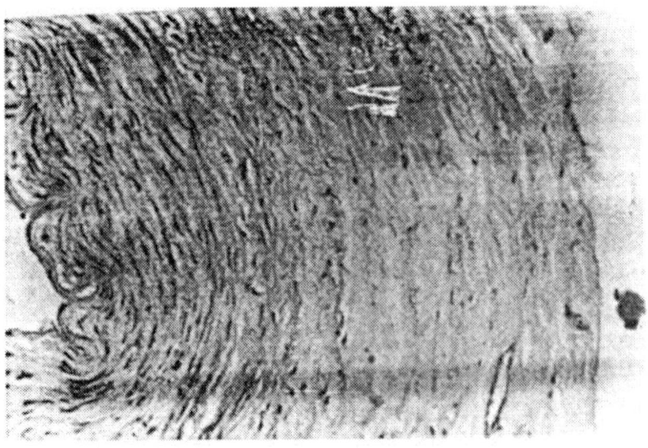

Figure 1. Femoral artery of 1 year old baby showing well defined thick internal elastic lamina, circularly arranged smooth muscle cells in media, tunica adventitia is thicker than media, sub-endothelial layer are scarcely perceptible. (HandE staining, ×100).

Group IV (15-40 years): Tunica media shows consistent increase in thickness and regional intimal hypertrophy. Even it was thicker than previous group, average thickness of this layer was 347 micron and average regional intimal hypertrophy was 115 micron. There was consistent increase in amount of collagen tissue. Tunica media also shown increasing amount of collagen tissue and elastic fibers. Tunica adventitia remains similar in appearance.

Group V (40-60 years): Tunica intima in this group shows consistent increases in thickness and diffuse intimal hypertrophy was notice. Internal elastic lamina was seen thick and showed replication. Sub-endothelial layer shows fibrocytes, collagen fibers, elastic fibers, and smooth muscle cells. Average thickness of tunica media in this group was 512 micron and localized intimal hypertrophy was 165 micron. Elastic laminae of outer media are seen like previous group. Tunica media shows increased amount of collagen and elastic tissue than the previous group. Average thickening of tunica media was 512 micron.

Group VI (Above 60 Years): Tunica intima shows increased thickening as well as reduplication of internal elastic lamina. At this stage, regional intimal hypertrophy becomes diffused in nature.

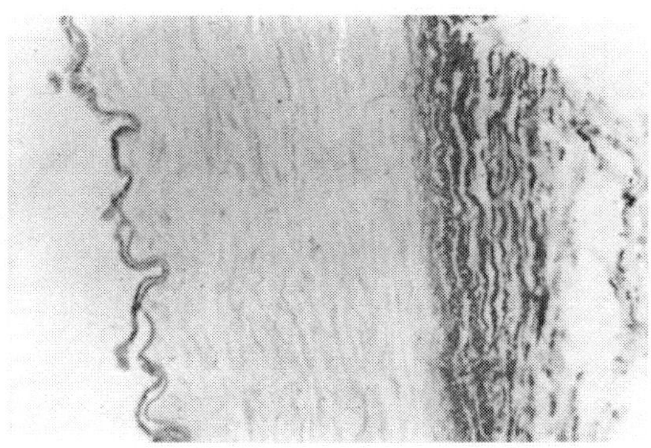

Figure 2. Femoral artery of 14 year old men showing localized splitting of internal elastic lamina, localized thickening in intima and few elastic fibers in tunica media (Orcein staining, ×100).

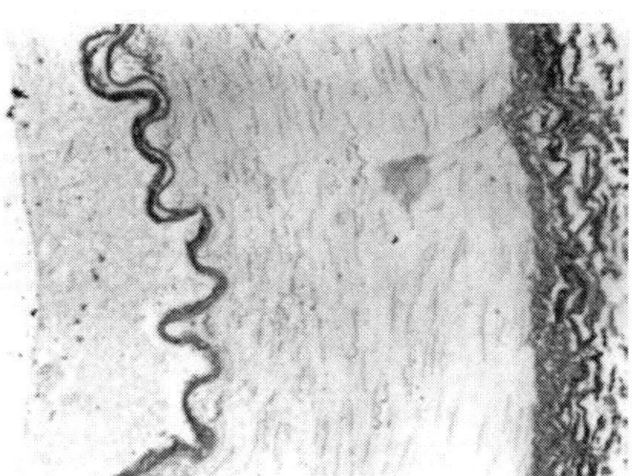

Figure 3. Femoral artery of 25 year old men showing localized splitting of internal elastic lamina, thickening in intima and more elastic fibers in tunica media (Orcein staning, ×100).

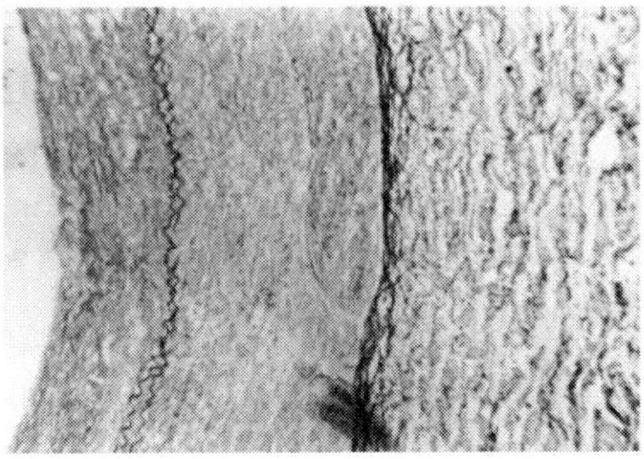

Figure 4. Femoral artery of 47 year old men showing localized splitting of internal elastic lamina, thickening in intima and increase in elastic tissue in tunica media (Orcein staining, ×100).

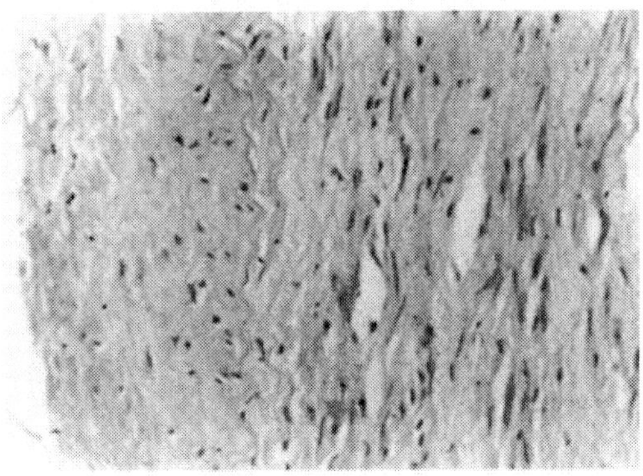

Figure 5. Femoral artery of 47 year old men showing replication of internal elastic lamina few places, thickening in intima and smooth muscle cell are few in tunica media (HandE staining, ×100).

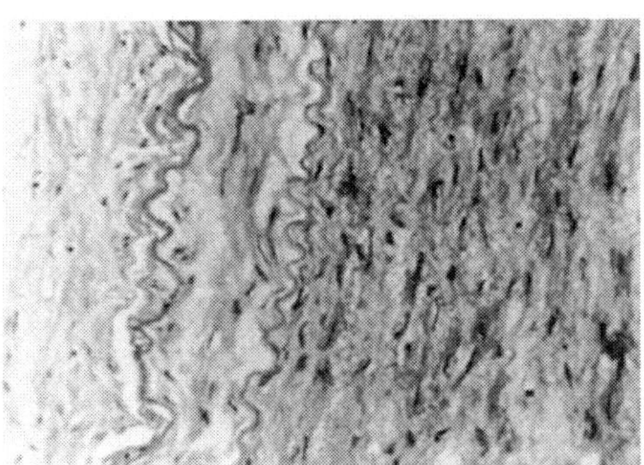

Figure 6. Femoral artery of 70 year old men showing replication of internal elastic lamina, thickening in intima and increase collagen tissue in tunica media (Masson's trichrome staining, ×100).

Average thickness of regional intimal hypertrophy was 195 micron and average thickening of tunica intima and 165 micron. Tunica media of femoral artery become more fibrosed and there is increase in elastic tissue content. Average thickness (428 micron) of this layer is less than the previous group.

Discussion

Arteries are so continually at work that it is not surprising that they show degenerative changes as age increases. These degenerative changes are more common in the arteries than anywhere else in the body and their degeneration is among the most important cause of disease and death.

There is indeed no system in the body in which the border line between normal and pathological state is so ill defined as in the vascular tree.

Table 2. Thickness range and mean thickness of tunica intima of femoral artery

Group	Age	Thickness Range (μ)	Mean Thickness (μ)
1	0-1	Not seen	Not seen
2	1-5	0-20	12
3	5-15	15-35	22
4	15-40	35-65	40
5	40-60	35-85	68
6	60+	100-285	165

The above table shows increase in thickness of tunica intima of femoral artery with age. During infancy thickness of tunica intima was minimal and it could not be measured, where as in the age group 60 years and above thickness of tunica intima was maximum, i.e., 166 (100-285).

Table 3. Thickness range and mean thickness of tunica media of femoral artery

Group	Age	Thickness Range (μ)	Mean Thickness (μ)
1	0-1	50-185	115
2	1-5	130-225	165
3	5-15	235-350	300
4	15-40	285-495	347
5	40-60	485-535	512
6	60+	415-435	428

The above table shows increase in thickness of tunica media of the femoral artery. During infancy thickness of tunica media was minimum, i.e., 115 (50-185).

Table 4. Localized thickness range and mean localized thickness of tunica intima of femoral artery

Group	Age	Thickness Range (μ)	Mean Thickness (μ)
1	0-1	Not seen	Not Seen
2	1-5	15-65	50
3	5-15	45-110	85
4	15-40	60-150	115
5	40-60	85-290	165
6	60+	100-135	195

The above table shows increase in localized thickness of tunica intima with age. During infancy localized thickness of tunica intima was minimal and it could not be measured, whereas in the age group 60 years and above, localized thickness of tunica intima was maximum, i.e., 195 (100-335).

The blood vessels undergo progressive and gradual changes from birth to death and it is difficult to say where the normal growth process ends and process of involution begins. There is also often difficulty in distinguishing changes in arteries that are merely part of aging process from the changes that are definitely pathologic in nature. Histological changes observed in this study can be discussed under following headings: 1) medial fibrosis 2) intimal thickening and 3) elastic degeneration.

In medial fibrosis, there is progressive increase in the collagen content of the tunica media this increase is partly an addition to the bulk of the media, but to an increasing extent in later life, it is at the expense of smooth muscle fibers and can be designated as fibrous replacement of muscle. The intimal thickening is the most important change due to aging, but to an increasing extent in later life, it is also at the expense of smooth muscle fibers and can be designated as fibrous replacement of muscle too. The intimal thickening is the most important change due to aging which carries the seeds of arterial diseases. Intima, which comprises of merely the endothelial cell layer and internal elastic lamina with a scarcely perceptible subendothelial connective tissue layer between two in infants, may be nearly equal in thickness to that of media in old age.

In our study, it was found that, in arteries of infants, all three layers were present. Tunica intima consists of only internal elastic laminae and one endothelium with scarcely perceptible sub-endothelial connective tissue layer. We have not found intact endothelium as it was lost during preparation but only few endothelial cells nuclei could be seen projecting in to lumen. So the internal elastic lamina appeared as inner most layer and was continuous, thick wavy in appearance.

The tunica media of infant femoral artery shows circularly arranged elastic laminae. These elastic laminae are long and run parallel to each other and are uniform in thickness. They appear thicker than those of adult aorta and are fairly compactly arranged. Similar findings also suggested by Boucek RJ et al. (1964), [18] when they compared the pulmonary trunk and artery at different ages. Between elastic fibers, smooth muscle cell layers, surrounded by collagen tissue, are present.

Lakatta EG [19] showed that elastic fibers in the media of infant femoral arteries were thick elastic membranes while tunica adventitia was smaller than media. He also found that sub-endothelium connective tissue was absent at birth. Crawford T [15] also described similar changes at birth.

Bloom and Fawcett (1967) [14] have stated that, in intima between endothelium and internal elastic lamina, sub-endothelial elastic muscular layer appears in early childhood. It arises partly by a splitting of internal elastic membrane and partly by the new formation of collagenous and elastic fibers. It gradually increases in thickness and produces progressive thickening. They have also shown that number and thickness of elastic membranes in media gradually increase in early childhood.

Crawford T [15] has also shown that, in early childhood, internal elastic lamina and endothelium layers become separated by loose mesh of connective tissue and it tend to increase in among and density with passage of years.

We have also found that in 1-5 years age group, sub-endothelial connective tissue becomes clearer between internal elastic lamina and endothelium. Elastic laminae in the femoral artery were long and parallel to each other, and similar to the elastic laminae of infant, with exception of occasional anastomoses of elastic laminae. There is a localized splitting of internal elastic lamina with some intimal thickening over elastic splitting. An increase in thickness of tunica media was observed in comparison to the previous group.

Ross R [20] has shown that, in older children, more definite intimal thickening appears and almost invariably there is some stretching, splitting or fraying of underlying internal elastic lamina. With increasing age, these features become more pronounced.

According to Homan R et al., [21] intimal thickening beings to appears in 1st decade of life and covers increasing proportion of intimal surface. Prior and Hutter [22] has shown that continued deposition of sub-endothelial connective tissue throughout life produces progressive thickening of this layer. The elastic fibers deteriorate and are replaced by fibrous tissue; with this loss of elasticity, these vessels expand less readily. Tyagi SP et al. [23] have shown that the collagen fibers are less numerous and adventitia is rich in cells in young animal than older one. Lakatta EG [19] also found that, in childhood, intima of femoral artery has highly developed layer of circular elastic fibers.

According to Crawford T, [15] arteries of the arm in early adult life show scarcely perceptible amount of sub-endothelial connective tissue and abdominal aorta and large arteries of lower limb shows intimal thickening even in early childhood.

In the present study, we also found the gradual increase in thickening of intima in 5-15 years age group. Along with this, internal elastic lamina showed localized splitting. The elastic laminae of tunica media of the femoral artery are similar to earlier age group. The laminae are still long uniform and parallel and only occasional laminae anastomose. Movat HZ et al. [24] have also found similar structure in early adult life. We have also found that elastic laminae of media of femoral artery increased in number. Apart from increased thickness of intima as a whole, localized or regional intimal thickening is also seen with increasing age. Boucek RJ et al. (1964) [25] have also shown that this regional intimal hypertrophy is present by 12 years of age and it increases in amount and diffuseness with age. This regional or localized intimal hypertrophy is present only up to 40 years of age. Boucek RJ et al. (1963) [26] have shown that after 40 years there is diffuse hypertrophy of tunica intima and it is associated with atrophy of media and adventitia. They have also shown that this regional intimal hypertrophy appears at sites where unusually physical stresses can be predicted to occur within the arterial wall.

According to Velican and Velican, [27] elastic laminae of arteries in adult life have shown some branching and irregularity of elastic fibers and the laminae are crenated. We also found similar changes in elastic laminae between 15-40 years of age.

In present study, we have measured the thickness of normal intima and media and found that thickness of intima increases with age, whereas the thickness of the media increases until the 6th decade and thereafter it decreases, possibly due to atrophy of smooth muscle cells and deposition of collagen tissue. The relative measurements of the intima and media of the arteries show that localized thickening of intima is usually associated with thinning of the media. Similar result was also shown by Crawford T (1961). [28] The possible explanation of thinning of media over regional intimal hypertrophy is that, at parts where intima is thickened, the diastolic rebound of the vessel wall will be reduced and it will tend to remain in a more or less distended position after death. But there is also the possibility that the thinning may represent a basic weakness of the muscular coat.

According to Schlatmann and Becker, [16] in middle and old age, degenerative changes develop in the elastic fibers of internal and external elastic laminae and the media. These changes are most noticeable in the internal elastic lamina and there was often some reparative process shown by the development of thinner fibers irregularly oriented around the original membrane. Studies have also shown that concentrations of both collagen and elastin increased significantly with age, substantially after the age of 45. [29]

In present study, it was observed that, between 41-60 years, thickness of tunica intima continues to increase an average thickness of tunica intima may reach upto 30-40 percent of the thickness of the tunica media.

There was a reduplication of the internal elastic lamina. The intimal thickening is irregular and causes asymmetrical lumen, collagen tissue in media also increases in quantity. This increase in collagen tissue is more noticeable near junction of media and adventitia. These possibly correspond to the course of vasa vasorum.

Increase in collagen tissue causes atrophic changes in smooth muscles of media and bring tortuosity of vessel. The increase in collagen tissue and its lack of association with any inflammatory process seem to favor the phenomenon of physical aging. Boucek RJ et al. (1963) [26] have also shown that after 40 years of age muscle bundles become atrophied in certain region and adventitia become thinned.

All these changes suggest that, with age, arteries lose their active function and assume a more passive role as conduit for transport of blood. Tunica adventitia also becomes thinned. Heath D, et al. (1959) [30] have shown that elastic laminae become thinned and show much fragmentation by the age of 45 with occasional clumping. The thinning is much more prevalent to the inner media, where there is much fibrosis. The elastic laminas of the outer media were thicker and appear like the lamina of younger age. We also found the similar structure of elastic lamina after 40 years.

According to Crawford T et al. (1961) [28] intima by the age of 68 years may equal or even exceed the media in thickness and internal elastic lamina become thickened and there may be second incomplete ring inside the original lamina. The thickened intima shows the presence of smooth muscle cells, collagen and elastic fibers and round cells.

Clark JM, Glagov S [31, 32] has shown that, with more advanced age, collagen fibers tend to undergo hyaline degeneration and appear swollen, glassy and increasingly eosinophilic. In the present study, in 60 years and above age group, decrease in thickness of tunica media of femoral artery was recorded, while thickness of tunica intima continues to increase. It becomes half or two third of tunica media. As explained earlier it could be due to atrophy of smooth muscle cells and deposition of connective tissue in media. Elastic laminae of tunica media become thinner and fragmented.

Conclusion

Light microscopy of femoral arteries revealed different characteristics in various age groups. A detailed microscopic study was carried out to seen the changes in histological pattern of femoral arteries. All three layers of the arterial wall can be identified in infants. Tunica intima consists of internal elastic lamina and endothelium. Sub-endothelial layer is scarcely perceptible. Elastic fibers on tunica media of the arteries are long and parallel, uniformly arranged and thicker than adult femoral arteries.

In the earlier childhood, sub-endothelial connective tissue layer becomes more apparent. There is occasional localized splitting of internal elastic laminate. Adventitia is thinner than media. In adults, thickness of tunica intima increases while tunica media appears thin over regional intimal hypertrophy, but in the old age, diffuse intimal hypertrophy is seen. Tunica intima become half, to two third of tunica media. Lumen becomes asymmetrical. There is complete reduplication of internal elastic lamina.

Hyalination of collagen tissue of media occurs with loss of normal staining properties of elastic tissue. Elastic laminae become thinned and show fragmentation with occasional clumping. These changes are more prominent near inner media. Elastic laminae of inner media appear liked adult. From the finding in the present study, one might conclude that there is definite intimal hypertrophy and media fibrosis with increasing age. These degenerative changes, which develop gradually with the age, lead to arteriosclerosis or hardening of the arteries. The gross morphological effect of all these changes is that the arteries tend to dilate and elongate. These inevitably lead to tortuosity of the artery. The arteries become harder and easily palpable in such sites as the wrist and neck. They loss some of their active function and assume a more passive role as a conduit for transport of blood.

References

[1] Medawar, P. B. An Unsolved Problem in Biology. London: HK Lewis; 1952. Available at "http://science-of-aging.healthaliciousness.com/ timelines/medawar-theory-mutation-accumulation.php"

[2] Williams, G. C. Pleiotropy, Natural Selection, and the Evolution of Senescence. *Sci. Aging Knowl. Environ*. 2001; l: 1.

[3] Hamilton, W. D. The moulding of senescence by natural selection. *J. Theor. Biol.* 1966; 12(1): 12-45.

[4] Kirkwood, T. B. Evolution of ageing. *Nature*. 1977; 270(5635): 301-4.

[5] Bass, T. M., Weinkove, D., Houthoofd, K., Gems, D., Partridge, L. Effects of resveratrol on lifespan in Drosophila melanogaster and Caenorhabditis elegans. *Mech. Ageing Dev*. 2007; 128(10): 546-52.

[6] Ryley, J., Pereira-Smith, O. M. Microfluidics device for single cell gene expression analysis in Saccharomyces cerevisiae. *Yeast*. 2006; 23(14-15): 1065-73.

[7] Gavrilov, L. A., Gavrilova, N. S. The reliability theory of aging and longevity. *J. Theor. Biol.* 2001; 213(4): 527-45.

[8] Goldman, L., Pearl, R. Smoking and Longevity. *Genetics*. 2002; 162: 997-1001.

[9] Brunet-Rossinni, A. K., Austad, S. N. Ageing studies on bats: A review. *Biogerontology*. 2004; 5(4): 211-22.

[10] Magalhães, J. P., Costa, J., Church, G. M. An analysis of the relationship between metabolism, developmental schedules, and longevity using phylogenetic independent contrasts. *J. Gerontol. A Biol. Sci. Med. Sci*. 2007; 62(2): 149-60.

[11] Macintosh, K. L. *Illegal Beings: Human Clones and the Law*. Cambridge: Cambridge University Press; 2005.

[12] Swynghedauw, B., Besse, S., Assayag, P. Biology of cardiac and vascular senescence. *Bull. Acad. Natl. Med*. 2006; 190(4-5): 783-92.

[13] Le Couteur, D. G., Lakatta, E. G. A vascular theory of aging. *J. Gerontol. A Biol. Sci. Med. Sci.* 2010; 65(10):1025-7.

[14] Fawcett, D. W., editor. Bloom and Fawcett, a *Textbook of Histology*. 12th edition. A Hodder Arnold Publication, Chapman and Hall, 1994: 359-582.

[15] Crawford, T. Some pathological aspects of arterial disease. *Postgrad. Med. J.* 1955; 31(353): 108-12.

[16] Schlatmann, T. J., Becker, A. E. Histologic changes in the normal aging aorta: implications for dissecting aortic aneurysm. *Am. J. Cardiol.* 1977; 39(1): 13-20.

[17] Francis, G., John, R., Thomas, J. Biosynthetic pathway of desmosines in elastin. *Biochem. J.* 1973; 136(1): 45-55.

[18] Boucek, R. J., Takashita, R., Fojaco, R. Functional anatomy of the ascending aorta and the coronary ostia (dog). *Am. J. Anat.* 1964; 14: 273-482.

[19] Lakatta, E. G. Arterial and cardiac aging: major shareholders in cardiovascular disease enterprises: Part III: cellular and molecular clues to heart and arterial aging. *Circulation.* 2003; 107(3): 490-7.

[20] Ross, R. The arterial wall and atherosclerosis. *Annu. Rev. Med.* 1979; 30: 1-15.

[21] Homan, R., Hanselman, J. C., Bak-Mueller, S., Washburn, M., Lester, P., Jensen, H. E., Pinkosky, S. L., Castle, C., Taylor, B. Atherosclerosis in Octodon degus (degu) as a model for human disease. *Atherosclerosis.* 2010; 212(1): 48-54.

[22] Prior, J. T., Hutter, R. V. Intimal repair of the aorta of the rabbit following experimental trauma. *Am. J. Pathol.* 1955; 31(1): 107-23.

[23] Tyagi, S. P., Dadgar, S. K., Hameed, S., Singh, R. P. Coronary atherosclerosis in Aligarh district, Uttar Pradesh. A morphological study. *Jpn. Circ. J.* 1978; 42(9): 1033-40.

[24] Movat, H. Z., More, R. H., Haust, M. D. The diffuse intimal thickening of the human aorta with aging. *Am. J. Pathol.*1958; 34(6): 1023-31.

[25] Boucek, R. J., Fojaco, R., Takashita, R. Anatomic considerations for regional intimal changes in the coronary arteries (dog). *Anat. Rec.* 1964; 148: 161-9.

[26] Boucek, R. J., Takeshita, R., Brady, A. H. Microanatomy and intramural physical forces within the coronary arteries (man). *Anat. Rec.* 1965;153(3):233-41.

[27] Velican, C., Velican, D. Discrepancies between data on atherosclerotic involvement of human coronary arteries furnished by gross inspection and by light microscopy. *Atherosclerosis*. 1982; 43(1): 39-49.

[28] Crawford, T. Morphological aspects in the pathogenesis of atherosclerosis. *J. Atheroscler. Res.* 1961; 1: 3-25.

[29] Cattell, M. A., Anderson, J. C., Hasleton, P. S. Age-related changes in amounts and concentrations of collagen and elastin in normotensive human thoracic aorta. *Clin. Chim. Acta.*1996; 245: 73-84.

[30] Heath, D., Wood, E. H., Dushane, J. W., Edwards, J. E. The structure of the pulmonary trunk at different ages and in cases of pulmonary hypertension and pulmonary stenosis. *J. Pathol. Bacteriol.* 1959;77(2):443-56.

[31] Clark, J. M., Glagov, S.Transmural organization of the arterial media. The lamellar unit revisited. *Arterioscler. Thromb. Vasc. Biol.*1985; 5:19-34.

[32] Majack, R. A., Bornstein, P. Biosynthesis and modulation of extracellular component by cultured vascular smooth muscle cell. In: Campball, J. H., Campball, G. R., editors. *Vascular smooth muscle culture in boca raton*, fla 1986, CRC Press.

In: Asia Pacific Life Sciences Compendium
Editor: Bandit Chumworathayi

ISBN: 978-1-62257-338-7

Chapter 18

SAFETY AND EFFICACY OF LOCAL ANESTHESIA FOR LICHTENSTEIN HERNIOPLASTY IN A GENERAL HOSPITAL, THAILAND

Wibun Phanthabordeekorn*
General surgeon, Paholpolpayuhasaena General Hospital, Kanchanaburi, Thailand

ABSTRACT

Objective: To evaluate safety and efficacy of local anesthesia for Lichtenstein hernioplasty in a general hospital. *Method:* After having permission from the Ethical Review Board of the hospital, any patients who had this procedure done by author, from July 2007 to February 2011, were prospectively assessed and recorded about their demographic data, intra-operative pain scores, complications, and additional anesthesia/ analgesia. *Result:* There were 170 male patients aged 18-97 year-old (mean 60.5) having this procedure done by the author during the mentioned period. Most of the patients had only initial pain in the first infiltration, with low intra-operative pain score (mean 2.3, range 1-4). No one needed additional anesthesia, neither regional nor general, or pre-operative/ intra-operative analgesia/ sedative, neither oral nor injected. Also, no one needed to be observed in the recovery room. Every patient was able to return to their residences immediately after surgery. No one needed re-admission from either surgical complications or post-operative pain. There was neither post-operative acute urinary retention nor wound infection found in all of the 170 patients. Only one complication, that author found, was minute wound hematoma, which could be treated by only supportive approach until it was completely healed. *Conclusion:* Local anesthesia for Lichtenstein hernioplasty can be done efficaciously and safely in a general hospital, although most patients in this study had one or more underlying medical diseases such as; coronary heart disease, diabetes mellitus, hypertension, which are commonly found in the elderly.

Keywords: local anesthesia; Lichtenstein; inguinal hernioplasty; day-case surgery

---

* E-mail: wibunphantha@yahoo.co.th

BACKGROUND

Inguinal hernioplasty is a common general surgical procedure. In 2003, there were 800,000 procedures done in the US [1]. Most are elderly patients with underlying diseases such as; myocardial infarction, diabetes mellitus, hypertension, COPD, etc. If regional or general anesthesia were used, there would be high risks of post-operative complications for these patients. Kehlet et al found that only 18.1% of these procedures were done under local anesthesia [2]. Inguinal hernioplasty under local anesthesia has not also been popularized by most Thai surgeons although it was found to be safe and save [2-4] compared with either regional or general anesthesia. Importantly, no admission is needed. However, the techniques of injecting local anesthesia for this procedure vary between the surgeons. The procedure described below, as developed by the author, have been used for Lichtenstein hernioplasty for more than 10 years and was used in this study. The study's objectives were to evaluate safety and efficacy of local anesthesia for Lichtenstein hernioplasty in a general hospital.

METHOD

After having permission from the Ethical Review Board of the hospital, any patients who had this procedure done by author, from July 2007 to February 2011, were prospectively assessed and recorded about their demographic data, intra-operative pain scores, complications, and additional anesthesia/ analgesia. Only patients who gave informed consents were included. Pain scores were achieved by visual analog scale (VAS) assessment by patients during having the Lichtenstein tension-free hernioplasty. 0 meant no pain and 10 meant most severe pain. There was no need for preoperative gastric emptying, laboratory investigation, or intravenous fluid replacement. After operation, patients were allowed to return to their residences and appointed for 1-week follow-up. Exclusion criteria were history of allergy to local anesthesia and bleeding tendency. Infection prophylaxis was done by intravenous Cefazolin injection, 30 minutes before doing surgical incision. During operation, if there was breakthrough pain, additional analgesia might be given. After returning home, patients' personal daily activities could be done without any limitations. Oral analgesia would be used if needed.

Local Anesthesia Injection Technique Developed by the Author

1) Prepare 2 syringes of 10 ml 1% xylocaine and 10 ml 0.5% bupivacaine solutions, separately, then mark the injection site.

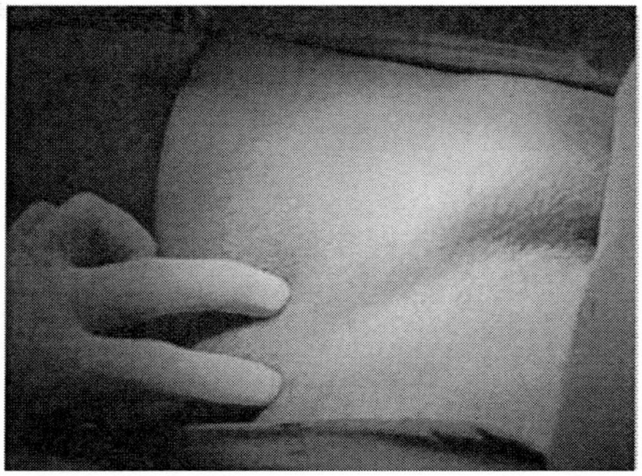

2) Infiltrate the 1% xylocaine solution subcutaneously/intradermally at 2 cm medial to anterosuperior iliac spine (ASIS).

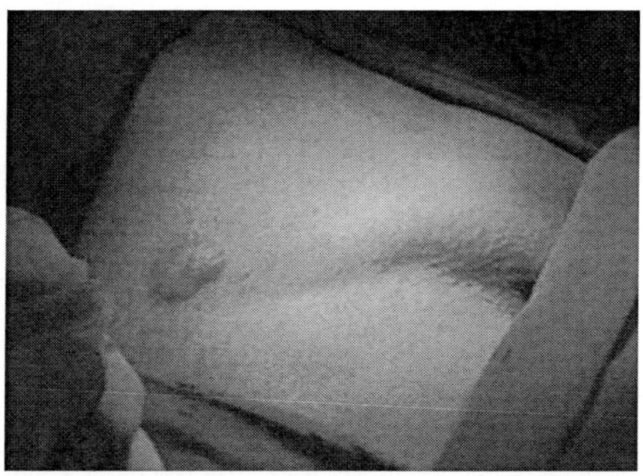

3) Infiltrate the solution into the muscular layer pointing to the umbilicus.

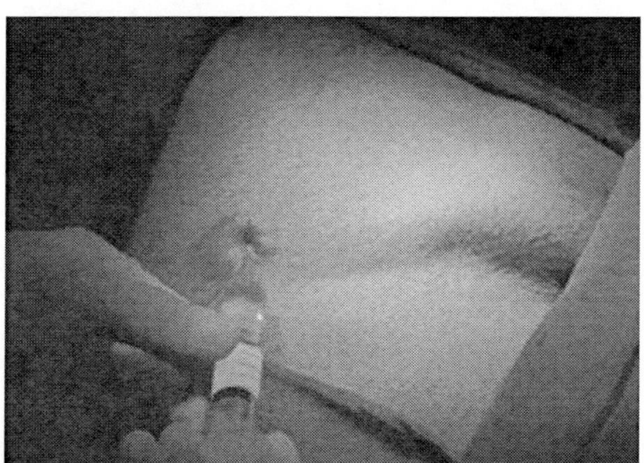

4) Infiltrate the solution into the muscular layer pointing downward.

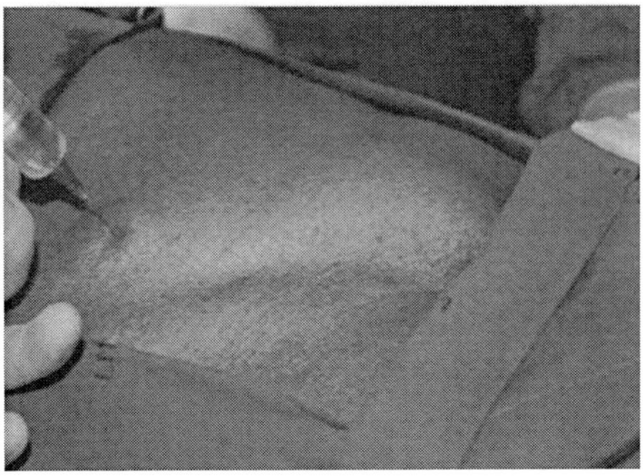

5) Infiltrate the solution into the muscular layer pointing to the inguinal canal. Then, use the 0.5% bupivacaine solution and repeat steps 2-5.

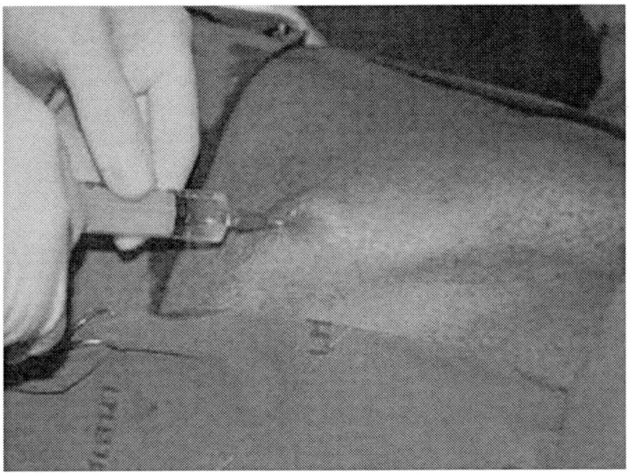

6) Wait for 5 minutes. This is to completely block the ilioinguinal and iliohypogastric nerves.

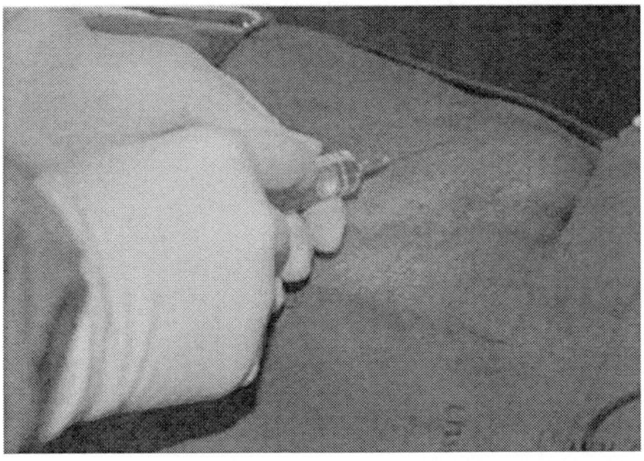

7) Infiltrate 10 ml 1% xylocaine at the incision site (local infiltration).

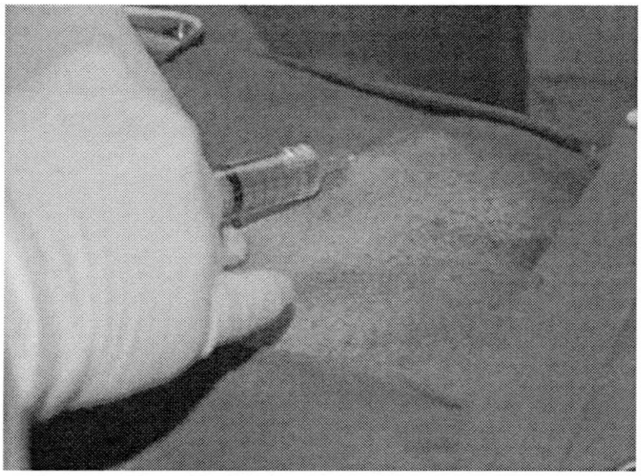

8) Do the incision.

9) When external oblique aponeurosis is reached, 1-2 ml 1% xylocaine was infiltrated underneath this aponeurosis, after then external oblique aponeurosis was opened.

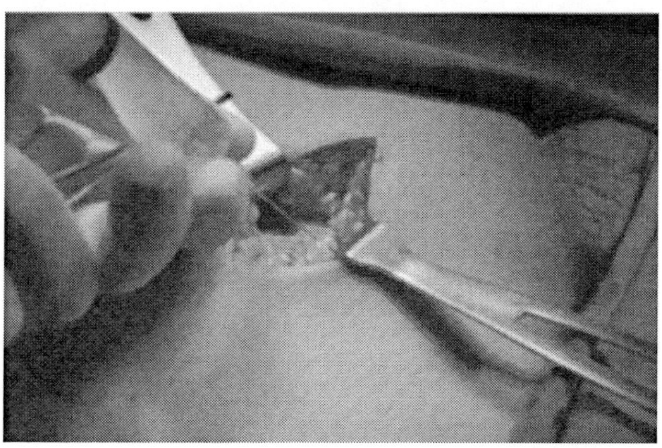

10) When reaching where mesh prosthesis must be sutured with the tissue at pubic tubercle, 1 ml 1% xylocaine was infiltrated here then continue the hernioplasty.

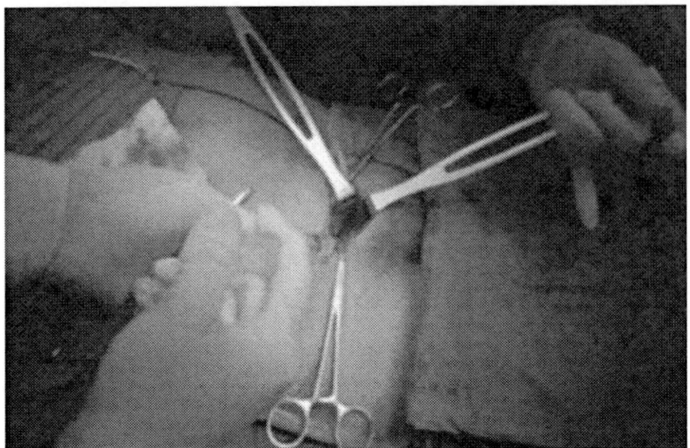

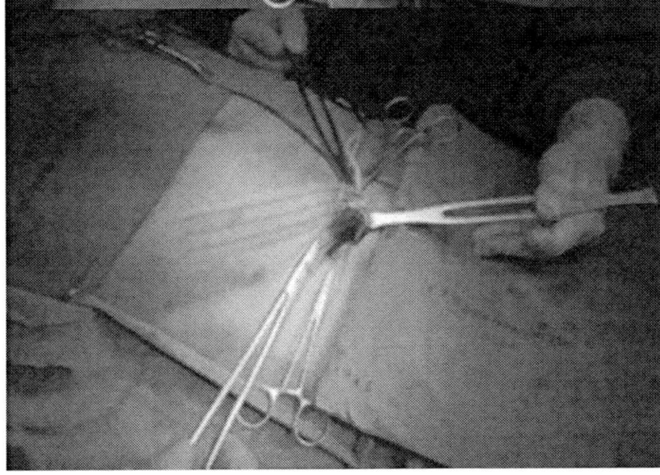

11) When finished, infiltrate 10 ml 0.5% bupivacaine around the incision just before closing the incision.

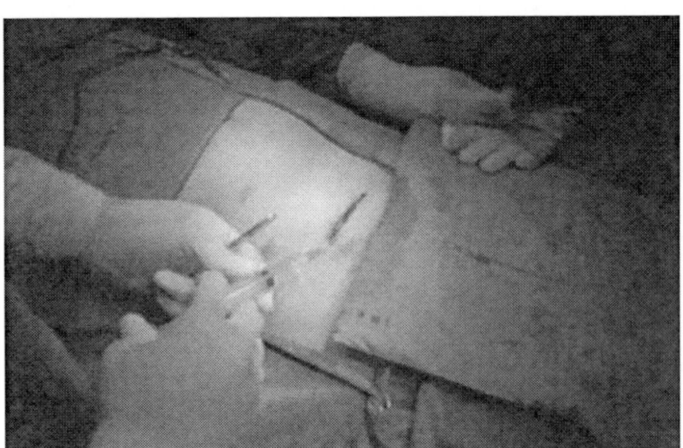

RESULT

From July 2007 to February 2011, there were 170 male patients aged 18-97 year-old (mean 60.5) having this procedure done by the author. Most of them had an underlying disease; 82 myocardial infarctions, 76 hypertensions, 70 dyslipidemias, 50 diabetes mellituses, 20 chronic kidney diseases, 18 COPDs, and 2 epilepsies. 106/170 had right inguinal hernioplasties. 59/170 had left inguinal hernioplasties. 5/170 had bilateral inguinal hernioplasties. Importantly, 6/170 were the recurrent cases.

Most of the patients had only initial pain in the first infiltration, with low intra-operative pain score (mean 2.3, range 1-4). When 0 and 10 assumed no pain and most severe pain, respectively, one had 1, 120 had 2, 47 had 3, and two had 4. Importantly, no one needed additional anesthesia, neither regional nor general, or pre-operative/ intra-operative analgesia/ sedative, neither oral nor injected.

No any patients needed to be observed in the recovery room. Every patient was able to return to their residences immediately after surgery. No one needed re-admission from either surgical complications or post-operative pain. There was neither post-operative acute urinary retention nor wound infection found in all of 170 patients. Only one complication, that author found, was minimal wound hematoma, which could be cured by only supportive approach.

DISCUSSION

Standard Lichtenstein hernioplasty can be done efficaciously and safely under local anesthesia [3-7]. In this study conducted in a general hospital, although most patients in this study had one or more underlying medical diseases which contribute more risks to regional or general anesthesia, it is still also safe and save. Right inguinal hernias were found more than left inguinal hernias. Indirect inguinal hernias were found more than direct inguinal hernias. Likewise, 5 bilateral inguinal hernias found in this study could be safely repaired in the same operation as previous reports [10-12]. All 5 were able to withstand the operation without any additional complications. The 6 recurrent cases were also able to withstand the operation and had pain scores not different from the new ones. In this study, there was no one needing additional anesthesia, neither regional nor general. Practical simple evaluation, by letting patient sat and got down the bed, could predict patient's readiness for self caring at home.

One of the most important advantages of local anesthesia over general anesthesia is that it has no effect on the cardiovascular and respiratory systems. Therefore, it can be used safely in COPDs and heart diseases. However, some of disadvantages are;

1) Patients are awake during the procedure causing some anxiousness. This feeling could be relieved by light talking with surgeon or accompanying health personnel. This is to reassure the patient that he does not need to be patient. If there is some pain or discomfort, he can ask the care team immediately for help. The team should also warn the patient if there will be some discomfort, such as traction on a spermatic cord, to help him prepared. Local anesthesia can help relief only pain not anxiousness (anxiolytic), therefore, an attention must be paid to patients' feelings. This can help increase patients' satisfaction to the local-anesthesia-based procedure.

2) When doing hernioplasty, there will be some local edema from infiltration of fluid. There will also be slightly bleeding because the local anesthesia had a vasodilator effect which can mostly be spontaneously stopped. If this bleeding continues, electrocauterization may be used.

To help prevent the initial pain during local infiltration, the author used;

1) 1% xylocaine without epinephrine because it has less acidity compared with the one with epinephrine.
2) Subcutaneous infiltration before intradermal infiltration

This local anesthesia for inguinal hernioplasty in this study is to block the ilioinguinal and iliohypogastric nerve helping decreasing pain from both during and after the surgery like Andersen et al's study [13] which was a double-blind randomized clinical trial in 160 patients. Pre-operative ilioinguinal nerve block could significantly decrease intra-operative pain and post-operative complications compared with regional or general anesthesia. Acute urinary retention, which could be found in previous studies [4,8,9] was not found in this study. Likewise, wound infection was not found. Therefore, the patients could leave the hospital for self caring without admission.

CONCLUSION

Local anesthesia for Lichtenstein hernioplasty can be done efficaciously and safely in a general hospital, although the patients had one or more underlying medical diseases. This can help decreasing both hospital's burden and patient's expenses by eliminating the needs of pre-operative investigation (blood chemistry, CXR, EKG), intraveneous fluids, analgesias, sedatives, regional or general anesthesias, post-operative recovery room, admissions, and complications-related costs such as expenses for acute urinary retention and cardio-pulmonary complications from regional and general anesthesias [14].

REFERENCES

[1] Rutkow IM. Demographic and socioeconomic aspects of hernia repair in the United States in 2003. *Surg. Clin. North Am.* 2003; 83:1045-51.
[2] Kehlet H, Bay-Nielsen M. Anaesthetic practice for groin hernia repair-a nation-wide study in Denmark 1998-2003. *Acta Anaesth Scand* 2005; 49: 143-6.
[3] Kehlet H, Aasvang E. Groin hernia repair:anesthesia. *World J. Surg.* 2005; 29: 1058-61.
[4] Sanjay P, Woodward A. Inguinal Hernia Repair: Local or General Anesthesia? *Ann. R. Coll Surg. Engl.* 2007; 89: 497-503.
[5] Milone F, Salvatore G, Leongito M. Hernia repair and local anesthesia: Results of a randomized clinical trial. *G. Chir* 2010; 31: 552-5.
[6] Yang B, Liang MJ, Zhang YC. Use of local or epidural anesthesia in inguinal hernia repair: a randomized trial. *Zhonghua Wai Ke Za Zhi* 2008; 15: 1234-6.

[7] Callesen T, Bech K, Kehlet H. One-thousand consecutive inguinal hernia repairs under unmonitored local anesthesia. *Anesth. Analg*. 2001; 93: 1373-6.

[8] Jessen P, Mikkelsen T. Postherniorrhaphy urinary retention-effect of local, regional, and general anesthesia: A review. *Regional Anesthesia and Pain Medicine* 2002; 27: 612-617.

[9] Bay-Nielsen M, Kehlet H. Anesthesia and post-operative morbidity after elective groin hernia repair: a nation-wide study. *Acta Anaesthesiol. Scand* 2008; 52: 169-74.

[10] Amid PK, Shulman AG, Lichtenstein IL. Simultaneous repair of bilateral inguinal hernias under local anesthesia. *Ann. Surg*. 1996; 223: 249-52.

[11] Dakkuri RA, Ludwig DJ, Traverso LW. Should bilateral hernias be repaired during one operation? *Am. J. Surg*. 2002; 183: 554-7.

[12] Kark AE, Belsham PA, Kurzer MN. Simultaneous repair of bilateral groin hernias using local anaesthesia: a review of 199 cases with a five-year follow-up. *Hernia* 2005; 9: 131-3.

[13] Andersen FH, Nielsen K, Kehlet H. Combined ilioinguinal blockade and local infiltration anaesthesia for groin hernia repair-a double-blind randomized study. *Br. J. Anaest*h. 2005; 94: 520-3.

[14] Gonullu N, Cubukcu A, Alponat A. Comparison of local and general anesthesia in tension-free (Lichtenstein) hernioplasty: a prospective randomized trial. *Hernia* 2002; 6: 29-32.

In: Asia Pacific Life Sciences Compendium
Editor: Bandit Chumworathayi

ISBN: 978-1-62257-338-7

Chapter 19

PHYTOCHEMICAL FINGERPRINTING, CYTOTOXIC, ANTIMICROBIAL, ANTITUBERCULAR, ANTIMYCOTIC POTENTIALS OF *SIDA RHOMBIFOLIA* SUBSP. *RETUSA* AND *EMBELIA TSJERIAM-COTTAM*

Radhika Poojari*
Department of Biological Sciences, Ramnarain Ruia College, Matunga, Mumbai, India

ABSTRACT

Objective: *Sida rhombifolia* subsp. *retusa* and *Embelia tsjeriam-cottam* were investigated for their potentials to exhibit cytotoxic, antibacterial, antitubercular and antimycotic activities *in vitro* for the first time. *Method:* Plants extracts were analyzed for phytochemical fingerprinting by High Performance Thin Layer Chromatography and High Pressure Liquid Chromatography, cytotoxic activity against Human Non-Small-Cell Lung Cancer (HOP 62) and Hepatocellular Carcinoma (Hep G2) cell lines by Sulforhodamine B assay, antimicrobial, antimycotic activity by agar well-diffusion method and antitubercular activity on Lowenstein-Jensen medium. *Result:* Phytochemical fingerprinting revealed Ecdysterone in *S. rhombifolia* subsp. *retusa,* Embelin in *E. tsjeriam-cottam* as chemotaxonomic markers and for routine quality control. Methanolic seed extract of *S. rhombifolia* and Embelin in fruits of *E. tsjeriam-cottam* exhibited marked cytotoxic activity against HOP 62, Hep G2 cells at $GI_{50} \leq 20$ µg/ml. They exhibited potent inhibitory effects against *Staphylococcus aureus, Diptheriae* sps, *Candida krusei, Aspergillus flavus* with inhibitory zones ranging from 18-26 mm and were also active to *Mycobacterium tuberculosis* strains. *Conclusion:* Present findings demonstrate potent cytotoxic, antibacterial, antitubercular and antimycotic activities in methanolic seed extract of *S. rhombifolia* subsp. *retusa* and Embelin in fruits of *E. tsjeriam-cottam.* This can be regarded as potential source of alternative natural products.

Keywords: Sida rhombifolia subsp. retusa, Embelia tsjeriam-cottam, phytochemical fingerprinting, cytotoxicity, antibacterial, antitubercular, antimycotic

---

* Corresponding author: Radhika Poojari. Department of Biological Sciences, Ramnarain Ruia College, Matunga, Mumbai, 400019, India. Tel: 022-26862684, E-mail: drradhikapoojari@gmail.com.

BACKGROUND

Bacterial infections are primarily associated with diarrhoeal diseases, enteric fever, urino-genital and respiratory tract of which 50% affects the pediatrics. Added to this burden is more than a quarter of the world population plagued with *Mycobacterium tuberculosis* and HIV infections---a risk factor. Every year, there are about 8 million new cases of TB and 2 million TB deaths. Multidrug resistance and adverse drug reactions of high dose of chemotherapeutic drugs, remains a major cause of public health concern worldwide. This has prompted 80% masses towards resurgence of traditional medicine globally to meet the primary health care---a WHO report. [1]

There is a constant quest for new therapeutic phytocomplexes to explore a variety of lead structures for new drug development.

Sida rhombifolia Linn. subsp. r*etusa* (Linn.) Borss (Malvaceae), syn. *Sida retusa,* (Bala, Atibala) is an annual or perennial branched undershrub found throughout the tropical, subtropical parts of India growing wild along road sides gregariously as weeds. *Embelia tsjeriam-cottam* A.D.C. (Myrsinaceae), syn. *Embelia robusta,* (Vidang) is a large scandent shrub found throughout India, ascending upto 1600 m in the hills.

Roots, seeds of *Sida* and fruits of *Embelia* species are being used in the Ayurvedic system of medicine and form the chief ingredient of several important formulations. *Sida* species is widely used in ethnopharmacology as a remedy for rheumatism, arthritis and other neurological disorders. It possesses astringent, cooling, stomachic, tonic properties and is a very good drug for asthma. They are considered to be aromatic bitters having febrifuge, demulcent and diuretic properties. Seeds are colic, aphrodisiac, spermatic tonics; have fertility regulating properties, useful in gonorrhoea and cystitis. [2,3] Fruits of *Embelia* species possess potent analgesic, anti-inflammatory, antihelminthic and contraceptive properties. They are astringent, bitter, carminative, antiseptic, thermogenic, useful in gastrointestinal disorders, bronchitis, chest ailments, skin diseases and leprosy. [4,5]

Being Rasayana drugs both *Sida* and *Embelia* species are rejuvenating and age-sustaining tonics for promoting vitality and longevity. Phytochemically, *Sida* species possesses alkaloids (β-phenylthylamine, ephedrine, Ψ-ephedrine, vasicine, vasicinol, vasicinone), β-sitosterol, phenolics, steroids, sugars, glycosides, amino acids, gums and mucilages. Seed oil consists of cyclopropenoid fatty acids viz; sterculic acid, malvalic acid, linoleic acid, myristic acid, palmitic acid, stearic acid and oleic acid. [3] Fruits of *Embelia* species possess embelin, vilangin, quercitol, tannins, christembine, fatty ingredients, resinoids and volatile oil present. [6]

Presence of phytoecdysteroids in *Sida* species, [7-9] quantitative estimation of Embelin in *E.ribes* spectrophotometrically [10] and HPTLC in the formulations have been reported by earlier workers. [11,12]

Previous studies have shown *S. rhombifolia* leaves exhibited antitumor and cytotoxic activity against 60 human cell lines. [13,14] Embelin exhibited promising cytotoxic activity on polymorphonuclear leucocytes [15] and Methycholanthrene-induced fibrosarcoma cancer cells. [16] Moreover, antimicrobial activity has been detected in some *Sida, Embelia* species, [17-19] but at present no data on this regard are available on *S. retusa* and *E. tsjeriam-cottam* species.

The present research work describes a normal phase High Performance Thin Layer Chromatographic (HPTLC) and reversed phase High Pressure Liquid Chromatographic (HPLC) method for establishing the fingerprint for *S. rhombifolia* subsp. *retusa* and *E. tsjeriam-cottam* for the first time, which is qualitative in nature.

In context of the above claims it was, therefore of, interest to investigate their cytotoxic, antimicrobial, antitubercular as well as antimycotic activities *in vitro*.

METHOD

Plant Materials and Extraction

Whole plant of *Sida rhombifolia* subsp. *retusa* (Linn.) Borss (Malvaceae) and its seeds were collected by the author locally from hilly forest areas of Borivali National Park, Mumbai, whereas whole plant of *Embelia tsjeriam-cottam* A.D.C. (Myrsinaceae) and its fruits was collected by the author from Lonavala, Pune, during the months of October, November, and May, respectively. Authentication of these plants was established at National Institute of Science Communication and Information Resources (NISCAIR), New Delhi. A voucher specimen of *S. rhombifolia* subsp. *retusa* (NISCAIR / RHM / F-3./ 2004 / Consult./ 512 / 88) and *E. tsjeriam-cottam* (NISCAIR / RHM / F-3/ Consult./ 06 / 706 / 23-R07107) was deposited in the laboratory at R. Ruia College, Mumbai.

Chemical authenticity of the plants was established by HPTLC and HPLC fingerprint profiles. Powdered plant materials viz; roots, stem, leaves, stem+leaves (1:1) (aerial part), root+stem (1:1), whole plant (root+stem+leaves) (1:1:1), seeds, flowers, fruits of *S. rhombifolia* subsp. *retusa* and *E. tsjeriam-cottam,* of each plant were individually extracted in a Soxhlet with 80% methanol (twice) for 36 h. The individual extracts were filtered and concentrated at 50°C under reduced pressure on a rotary evaporator. The residues were dissolved in 5 ml methanol.

Embelin (2,5-dihydroxy-3-undecyl-1,4-benzoquinone), an active constituent in fruits of *E. tsjeriam-cottam* was extracted in a Soxhlet using n-hexane. The concentrated extract was kept in refrigerator (0°C) on cooling deposited crude embelin residue. This, on recrystallization twice with cold ethanol, gave orange glistening crystals of embelin. The crystals were dissolved in 5 ml of chloroform and used for HPTLC/HPLC analysis except in case of HPLC; it was diluted in methanol (1:1).

Standards for *S. rhombifolia* subsp. *retusa;* Ecdysterone (gifted by Dr. Laddha, University Institute of Chemical Technology, Mumbai), 1 mg/ml was prepared in methanol whereas for *E. tsjeriam-cottam;* Embelin (gifted by Dr. N.D. Grampurohit, Dr. Bhanuben Nanavati College of Pharmacy, Mumbai) 0.2 mg/ml in chloroform. For cytotoxic, anti-infective bioassays methanolic seed extract of *S. rhombifolia* subsp. *retusa* and Embelin in the fruits of *E. tsjeriam-cottam* were dissolved in Dimethylsulfoxide (0.1% DMSO) before use to obtain a concentration of 100 mg/ml constituting the stock solutions. Further dilutions were prepared in sterile distilled water.

Phytochemical Fingerprinting

a) Chromatographic Conditions for HPTLC Analysis

HPTLC was performed on 20 cm X 10 cm, pre-coated Silica gel 60 F_{254} TLC plates (0.2 mm) (E. Merck, Germany). Standard and sample solutions 10 μl for methanolic extracts of root, stem, leaves, aerial part, whole plant, seeds and flowers of *S. rhombifolia* subsp. *retusa* and 2 μl for *Embelia* (methanolic extracts of roots, stem, leaves, aerial part, root+stem, whole plant, and fruits) were applied as narrow bands (8 mm width) by CAMAG Linomat V applicator. Plates were developed in a saturated CAMAG Twin Trough Development Chamber (10 min), to a distance of 80 mm, with Chloroform: Methanol (9:1 v/v) for *Sida* whereas n-Propanol: Butanol: 2N Ammonia (6:1:3 v/v) for *Embelia* as the mobile phase. The chromatograms were subjected to densitometric scanning on CAMAG TLC Scanner 3 equipped with winCATS software at λ=254 nm (*Sida*), 333 nm (*Embelia*) absorbance modes. Post-chromatographic derivatisation with Vanillin-Sulphuric acid reagent gave a good resolution on scanning at λ=580 nm (*Sida*), 480 nm (*Embelia*) visible modes. Photodocumentation was carried out using CAMAG Reprostar 3.

b) Chromatographic Conditions for HPLC Analysis

Sample solutions for *S. rhombifolia* subsp. *retusa* and *E. tsjeriam-cottam* are similar as mentioned above in HPTLC analysis. HPLC analysis was carried out on Dionex-3000 pump (*Sida),* Dionex P-680 Low pressure gradient (*Embelia*) equipped with autosampler Dionex-3000 (*Sida),* Dionex ASI 100 T (*Embelia*) and Dionex-3000 PDA detector running Chromelon (c) Dionex software. Column used was Acclaim C_{18} (Make-Dionex, US), 25cm X 4.6mm i.d., 5 μm particle size, 120 A^{0}. Column oven temperature (Dionex-TCC 100) was 25°C. In case of *S. rhombifolia* subsp. *retusa,* this was initially equilibrated with Methanol:Water (HPLC grade, Merck) in the ratio 30:70 at a flow rate of 1 ml/min, but following sample injection of 20μl, a linear gradient to 100 % MeOH for a further 35 min was performed and the final mobile phase was maintained for a further 7 min. In case of *E. tsjeriam-cottam,* initial equilibration was carried out using solvent A [0.01M KH_2PO_4: MeOH (90:10)]: solvent B [MeOH: 0.01M KH_2PO_4 (90:10)] in 70:30 (A:B) ratio at a flow rate of 1 ml/min, but following injection of the sample 20μl, a low pressure gradient system of A:B in a ratio of 20:80 for 40 min was performed and the final mobile phase was maintained in a ratio of 70:30 (A:B) for a further 15 min. Separation was monitored at 245 nm (*Sida)* and 225 nm (*Embelia*). Compounds which gave UV-absorbing peaks were compared to reference standards with respect to retention times (Rt) and UV spectra's in the same system.

Cytotoxic Activity

a) Cancer Cell Lines

Human Non-Small-Cell Lung Cancer (NSCLC-HOP 62) and Human Hepatocellular Carcinoma (HEP G2) cell lines (NCI, US) were grown as monolayer in RMPI 1640 medium (GIBCO, BRL) supplemented with 10% fetal bovine serum (GIBCO, BRL) and 2 mM L-glutamine and Antibiotic cocktail (Streptomycin 100 μg/ml, Penicillin 100 U/ml, Mycostatin 25 U/ml). Cells were grown at 37°C in a humidified 5% CO_2 atmosphere.

b) Cytotoxicity Bioassay

HOP 62 and HEP G2 cells in the exponential growth phase were harvested using 0.25 mM Trypsin-0.025% EDTA (GIBCO, BRL) for three minutes and resuspended in medium. They were seeded at $5x10^3$ and $6x10^3$ cells/well---lung and liver cancer cells, respectively, in 96 well microtitre plates in 100 μl plating densities (Nunc, Denmark) and allowed to attach for 24 h. Methanolic seed extract of *S. rhombifolia* subsp. *retusa* and embelin in the fruits of *E. tsjeriam-cottam* were added to the cell culture plates at 10, 20, 40, 80, 200 μg/ml concentrations and incubated at standard conditions for 48 h. 0.1% DMSO was used as the vehicle control. Doxorubicin (Adriamycin ADR, Pharmacia) 1 mg/ml stock solution was prepared in distilled water and further dilutions at 10, 20, 40, 80 μg/ml were prepared. This was used as the standard lung and liver anticancer drug for comparison. The Sulforhodamine B (SRB) assay was carried out as per National Cancer Institute (NCI, USA) recommendations. [20,21] The cells were stained with 0.4% (w/v) SRB (Sigma Co., US), bound protein-dye was extracted and the absorbance was read at 540 nm with 690 nm reference wavelength using a microtitre ELISA plate reader (TECAN). Three independent experiments for each cell line and each concentration was performed. Percent growth inhibition was calculated on a plate-by-plate basis and expressed as the ratio of average absorbance of the test wells to the average absorbance of the control wells *100 (*DMSO). Percentage growth inhibition was calculated as: [(Ti-Tz)/(C-Tz)]x100 for concentrations for which Ti >/= Tz (Ti-Tz) positive or zero, [(Ti-Tz)/Tz]x100 for concentrations for which Ti < Tz (Ti-Tz) negative [time zero (Tz), control growth (C) and test growth in the presence of drug at the five concentration levels (Ti)]. Growth inhibition of 50% (GI_{50}) was calculated from [(Ti-Tz)/(C-Tz)]x100 = 50, total growth inhibition (TGI): Ti = Tz and LC_{50}: [(Ti-Tz)/Tz]x100 = -50.

Antimicrobial, Anti-Tubercular and Antimycotic Activities

a) Clinical Isolates

Bacterial strains Staphylococcus aureus ATCC 25923, Bacillus subtilis NCIM 2921, Salmonella typhi NCTC T786, Diptheriae sps. (skin), Streptococcus pneumoniae (sputum), Staphylococcus albus, Staphylococcus citreus, Vibrio cholerae (patient stool isolate), Klebsiella pneumoniae (pus), mycobacterial strains; Mycobacterium tuberculosis $H_{37}Rv$ (Standard strain), clinical isolates 1 (sputum culture of TB patient 1) and 2 (sputum culture of TB patient 2), fungal strains; Candida albicans, Candida krusei, Candida tropicalis, Aspergillus flavus, Aspergillus niger and Mucor (K.E.M. Hospital, Haffkine Institute, Mumbai), were used for the assay.

b) Anti-Infectives Bioassay

Antimicrobial and antimycotic activities against different bacterial or fungal strains were determined by agar well-diffusion method [22] using Nutrient agar and Sabouraud's glucose agar (HiMedia Laboratories, Mumbai) plates, respectively, except *S. pneumoniae* was grown on blood agar supplemented with defibrinated sheep blood. Overnight cultures in nutrient broth were diluted 1:100 with sterile saline such that resultant density of bacterial suspensions was 10^6-10^7cfu/ml by comparing its turbidity with that 0.5 McFarland opacity standard. In case of fungal strains inoculum was prepared from 48 hr old culture by picking up a loopful of colonies/conidial mass suspended in Sabouraud's glucose broth. The resultant suspension, after 48 hr, was diluted as above to obtain a count of $1x10^6$ cells/ml. Agar plates were seeded with 0.1 ml of the culture suspension. After solidification, wells were punched (8 mm diameter) on the agar medium.

Concentrations of 0.75, 1.25, and 2.5 mg/50µl of methanolic seed extract of *S. rhombifolia* subsp. *retusa* and embelin in the fruits of *E. tsjeriam-cottam* corresponding to 15, 25, and 50 mg/ml, respectively, were loaded from the stock solutions into each well. Standard reference drugs, Ofloxacin and Fluconazole (Cipla Ltd, Mumbai) were used for comparative evaluation. A stock solution of 1mg/ml was prepared in sterile distilled water. Further antibiotic dilution range 2-5 µg/ml was tested. 50µl of each drug was added per well. The inoculated plates were kept for pre-diffusion time of 1 hr before incubation at 37°C for 24 hr (except *S. pneumoniae* at 37°C, CO_2 atmosphere) and in case of fungi, 28°C for 48 hr. The diameter of the zone of inhibition (ZOI) was measured and expressed in millimeters. For each data, an average of three independent determinations for each strain and each concentration was recorded.

For testing anti-tubercular activity, mycobacterial strains were grown on Lowenstein-Jensen medium. [23] 7 ml of the medium containing methanolic seed extract of *S. rhombifolia* subsp. *retusa* and embelin in the fruits of *E. tsjeriam-cottam* were dispensed into the medium at concentrations of 0.5, 1.25, 2.5, and 5.0 mg/ml prior to inspissation into slants at 85°C for 45 minutes for three consecutive days. Standard drugs; Isoniazid (1 µg/ml), Rifampicin (40 µg/ml) (provided by K.E.M. Hospital, Mumbai) and drug-free control slants were used for comparative evaluation. The inoculum was prepared by picking up 2/3rd loopful of growth of colonies of each of the strains of mycobacterium, suspended in 1ml of sterile distilled water and vortexed for 30 seconds. The contents of each tube were diluted with sterile distilled water to a density equivalent to 1 McFarland standard. Two loopfuls of this were streaked on the Lowenstein-Jensen medium slants containing the drug extracts and incubated at 37°C for 4 and 8 weeks. The results were evaluated as sign '+' inactive if growth was detected and sign '–' active if not detected. Three independent experiments for each mycobacterium strain and each concentration was performed. Control wells or slants were loaded with 0.1% DMSO and sterile distilled water.

RESULT

Phytochemical fingerprinting of *S. rhombifolia* subsp. *retusa* and *E. tsjeriam-cottam* by HPTLC and HPLC is a simple, reproducible and cost-effective method useful for routine analysis of large number of samples with a short analysis time. HPTLC chemoprofiling-assisted characterization and presence of two markers viz; Ecdysterone a major phytoecdysteroid, an insect moulting hormone (Rf 0.30), a turquoise green band (arrow) in *S. rhombifolia* subsp. *retusa* of chemotaxonomic value (Figure 1A) and Embelin, a naturally occurring benzoquinone (Rf 0.50), a orange band (arrow) in *E. tsjeriam-cottam* fruits (Figure 2A) were detected in visible light on comparison with known standards. Seed and root extracts of *S. rhombifolia* subsp. *retusa* showed presence of Ecdysterone to a maximum extent. Embelin having chemotaxonomic as well as therapeutic efficacy to which clinical effects are attributed was detected highest in the whole fruit extract and next in the root extract of *E. tsjeriam-cottam*. The RP-HPLC fingerprints of seed and root extracts of *S. rhombifolia* subsp. *retusa,* showed the presence of Ecdysterone at Rt. 20.80, 20.82 minutes to a maximum extent was ascertained by comparing the retention time and UV-spectra of the reference standard Rt 20.65 minutes at λ_{max} 245 nm the peak was superimposable (Figure 1B).

Extracts of whole plant and leaves showed its presence in negligible amount with stem being at the undetectable level (data not shown). In case of *E. tsjeriam-cottam,* RP-HPLC fingerprints showed the presence of Embelin to a maximum extent in the whole fruit extract and next in the root extract at Rt 6.72, 6.80 minutes on comparison with UV-spectra of the reference standard Rt 7.08 minutes at λ_{max} 225 nm the peak was superimposable (Figure 2B).

Other extracts viz. stem, leaves, and whole plant, showed the presence of Embelin to a lesser extent (data not shown).

a)

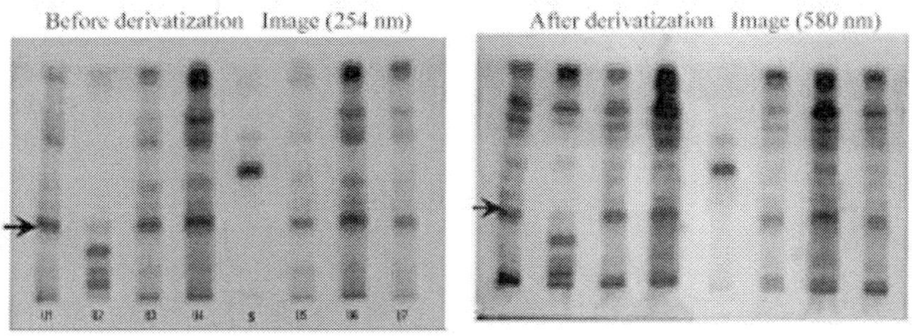

b)

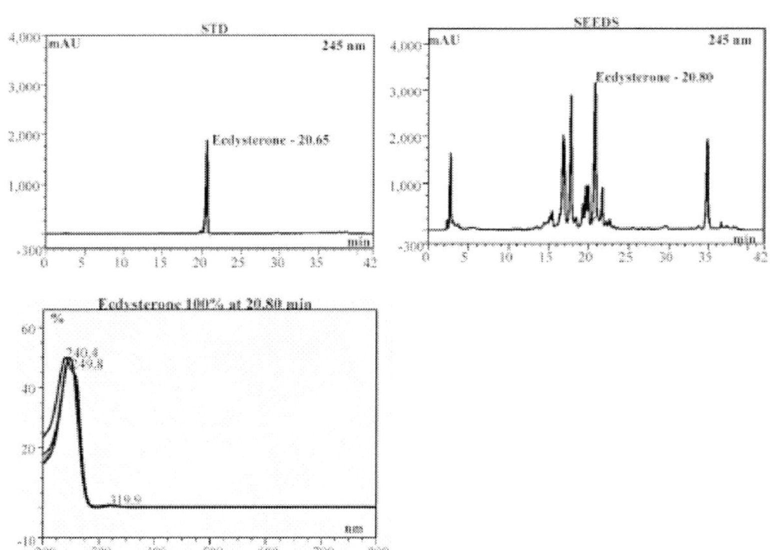

U1: Roots, U2: Seeds, U3: Stem, U4: Leaves, S: Standard Ecdysterone, U5: Flowers, U6: Stem+Leaves, U7: Whole plant.

Figure 1. *S. rhombifolia* subsp. *retusa;* A. HPTLC fingerprint (arrow indicates presence of marker Ecdysterone), B. HPLC fingerprints of standard Ecdysterone, methanolic seed extract and its UV spectra for Ecdysterone.

a)

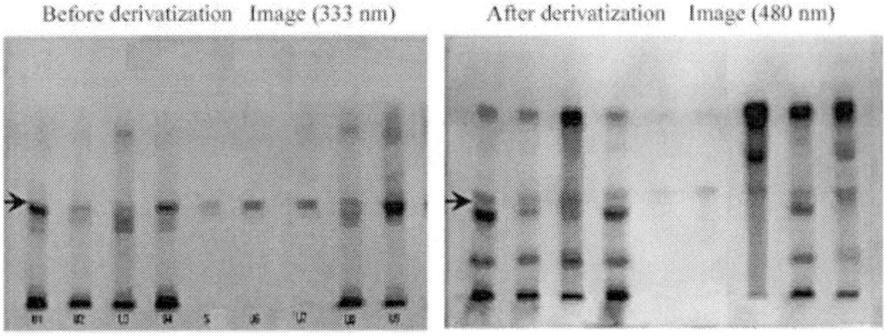

b)

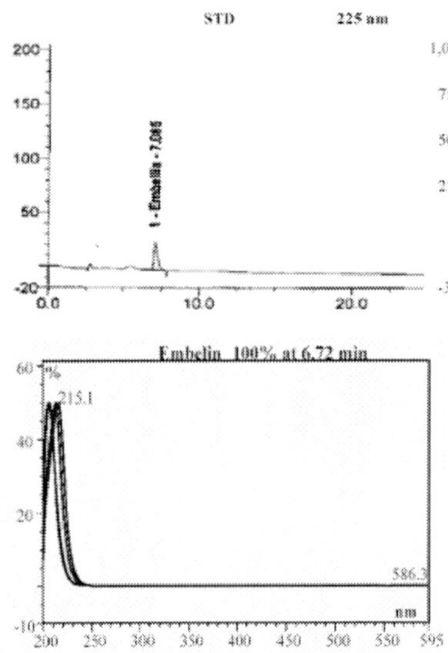

U1: Root, U2: Stem, U3: Leaves, U4: Root+Stem, S: Standard Embelin, U5: Embelin Isolated, U6: Stem+Leaves, U7: Whole Plant, U8: Whole Fruits.

Figure 2. *E. tsjeriam-cottam*; A. HPTLC fingerprint (arrow indicates presence of marker Embelin), B. HPLC fingerprints of standard Embelin, Embelin isolated from the fruits, and its UV spectra for Embelin.

For the cytotoxic and anti-infective assays, except methanolic seed extract of *S. rhombifolia* subsp. *retusa* and Embelin isolated from the fruits of *E. tsjeriam-cottam;* other parts of plant extracts showed feeble or no growth inhibition which was considered insignificant (data not shown). In HOP 62 and HEP G2 cell lines methanolic seed extract of *S. rhombifolia* subsp. *retusa* and Embelin in the fruits of *E. tsjeriam-cottam* exhibited marked dose-dependent cytotoxic effect ($GI_{50} \leq 20$ μg/ml) on the growth of cancer cells after 48 hr incubation. Vehicle control 0.1 % DMSO showed no inhibition with 100% growth for both the cell lines. Standard drug, adriamycin ($GI_{50} \leq 10$ μg/ml), was also active to both the cell lines (Figure 3).

The agar well plate method based on the diffusion of the extracts from the well into the surrounding agar medium and antituberculosis on Lowenstein-Jensen medium measured the growth inhibition indicating its effectiveness against the test organisms. All the extracts at their respective concentrations depicted varying diameter of zone of inhibition (ZOI). Table 1 and 2 depicts the results of the experiment at 0.75, 1.25, 2.5 mg/50µl with antimicrobial and antimycotic activities in the range of 15-26 mm. Methanolic seed extract of *S. rhombifolia* subsp. *retusa* and Embelin in the fruits of *E. tsjeriam-cottam* were found to be active against most strains. Potent inhibitory activities were observed against the growth of *Diptheriae* sps, *S. aureus* ATCC 25923, *A. flavus* and *C. krusei* (Figure 4).

Inhibitory effects against *C. albicans* and *Mucor* were moderate. Methanolic seed extract of *S. rhombifolia* subsp. *retusa* and embelin in the fruits of *E. tsjeriam-cottam* were also active to all the three *M. tuberculosis* strains at 0.5-5.0 mg/ml (Table 3). Control wells or slants with solvents alone showed no inhibitory zones or was inactive to the mycobacterial strains. Ofloxacin (2-5 µg/ml) was active against both gram-positive and gram-negative bacteria with inhibitory zones ranging from 20-32 mm. Fluconazole (5 µg/ml) exhibited inhibitory effect in the range of 14-17.5 mm against the fungal strains. Isoniazid (1 µg/ml) was active whereas Rifampicin (40 µg/ml) was inactive to the mycobacterial strains.

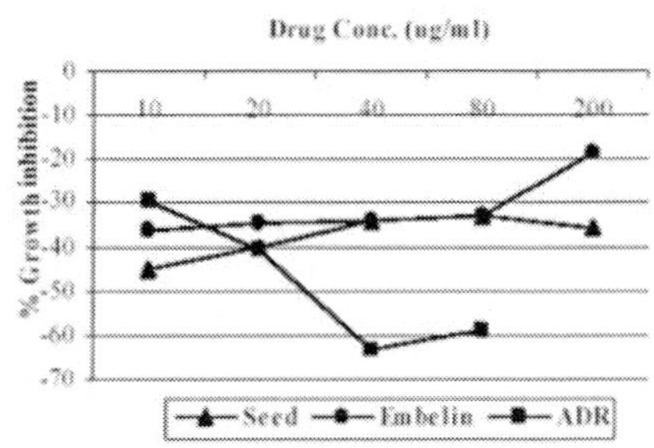

IC-50 graph of Human Non Small Cell Lung Cancer (NSCLC) HOP- 62

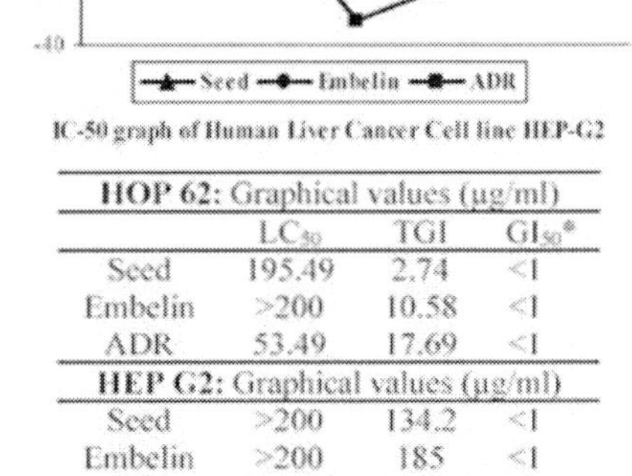

IC-50 graph of Human Liver Cancer Cell line HEP-G2

HOP 62: Graphical values (µg/ml)			
	LC_{50}	TGI	GI_{50}*
Seed	195.49	2.74	<1
Embelin	>200	10.58	<1
ADR	53.49	17.69	<1
HEP G2: Graphical values (µg/ml)			
Seed	>200	134.2	<1
Embelin	>200	185	<1
ADR	77.871	35.09	<1

* represent average values of [illegible] experiments. Growth inhibition of 50% (GI_{50}) is the drug concentration resulting in a 50% reduction in the net protein increase in control cells during the drug incubation. ADR Adriamycin

Figure 3. Cytotoxicities of methanolic seed extract of *S. rhombifolia* subsp. *retusa* and Embelin from fruits of *E. tsjeriam-cottam* in the HOP 62 and HEP G2 cell culture assay.

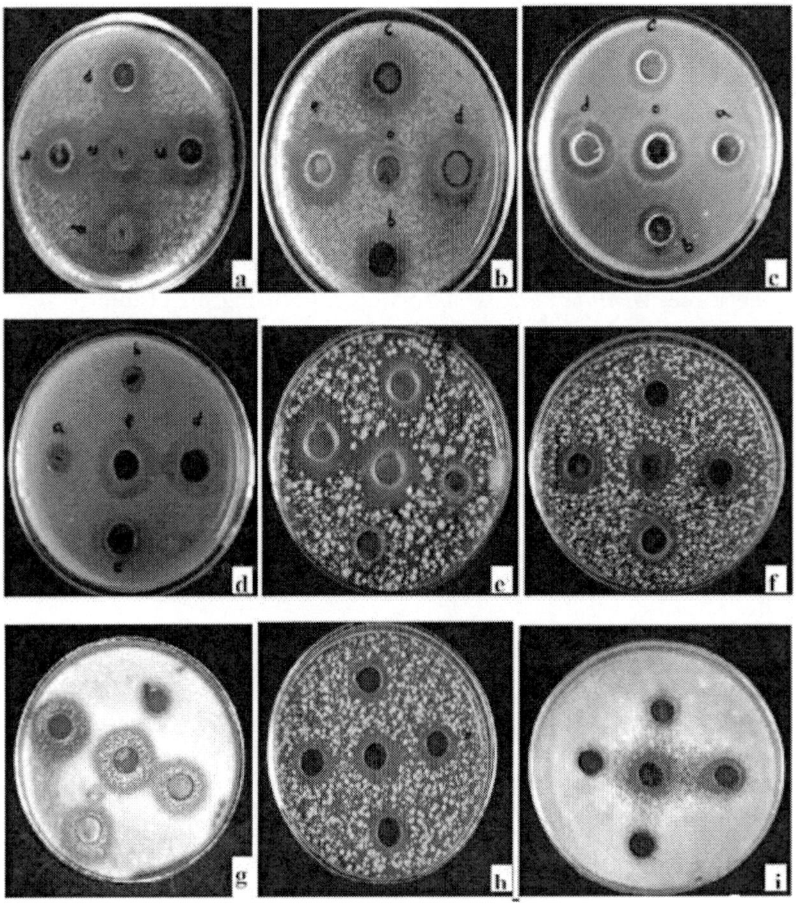

Figure 4. Antimicrobial and antimycotic activities of *S. rhombifolia* subsp. *retusa* and *E. tsjeriam-cottam* showing Zones of inhibition exhibited by; (a) embelin and (b) methanolic seed extract on *Diptheriae*, (c) methanolic seed extract and (d) embelin on *S. aureus*; (e) embelin on *C. krusei,* (f) *C. tropicalis* and *(g) A. flavus, (*h*)* methanolic seed extract on *C. tropicalis* and *(*i*) A. niger.*

DISCUSSION

The phytochemical profiling of the plant extracts by normal phase HPTLC and reversed phase HPLC fingerprints established a unique characteristic chemical pattern. It revealed the presence of Ecdysterone in *S. rhombifolia* subsp. *retusa* and Embelin in *E. tsjeriam-cottam* which can serve as chemotaxonomic markers and routine quality control (detecting adulterations and substitutions) of raw materials in crude form entering into formulation, an important step towards standardization of herbal medicines. The results demonstrate potent cytotoxic, antibacterial, antitubercular, and antimycotic activities presented in the methanolic seed extract of *S. rhombifolia* subsp. *retusa* and Embelin, the active constituent in the fruits of *E. tsjeriam-cottam* at a lower and safe non-toxic dose.

Table 1. Antimicrobial activities of methanolic seed extract of S. rhombifolia subsp. retusa and Embelin from fruits of E. tsjeriam-cottam*

Micro-organism	Extract (mg/50μl)	Seed	Embelin	Ofloxacin (μg/ml)	ZOI	DMSO (0.1 %)
Staphylococcus aureus 25921	0.75	18.4	17	2	15	0
	1.25	19.5	22.5	3	20	
	2.5	21.5	22	5	31	
Bacillus subtilis 2921	0.75	17	17	2	24	0
	1.25	19	20	3	26	
	2.5	20	21.5	5	32	
Diptheriae sps.	0.75	18	19.5	2	15.5	0
	1.25	21.5	22	3	20	
	2.5	24	25	5	27	
Streptococcus pneumoniae	0.75	13	14	2	--	0
	1.25	15	15	3	--	
	2.5	16	16	5	--	
Micro-organism	Extract (mg/50μl)	Seed	Embelin	Ofloxacin (μg/ml)	ZOI	DMSO (0.1 %)
Staphylococcus albus	0.75	17.8	16.5	2	13	0
	1.25	18	18	3	16	
	2.5	18.1	18	5	20	
Staphylococcus citreus	0.75	13	14.2	2	15	0
	1.25	13.5	15.8	3	18	
	2.5	14	16	5	25	
Vibrio cholerae	0.75	14	15	2	16.5	0
	1.25	15.5	16.2	3	20	
	2.5	16	17	5	20.5	
Klebsiella pneumoniae	0.75	15.5	14.5	2	14	0
	1.25	16	15	3	16	
	2.5	16	15.5	5	21.5	
Salmonella typhi T 786	0.75	14.8	15	2	26	0
	1.25	15	15.2	3	29	
	2.5	15	16	5	30.2	

*Results are expressed as diameters of zone of inhibition (mm), are average of three independent experiments.

CONCLUSION

Methanolic seed extract of *S. rhombifolia* subsp. *retusa* and Embelin in fruits of *E. tsjeriam-cottam* have potent cytotoxic, antibacterial, antitubercular, and antimycotic activities, which can be regarded as potential source of alternative natural products or possibly synergistic compounds to the conventional chemotherapeutic drugs. Further studies in infected animal models, to determine its therapeutic efficacy, pharmacokinetics, and overall toxicity, remain to be evaluated.

Table 2. Antimycotic activities of methanolic seed extract of S. rhombifolia subsp. *retusa* and Embelin from the fruits of *E. tsjeriam-cottam*[*]

Fungi	Extract (mg/50 µl)							
	Seed			Embelin			Fluconazole	DMSO
	0.75	1.25	2.5	0.75	1.25	2.5	5 µg/ml	0.1 %
Candida albicans	--	13	14	--	13.5	14	17.5	0
Candida krusei	12	16	16	18	21.5	22	10	0
Candida tropicalis	13.5	15	16	15	16.5	17	16	0
Aspergillus flavus	13	14.5	15	19	22.5	26	15	0
Aspergillus niger	12	15.5	16	12	15.2	16	14	0
Mucor	--	--	15	--	14.5	16	--	0

[*]Results are expressed as diameters of zone of inhibition (mm), are average of three independent experiments.

[--]No inhibition.

Table 3. Antitubercular activities of methanolic seed extract of *S. rhombifolia* subsp. *retusa* and Embelin from fruits of *E. tsjeriam-cottam*[*]

Mycobacterium tuberculosis												
Extract (mg/ml)	$H_{37}Rv$				Clinical isolate 1				Clinical isolate 2			
	0.5	1.25	2.5	5.00	0.5	1.25	2.5	5.00	0.5	1.25	2.5	5.00
Seed	--	--	--	--	--	--	--	--	--	--	--	--
Embelin	--	--	--	--	--	--	--	--	--	--	--	--
Isoniazid (1 µg/ml) (positive control)	--				--				--			
Rifampicin (40 µg/ml) (positive control)	+				+				+			
DMSO (0.1 %) (negative control)	+				+				+			

[*]Results are expressed as growth response to Mycobacteria.

[--]Active, +Inactive of three independent experiments.

ACKNOWLEDGMENTS

Author is thankful to Mr. T.B. Tithe (Anchrom Pvt Ltd, Mumbai) for providing HPTLC and to Dr. V.R. Bhate, Mr Umakanth (Analytical Solutions, Navi, Mumbai) for HPLC instrumentation facilities and their assistance, to Dr. A. Juvekar and Mr. Subrata Sen (Advanced Centre for Treatment, Research, and Education in Cancer [ACTREC], Navi, Mumbai) for their help rendered in carrying out this work, to Dr G. Natarajan, Dr. S. Baweja (K.E.M. Hospital, Mumbai) for laboratory facilities and bacterial cultures, to Ms. Sheetal

Ghadi, Ms. Bhakti Iyer and Mr. Roshan Baste (Haffkine Institute, Mumbai) for their kind help in anti-infective studies. Thanks are also to Dr. S. Bhagwat (Ramnarain Ruia College, Mumbai) for his support.

REFERENCES

[1] World Health Organization. *Traditional Medicine Strategy* 2002-2005. Geneva: WHO; 2002.

[2] Kirtikar, K. R., Basu, B. D. *Indian Medicinal Plants* (Vol. I, II, 2nd edition). Allahabad, India: Lalit Mohan Basu; 1993, 1984.

[3] Khare, M., Srivastava, S. K., Singh, A. K. Chemistry and Pharmacology of genus Sida (Malvaceae): A review. *J. Med. Arom. Pl. Sci.* 2002; 24: 430-40.

[4] Sharma, P. C., Yelne, M. B., Dennis, T. J. *Database on Medicinal Plants used in Ayurveda* (Vol. 5). New Delhi: Central Council for Research in Ayurveda and Siddha, Govt of India; 2002.

[5] Murthy, G. K. Clinical trials with Embelia ribes. *Curr. Med. Pract.* 1960; 4: 413-5.

[6] Raghunathan, K., Mitra, R. *Pharmacognosy of Indigenous Drugs* (Vol. II). New Delhi: Central Council for Research in Ayurveda and Siddha, Govt of India; 1999.

[7] Pandit, S. S., Naik, S. D., Jathar, V. S., Kulkarni, A. B. Insect moulting hormone, ecdysterone from *Sida carpinifolia* Linn. *Indian J. Chem.* 1976; 14B: 907-8.

[8] Dinan, L., Bourne, P., Whiting, P. Phytoecdysteroid profiles in seeds of *Sida* spp. (Malvaceae). *Phytochem. Anal.* 2001; 12: 110-9.

[9] Jadhav, A. N., Pawar, R. S., Avula, B., Khan, I. A. Ecdysteroid Glycosides from *Sida rhombifolia* L. *Chemistry and Biodiversity* 2007; 4: 2225-30.

[10] Pal, S. K., Mukherjee, P. K., Saha, B. P., Pal, M. Estimation embelin in *Embelia ribes* Burm.F. (Myrsinaceae) and in formulation by spectrophotometric method. *Research and Industry* 1995; 40: 118-20.

[11] Anonymous. *Indian Herbal Pharmacopoeia* (Vol. II, Revised new edition). *Mumbai: A joint publication of Regional Research Laboratories* (CSIR) Jammu Tawi and Indian Drugs Manufacturer's Association; 1999, 2002.

[12] Eapen Saumy, M. S., Grampurohit, N. D. *Chemical evaluation of Navayasa Churna. Indian Drugs* 2002; 39: 101-5.

[13] Muanz, D. N., Euler, K. L., Willians, L., Newmann, D. J. Screening of antitumor and anti HIV activities of nine medicinal plants from Zaire. *Int. J. Pharmacognosy* 1995; 33: 98-106.

[14] Ekramul Islam, M., Ekramul Haque, M., Mosaddik, M. A. Cytotoxicity and antibacterial activity of *Sida rhombifolia* (Malvaceae) grown in Bangladesh. *Phytother. Res.* 2003; 17: 973-5.

[15] Kinoshita, K., Morikawa, K., Fujita, M., Natori, S. Inhibitory effects of plant secondary metabolites of cytotoxic activity on polymorphonuclear leucocytes. *Planta Med.* 1992; 58: 137-45.

[16] Chitra, M., Sukumar, E., Devi, C. S. [3H]-Thymidine uptake and lipid peroxidation by tumor cells on embelin treatment: An in vitro study. *Oncology* 1995; 52: 66-8.

[17] Kumar, S., Bagchi, G. D., Darokar, M. P. Antibacterial activity observed in the seeds of some corphophilous plants. *Int. J. Pharmacognosy* 1997; 35: 179-84.

[18] Namba, T., Tsunezuka, M., Dissanayake, D. M. R. B., Pilapitiya, U., Saito, K., Kokiuchi, N., Hattoni, M. Studies on dental carries prevention by traditional medicines (Part VII): Screening of Ayurvedic medicines for antiplaque action. *Shoyakugaku Zasshi* 1985; 39: 146-53.

[19] Ikram, M., Inam-Ul-Haq. Screening medicinal plants for antimicrobial activities: Part III. *Fitoterapia* 1984; 55: 62-4.

[20] Vichai, V., Kirtikara, K. Sulphorhodamine B colorimetric assay for cytotoxicity screening. *Nature Protocols* 2006; 1: 1112-16.

[21] Skehan, P., Storeng, R., Scudiero, D., Monks, J., McMohan, D., Vistica, D., Warren, J. T., Bokesh, H., Kenney, S., Boyd, M. R. (1990). New colorimetric assay for anticancer-drug screening. *J. Natl. Cancer Inst.* 1990; 82: 1107-12.

[22] Cruickshank, R. *Medical Microbiology* (11th edition). London: The English Language Society and E and S Livingstone Ltd; 1968.

[23] Canetti, G., Froman, S., Grosset, J. Mycobacteria: laboratory methods for testing drug sensitivity and resistance. *Bull. WHO* 1963; 29: 565-78.

In: Asia Pacific Life Sciences Compendium
Editor: Bandit Chumworathayi
ISBN: 978-1-62257-338-7

Chapter 20

ISOLATION OF IMPROVED STRAIN OF *BACILLUS PUMILUS* SG2 AND ENHANCED PRODUCTION OF PROTEASE AND PROTEOLYSIS-RESISTANT LIPASE

***R. Sangeetha,*[1] *A. Geetha*[2*] *and I. Arulpandi*[3†]**

[1]Department of Biochemistry, School of Life Sciences, Vels University, Chennai, India
[2]Department of Biochemistry, Bharathi Women's College, Chennai, India
[3]Department of Microbiology, Asan Memorial College, Chennai, India

ABSTRACT

Strains which could exhibit enhanced production of protease and lipase were successfully produced from the wild strain *Bacillus pumilus* SG2 and enzyme production by the mutants was enhanced by 50% more than that of the wild strain. UV irradiation and EMS treatment was adopted to improve the wild strain and mutants which over-produced both protease and lipase were isolated. SGM4 was identified as the best mutant following UV irradiation. SGM4 was subjected to mutagenesis by EMS treatment and SGMM8 was identified as the potent producer of both protease and lipase. The enzymes produced by SGMM8 were subjected to ammonium sulphate fractionation and dialysis and the partially purified enzymes were characterized. Thus a stable mutant was which produced proteolytic and lipolytic enzymes was isolated and the enzymes were found to possess all the prerequisites of a detergent additive.

Keywords: UV irradiation, EMS, Bacillus pumilus, protease, lipase

---

* Corresponding author: A. Geetha. Department of Biochemistry, Bharathi Women's College, Chennai, 600108, India. E-mail: geethav21@yahoo.co.in.

† R. Sangeetha: Department of Biochemistry, School of Life Sciences, Vels University, Chennai, 600117, India, Tel: 914424461616. I. Arulpandi: Department of Microbiology, Asan Memorial College, Chennai, 600100, India.

Introduction

Enzymes from microbes have a wide range of industrial applications. Proteases and lipases are important biocatalysts in industries like dairy, leather, detergent and food. Proteases and lipases have been used in a mixture for various processes like soaking the leather prior to dehairing, cleaning soiled equipment used in a slaughter house, as detergent additive used in laundry [1]. Many species of bacteria have been studied extensively for the production of either protease or lipase [2,3]. However, the studies on the concomitant production of protease and lipase are few. Nevertheless, there are no reports on studies which involve quantitative enhancement of co-produced protease and lipase. We had earlier reported on the concomitant production of protease and a proteolysis-resistant lipase by *Bacillus pumilus* SG2 [4]. In this investigation, we aimed at improving the production of those co-produced enzymes using both physical and chemical mutagenesis. The enzymes produced by the mutant were compared with those produced by the parent to understand the efficiency of the mutagenesis that has been performed.

Method

Bacterial Strain

Effluent from a food processing industry located near Chennai, Tamilnadu, India was collected, serially diluted and plated on sterile nutrient agar plates (0.5% peptone, 0.3% yeast extract. 0.5% sodium chloride and 2% agar). The isolated colonies were then screened for protease production on casein agar (0.5% casein in nutrient agar) and nutrient gelatin (0.5% gelatin in nutrient agar) and lipase production using tributyrin agar (0.5% tributyrin in nutrient agar). A promising strain which produced comparatively higher amounts of both the enzymes was selected. The strain was identified as *Bacillus pumilus*, designated as *Bacillus pumilus* SG2 [5] and maintained in agar slants at 4 °C in our laboratory.

Strain Improvement

UV Irradiation

The parent strain was subjected to physical mutagenesis using Ultraviolet (UV) irradiation as the first step in strain improvement. The organism maintained on agar slants was scraped off into sterile phosphate buffer, pH 7.0 and a uniform suspension was formed. Four milliliter of this suspension was pipetted out in to sterile Petri plates and exposed to UV light. The exposure was carried out at 10.0 cm away from the germicidal lamp for various time intervals (10 to 120 min) with intermittent shaking. The plates were kept in dark overnight to avoid photoreactivation. Hundred microlitre of the treated bacterial suspensions was plated on nutrient agar plates and incubated for 48 h at 30 °C and observed for bacterial colonies.

Thirty mutant colonies were identified and were maintained in agar slants for further study. The mutants were studied for their extracellular protease and lipase production. The best mutant strain designated as SGM4 was used for further studies.

Ethyl Methane Sulphonate (EMS) Treatment

The best UV mutant, SGM4 was subjected to chemical mutagenesis using EMS as the mutagen. The bacterial suspension was prepared as described for UV irradiation. The suspension was incubated overnight at 37 °C. The culture was centrifuged at 6,000 g for 15 min and the pellet obtained was washed twice with saline. Then 80 µl of EMS was added to 2.0 ml of washed cells and incubated at 37 °C. The culture was drawn from the mixture at 30 min interval for 2 h. The withdrawn culture was centrifuged at 6000 g for 15 min and the sediment was washed with saline and suspended in sterile phosphate buffer, pH 7.0. The EMS treated bacterial suspensions (0.2 ml) was plated on nutrient agar plates to identify and isolate mutants. 1% survival growth was observed with 1 h incubation after EMS treatment. Twelve mutant colonies were isolated and maintained on agar slants. The mutants (designated as SGMM1 to SGMM12) were studied for extracellular protease and lipase production.

Growth and Enzyme Production

The production media consisted of (w/v) 0.04% $CaCl_2$, 0.02% $MgCl_2$, 1% glucose, 0.5% NaCl, 0.3% yeast extract and 1% tributyrin (in sodium phosphate buffer, pH 9.0). 5% overnight culture of wild type *Bacillus pumilus* SG2 and its hyper-producing mutant SGMM 8 were inoculated into separate 100 ml media and incubated on a rotary shaker (180 rpm) for 48 h at 37 °C. During the growth, culture medium was withdrawn aseptically at 4 h intervals. The bacterial growth was monitored by measuring the optical density of the culture at 650 nm. The proteolytic and lipolytic efficiencies of the withdrawn samples were assayed.

Enzyme Assay

Caseinolytic activity was measured by the photometric method of Rahman et al., [6]. One unit (U) of protease activity is equivalent to 0.5 µg of tyrosine liberated by 1.0 ml of enzyme solution under the assay conditions. Lipase activity was assayed by the photometric method of Kordel et al. [7]. One unit (U) of lipase activity is equal to 1micromole of p-nitrophenol liberated per min under the assay conditions. The total protein content was assayed by the method of Bradford [8].

Stability of Mutant

Bacillus pumilus SGMM8 was transferred from an agar slant to a second agar slant and incubated for growth at 37 °C for 24 h. The bacteria were further subcultured in a similar manner ten times and the lipase and protease production potential of the mutant was assayed to determine the stability of the mutant.

Characterization of the Partially Purified Enzyme

All the purification steps were performed at 4 °C. The cell free supernatant was subjected to gradient ammonium sulphate precipitation (40%, 50% and 60%) and the precipitates were suspended in 0.5 M sodium phosphate buffer (pH 7.0) and dialysed extensively against the same buffer. The dialysates were analysed for their protease and lipase activities. The dialysates exhibiting maximum enzyme activities were pooled.

The partially purified enzymes were studied to investigate their thermal stability (30-70°C), alkaline stability (pH 7.0-12.0), susceptibility to inhibitors (10 mM), response to detergents and metal ions (2% v/v or w/v).

The zymogram study was performed for protease with 0.1% copolymerized gelatin as substrate according to De Azeredo et al. [9] and for lipase with chromogenic substrates according to the method of Singh et al. [10].

RESULT AND DISCUSSION

Bacillus pumilus SG2 was subjected to mutagenic treatment with UV-irradiation and EMS. The selected mutant was used in further studies to analyze the efficiency of physical and chemical mutagenesis in improving strains to meet industrial demands. Of the thirty mutant colonies identified after UV irradiation, 12 colonies showed increase in protease production, 15 colonies exhibited increased production of lipase and only 3 colonies exhibited increase in production of both protease and lipase.

Of the three colonies, SGM4 exhibited 30% increase in production of protease (91 U/mg) and 22% increase in production of lipase (67.1 U/mg). SGM4 was thus selected based on the zone of hydrolysis on gelatin, ~~and~~ casein agar and tributyrin agar plates and by assessing the enzyme production in shake culture flasks.

SGM4 was subjected to EMS mutation and 12 mutants were identified and analysed for enzyme production. SGMM8 was the best mutant exhibiting 52% increase in protease production (106.4 U/mg) and 50% increase in lipase production (82.5 U/mg).

The parent strain produced maximum protease at 24 h~~ours~~ of incubation and lipase production was maximal after 30 h~~ours~~ of incubation. Both these enzymes were produced at the late log phase of the growth curve. But the mutant strain SGMM8 produced both these enzymes during the mid-log phase of the growth curve.

Thus the time taken for maximal enzyme production was reduced after mutation, though the growth curves of both the parent and mutant were almost similar (Figures 1 and 2). This increase in enzyme production and the decrease in time required for the same can be attributed either to the enhancement in enzyme secretion into the medium by altered permeability of cell membrane or to the increase in enzyme productivity per cell [11].

The characterization of protease and lipase produced by both the parent and mutant was performed to analyse both quantitative and qualitative enhancement achieved by mutagenesis. The studies on the properties of the partially purified protease and lipase showed that the proteolytic and lipolytic enzymes produced by SGMM8 were stable between 30-65 °C.

The optimum temperature for protease was 45 °C and that of lipase was 55 °C. A 30% reduction in the production of protease and 34% reduction in lipase activities were observed at 75 °C.

Both the enzymes exhibited 90% of their activities at 50 °C. The protease and lipase produced by the parent had an optimal temperature of 45 °C and were stable between 30-55°C. Thus, the enzymes of the mutant had exhibited a moderate increase in thermal stability.

The protease produced by SGMM8 had an optimal pH of 10.5 whereas the lipase produced had an optimal pH of 9.5. Both the enzymes were stable between 7.0-11.0. The study on the enzymes produced by the parent strain also showed similar results.

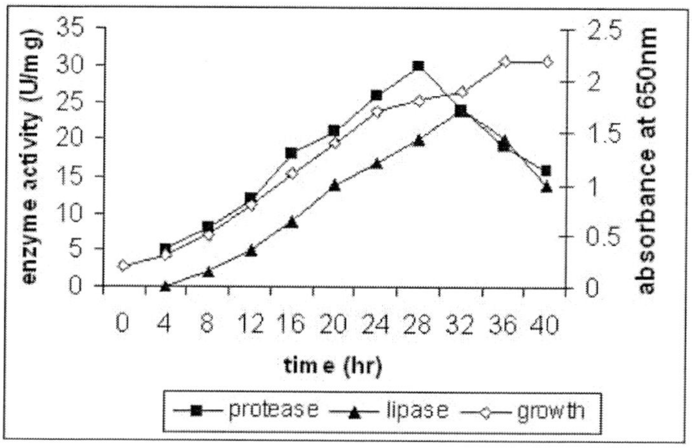

Figure 1. Growth and production of protease and lipase by Bacillus pumilus SG2.

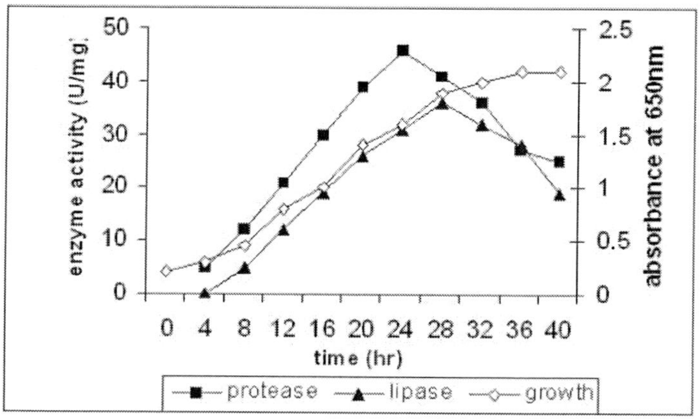

Figure 2. Growth and production of protease and lipase by *Bacillus pumilus* SGMM8.

The enzymes were pre-incubated with the inhibitors at 50 °C, pH 11.0 for 30 min and the residual activities were measured after incubation. The enzymes were considered to have 100% activities in the absence of any inhibitor. The protease produced by the parent and by the mutant SGMM8 was inhibited by 80% by PMSF (Phenyl methane sulphonyl fluoride) while no inhibition was observed with treatment with other inhibitors. Thus the protease had serine at or near the active site. The lipase of the parent and mutant was inhibited by 82% by both dithiothreitol and iodoacetate. Hence cysteine could be present at or near the active site of the coproduced lipase. EDTA was found to decrease the lipolytic efficiency of the enzyme produced by the wild strain by 24% whereas the lipase of the mutant strain sustained its original activity after incubation with lipase. Thus the lipase of the mutant was not dependent on any divalent cation for its activity. This is an important quality of an enzyme which can be considered for industrial exploitation.

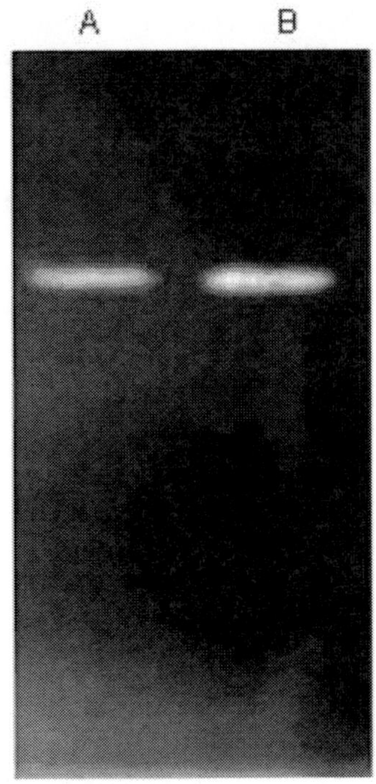

Figure 3. Zymogram study of protease produced by wild and mutant strain. Lane A: protease produced by *Bacillus pumilus* SG2; Lane B: protease produced by *Bacillus pumilus* SGMM8.

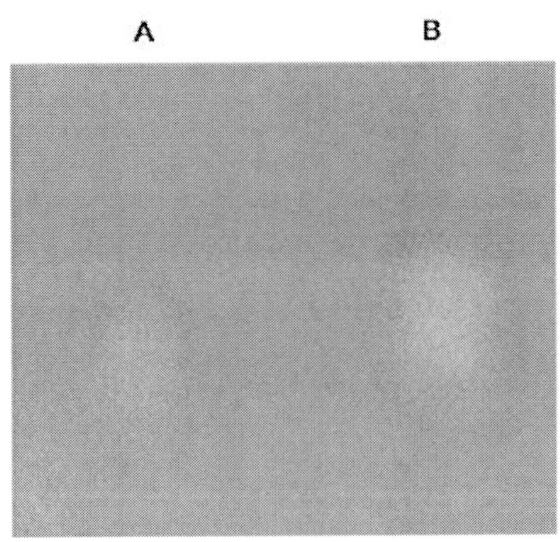

Figure 4. Zymogram study of lipase produced by wild and mutant strain. Lane A: lipase produced by wild strain, SG2; Lane B: lipase produced by mutant, SGMM8.

The enzymes were pre-incubated with various metal ions at 50 °C, pH 11.0 for 15 min and assayed to study the influence of divalent metal ions on the activity of protease and lipase. The effect of detergents (SDS, Triton X-100), oxidizing agents (H_2O_2, sodium hypochlorite) was studied by pre-incubating the enzymes with the additive at 50°C, pH 11.0 for 1 hr. The non-treated enzymes were considered to have 100% activity. All the cations, except for Ca^{2+} which had positive influence, had no effect on the catalytic efficiency of protease and lipase of the wild strain. Ca^{2+} enhanced the activity of protease by 20% while profoundly increased the activity of lipase by 65%. Thus the lipase produced by *Bacillus pumilus* SG2 was calcium dependent. Interestingly, both the enzymes of the mutant strain were not influenced by the presence of metal ions. This is an important prerequisite for enzymes to be used as detergent additives. The protease and lipase from the parent and the mutant strain were not influenced by the detergents and oxidizing agents. The desired properties in a detergent enzyme have been reported to be good temperature stability, stability at pH 9.0-10.5 and relative nondependence on metal ions, along with compatibility with detergent components, such as surface active reagents, builders, and bleaching agents [12]. The mutant strain, *Bacillus pumilus* SGMM8 thus produced proteolytic and lipolytic enzymes with all the desired properties of a detergent additive. Zymogram was performed to analyse the activities of the enzymes produced and to compare the efficiencies of the enzymes produced by the parent strain *Bacillus pumilus* SG2 and the mutant strain SGMM8. The activity staining of protease on a gelatin zymogram produces clear transparent bands against a blue background. The protease produced by the mutant produced a better transparent zone than that produced by the parent. This confirms the result obtained from spectrophotometric assay of protease activity (Figure 3). The zymogram analysis of lipase using chromogenic substrate shows yellow bands against a pink background. The band produced following the activity staining of lipase produced by the mutant SGMM8 was visibly darker than that produced by the parent (Figure 4). The improved strain *Bacillus pumilus* SGMM8 thus could produce significantly higher quantities of protease and lipase compared to the wild strain and the enzymes of the mutant were better suited for use in detergent industry.

REFERENCES

[1] Sangeetha, R., Geetha, A., Arulpandi, I. Concomitant production of protease and lipase by *Bacillus licheniformis* VSG1: Production, purification and characterization. *Braz. J. Microbiol.* 2010; 41: 179-85.

[2] Mala, B. Rao, Aparna, M. Tanksale, Mohini, S. Ghatge, Vasanthi, S. Deshpande. Molecular and biotechnological aspects of microbial proteases. *Microbiol. Mol. Biol. Rev.* 1998; 68: 597-635.

[3] Sharma, R., Yusuf Chisti., Uttam Chand Banerjee. Production, purification, characterization, and applications of lipases. *Biotechnol. Adv.* 2001; 19: 627-62.

[4] Sangeetha, R., Geetha, A., Arulpandi, I. Concomitant production, partial purification and characterization of a serine protease and a proteolysis-resistant metallolipase from *Bacillus pumilus* SG2. *Z. Naturforsch* 2010; 65: 61-5.

[5] Sangeetha, R., Geetha, A., Arulpandi, I. Optimisation of protease and lipase production by *Bacillus pumilis* SG 2 isolated from an industrial effluent. *The Internet J. Microbiol.* 2008; 5:2.

[6] Rahman, R. N. Z. A., Lee Poh Geok., Mahiran Basri. Physical factors affecting the production of organic solvent-tolerant protease by *Pseudomonas aeruginosa*. *Bioresour. Technol.* 2005; 96: 429-36.

[7] Kordel, M., Birgit Hofmann., Cietmar Schomburg., et al. Extracellular Lipase of *Pseudomonas* sp. Strain ATCC 21808: Purification, Characterization, Crystallization, and Preliminary X-Ray Diffraction Data. *J. Bacteriol.* 1991; 173: 4836-41.

[8] Bradford, M. A rapid and sensitive method for the quantification of microgram quantities of protein utilizing the principle of protein dye binding. *Anal. Biochem.* 1976; 72: 248–54.

[9] De Azeredo, L. A. I., Freire, D. M. G., Soares, R. M. A., et al. Production and partial characterization of thermophilic proteases from *Streptomyces* sp. isolated from Brazilian cerrado soil. *Enzyme Microb. Technol.* 2004; 34: 354–58.

[10] Singh, R., Namita Gupta., Vineet Kumar Goswami., et al. A simple activity staining protocol for lipases and esterases. *Appl Microbiol Biotechnol* 2006; 70: 679-82.

[11] Chen, J., Susumu Shimura., Kohtaro kirimura., et al. Enhancement of lipase production from hydrocarbons by mutation of *Trichosporon fermentans*. *Appl. Microbiol. Biotechnol* 1993; 38: 714-18.

[12] Jasvir, S., Navdeep Gill., Gina Devasahayam., et al. Studies on alkaline protease produced by *Bacillus* sp. NG312. *Appl. Biochem. Biotechnol.* 1999; 76: 57-63.

In: Asia Pacific Life Sciences Compendium
Editor: Bandit Chumworathayi

ISBN: 978-1-62257-338-7

Chapter 21

NOVEL FINDINGS OF AHI AND BMI IN SOME TAIWANESE SENIORS SUFFERING FROM SLEEP DISTURBANCE WITH RELATED LITERATURE REVIEW

Bing H. Tang*

Dr. Siu-T Chen's Systemic Biology Laboratory, Institute of Biological Chemistry, Academia Sinica, Nankang, Taipei, Taiwan
Researcher, Research and Ethics, Danville, CA, US

ABSTRACT

Objectives: 1) To explore the role of snoring affecting sleep disturbance 2) To examine the prevalence of abnormal body weight's influence on sleep disturbance 3) To examine the relationship between snoring and obstructive sleep apnea (OSA) (elevated AHI) in a community-dwelling older men and women as the studied subjects, with advancing age, age-related changes have been described for sleep-wakefulness and additional behavioral cycles. 4) To compare the trends in the relationship between elderly sleep disturbance and AHI 5) To signify the economic and cultural impacts on healthy weight and BMI maintenance with relationship to the prevalence of sleep disturbance 6) To review some of the related literature.

Methods: A retrograde single measures design was employed over nocturnal period. The study was conducted in a clinical sleep medicine laboratory of a Medical Center in Taiwan from January 2002 to January 2003. Totally 124 men and women were community-dwelling elders (mean age=71.85 years, SD=4.85 years) with complaint of sleep disturbance. Questionaires were asked. Polysomnography was conducted using a full 18-channel montage. Apnea was defined as a decrease in airflow of $\geq$ 90% for a minimum of 10 seconds. Hypopnea was defined as $\geq$ 30% decrease in airflow and desaturations required a $\geq$ 3% decrease in oxygen saturation for a minimum of 10 seconds. The apnea hypopnea index (AHI) was calculated as the sum of apneas and hypopneas divided by nocturnal hours of sleep.

---

* Dr. Siu-T Chen's Systemic Biology Laboratory, Institute of Biological Chemistry, Academia Sinica, Nankang, Taipei 115, Taiwan. Researcher, Research and Ethics, Danville, California, 94506, US.

Results: Sleeping apnea in this study was evaluated. AHI: Decile and the Quartile distributions were reported. A novel finding of AHI: a quantile-quantile plots (q-q plots) of AHI suggest deviations from normality, notwithstanding, none of the diagnostics of kurtosis and skewness on multiple linear regressions of AHI being significant or worrisome. For the correlation between AHI and BMI, after checking with Spearman's rank correlation coefficient (rs), it reveals 0.295. As to the relationship between AHI and Neck Circumference was noted to approximate with that between AHI and BMI, there was an insignificant effect on frequency of snoring by grouping factor of AHI. Spearman's rank correlation coefficient (rs) test reveals a mildly positive correlation between AHI and snoring (rs was 0.26). Characteristics of BMI in subjects with sleep disturbance were presented. Females had mean height, weight and cervical circumference compared with males. Note that the difference of mean BMI between non-snores and snorers is 2.02. AHI was noted as correlating with BMI, snoring, body height and weight, and cervical circumference respectively. Spearman's rank correlation test revealed that the relationship between snoring and BMI was highly significant. The effect of BMI grouping on the frequency of snoring as well as Asian BMI criteria was discussed.

Conclusion: The study of sleep disturbance in the elderly is a new area of scientific direction for fruitful areas of future research. Hopefully, this article provides some highlight in focusing on an individual patient-oriented approach to sleep disturbance. This is in contrast to a population-based approach in which the entire distribution of risk factors in total population in Taiwan will be expected in a very near future.

Keywords: AHI (Apnea-hypopnea Index), height; body mass index, sleep disturbance, older adults

INTRODUCTION

Population aging has materialized as an innovative demographic inclination with imperative insinuation for government programs, public health and education, and family restructuring. Among such changes, insomnia, snoring and sleep apnea, in conjunction with sleep hygiene have been usually ignored.

Changes in sleep pattern are part of the aging process. Nocturnal total sleep time can become more fragmented with age, with an increase in awaking early in the morning and nighttime awakenings.

In Taiwan, the prevalence of underweight is about less than ten percent, while approximately thirty percent of Taiwanese elderly are overweight. The main purpose of this article is to signify the economic and cultural impacts on healthy weight and BMI maintenance in potentially decreasing the prevalence of sleep disturbance and improving quality of the elderly life in Asian societies.

Sleep Disturbance in the elderly population is not uncommon. One of the most common complaints of sleep disturbance is insomnia. Whether it is a solitary disorder or merely a symptom is still unclear.

For example, in the Philippines, a country that is not far away from Taiwan, the prevalence of insomnia in a non-governmental community dwelling facility in Quezon City, where the elderly individuals had the prevalence of 43%, most of them were females [1]. The prevalence of chronic insomnia (> 4 weeks) in that facility was enormously female predominant. This was accredited to the much larger proportion of females in the dwelling association, where females were about 92% of 2,200 members at that time in that facility.

That mentioned, on the other hand, noticeably, during the decade from 1979 to 1989, most of East and South-East Asian countries, including Taiwan, have undergone a period of swift, communal, financial and demographic modifications. Population aging has since then materialized as an innovative demographic inclination with imperative insinuation for government programs, public health and education, and family restructuring.

However, changes that are usually ignored include sleep disturbance, sleep apnea, associated with sleep hygiene. In fact, changes in sleep are part of the aging process in the senior population. For example, nocturnal total sleep time can become more fragmented with age, with an increase in awaking early in the morning and nighttime awakenings [2]. Since nocturnal total sleep time is affected, daytime sleep may be prolonged. Sleep hygiene, beyond nocturnal and day time cycles may be warped. Such as daytime sleepiness and frequent sleep apnea and increases in nighttime awakenings, along with decreases in both slow-wave sleep and rapid eye movement (REM sleep) have been reported [3,4,5].

As to snoring, according to Sawit et al. [6], most physicians seldom inquire about snoring during their clinical encounter with patients. Both doctors and patients consider snoring to be a normal sleeping characteristic, without health implications. Thus, a closer study on snoring is warranted [6]. Snoring is a common phenomenon seen in approximately 20% of the adult population and in about 60% of men over 40 years of age. (Note that in this study, all the subjects are at the age of or older than 65.)

Hence, this article includes approaches to the understanding of sleep disturbance in Taiwanese seniors.

METHODS

[A]

Subjects Studied

From 1 January 2002 to 31 January 2003, the Sleep Medicine Laboratory, Changhua Christian Medical Center, Taiwan, admitted no subjects with heart failure and chronic obstructive lung disease, as there was no physician at night in the Sleep Medicine Laboratory.

The 124 participants aged from 65 to 88.5 years were studied. No pair of participants was related. No participant had more than one polysomnography (PSG). Among the 124 individuals, there were 117 with available BMI data. Written informed consents including personal and clinical data in the research database of all participants are obtained. The research protocol was approved by the Review Board of the Medical Center, and conformed to the Declaration of Helsinki.

Inclusion Criteria

1 The age of each subject is 65 or older.
2 All the subjects have complainted of some sort of sleep disturbance.
3 All the subjected accepted in-laboratory noctural PSG.

Exclusion Criteria

1. During the period from 1 January 2002 to 31 January 2003, at the Sleep Medicine Laboratory, Changhua Christian Hospital Medical Center admitted no individuals with heart failure or chronic obstructive lung disease, as there was no physician at night in the Sleep Medicine Laboratory.
2. People younger than 65 years were excluded.

AHI Definition and Classification Used in This Study

Apnea/ hypopnea Index (AHI) is defined as the number of apneic episodes (obstructive, central, and mixed) plus hypopnea per hour of sleep as determined by all-night PSG. AHI is synonymous with respiratory disturbance index. AHI was measured for all the subjects.

All AHI values were classified as first, second and third degrees of sleep-apnea severity. The gold standard for diagnosis of sleep disturbance is in-laboratory nocturnal PSG. Participants were required to have in-laboratory nocturnal PSG at sleep medicine laboratory. PSG was used not only to measure AHI, but also to evaluate sleep efficiency and snoring.

Obstructive sleep apnea/ hypopnea syndrome (OSAHS) was defined as AHI > 15 per hour. A computerized PSG system (Alice 4 Sleep Diagnostic System, Respironics) was used in this study. No second PSG study was conducted after the single-night PSG for any individual subject studied. The obstructive sleep apnea-hypopnea syndrome was classified according to the criteria of severity of sleep apnea. The degrees in severity of sleep apnea were defined based on the protocol of American Academy of Sleep Medicine (AASM) Task Force [8]. These include;

1. Zero (0) degree, apnea/ hypopnea index (AHI) < 5,
2. First degree, 5 < AHI < 15,
3. Second degree, 15 < AHI < 30, and
4. Third degree, AHI > 30.

Criteria Commonly Used in Applying Apnea-Hypopnea-Index (AHI)

For the indication for clinical treatment or not, the criteria include;

1. AHI < 5, SDB can probably be ruled out,
2. 10 < AHI < 15 gray zone,
3. AHI = 15 warranted treatment, and
4. 30 < AHI < 50 definitely needed to be treated clinically.

Sleep in the PSG findings were staged according to Rechtschaffen et al.'s criteria [10].

Clinical Classification of Obstructive Sleep Apnea Syndrome

Apnea was defined as a decrease in airflow of ≥ 90% for a minimum of 10 seconds. Hypopnea was defined as ≥ 30% decrease in airflow and desaturations required a ≥ 3% decrease in oxygen saturation for a minimum of 10 seconds. The apnea hypopnea index (AHI) was calculated as the sum of apneas and hypopneas divided by nocturnal hours of sleep.

Based on the protocol of American Academy of Sleep Medicine Task Force (1999), the degree in severity of sleep apnea is defined as following;

Degrees of Severity of Sleep Apnea (elevated AHI)

Sleep variables with Apnea (changes of AHI)	The Degree
Apnea (AHI<5)	Zero degree
Apnea (AHI 5~15)	First degree
Apnea (AHI 15~30)	Second degree
Apnea (AHI>30)	Third degree

BMI

BMI was calculated as weight (kilograms) divided by height (squared meters).

[B]

Control Data

A second data set consisting of 1,014 (aged 65 and over) control subjects came from 15,798 subjects who had participated in the Nutrition and Health Survey in Taiwan (decoded with permission NSC Council 93WFD2000205), between 1993 and 1996. The data of age and blood pressure of those 1,014 subjects were cited and analyzed. Among them, merely 665 subjects had complete data consisting of their age, blood pressure, weight, height, and BMI, while 95.79% (15,133/15,798) of the control subjects lacked one or more of the above listed information. As the subjects' data studied in this article covering the period from 2002 to 2003, in contrast, another report entitled "Elderly Nutrition and Health Survey in Taiwan (1999-2000): research design, methodology and content" was not published until 2005.

The Taiwanese Census Data Not Found in WHO Profile

Taiwan has been excluded from the United Nations and its related organizations since 1971. Taiwan has not been privy to the health related work of the World Health Organization (WHO). There is no Taiwanese census data available from WHO file.

[C]

PSG System

A computerized PSG system (Alice 4 Sleep Diagnostic System, Respironics) was used in this study. As previously mentioned, no second PSG study was conducted after the single-night PSG. Nocturnal polysomnography (PSG) (Alice 4 Sleep Diagnostic System, Respironics, Carlsbad, California, US) was done from about 9:30 pm to 6:30 am next morning in the sleep laboratory.

The following parameters were measured and recorded by the PSG:

a chest and abdominal wall motion by uncalibrated respiratory inductance plethysmography;

b heart rate by ECG;
c inspired and end-tidal carbon dioxide pressure (PETCO2), sampled at the nose or mouth at a rate of 60 mL/min by mass spectrometry (model 1100 Medical Gas Analyzer, Perkin Elmer; Pomona, CA) or by capnography (model 1000 Capnograph, Nellcor, Hayward, Calif. US); (iv) combined oral nasal air flow, sampled with a three-pronged thermistor placed at the upper lip;
d arterial oxygen saturation by pulse oximetry (model N 200, Nellcor, Hayward, Calif., US);
e oximeter pulse wave form;
f electro-oculogram;
g EEG in overnight PSG;
h chin electromyogram;
i actigraphy (placed on the hand); and microphone placed over the neck to monitor snoring. The transducers and lead wires permitted normal positional changes during sleep. Bedtime and awakening time were at each subject's discretion; the PSG was terminated after the final wakening.

Statistical Analysis

Mean, standard deviation, student t-test and correlation were measured with SPSS version 10. A two tailed p-value of < 0.05 was accepted as statistically significant. The mean age of male subjects was 71.6 + 4.5 years, while that of females was 72.3 + 5.5 years. All 124 subjects had chief complaints of sleeping disturbance. The ratio of female to male subjects was 1: 1.7 (the total number of subjects was 124). Pearson or Spearman correlation test was used. Multiple linear regression of AHI was used in Figures 1-4. Mann-Whitney test was used for three group comparison. [59]

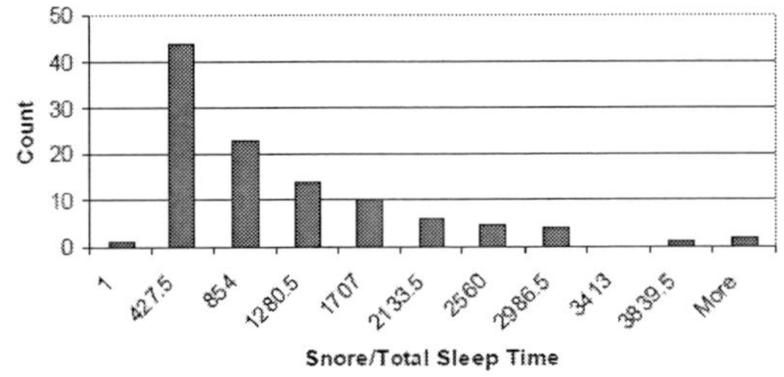

Figure 1. The Histogram of Snore.

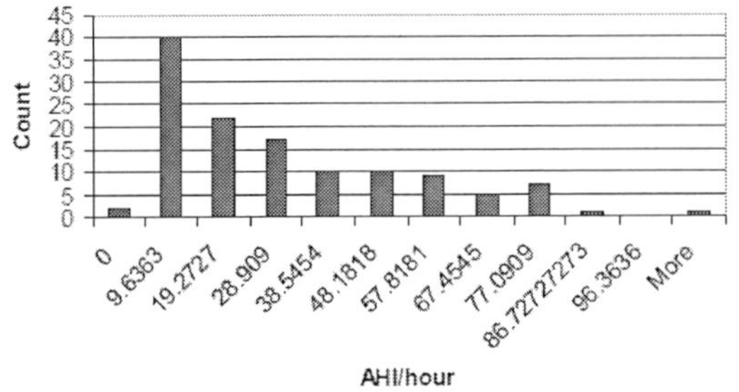

Figure 2. The Histogram of AHI.

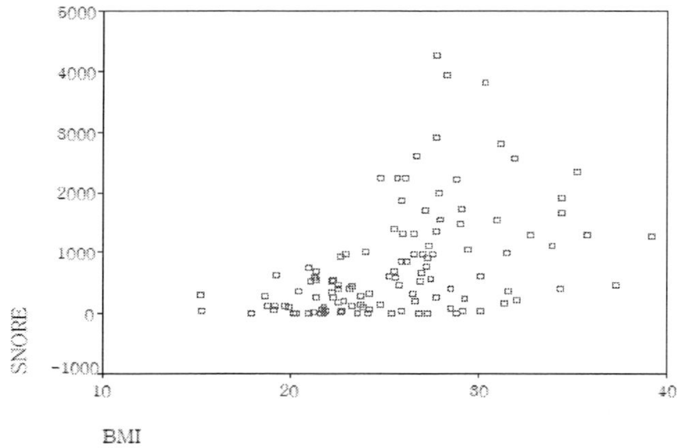

Figure 3. Spearman's Rank Correlation between BMI and snore.

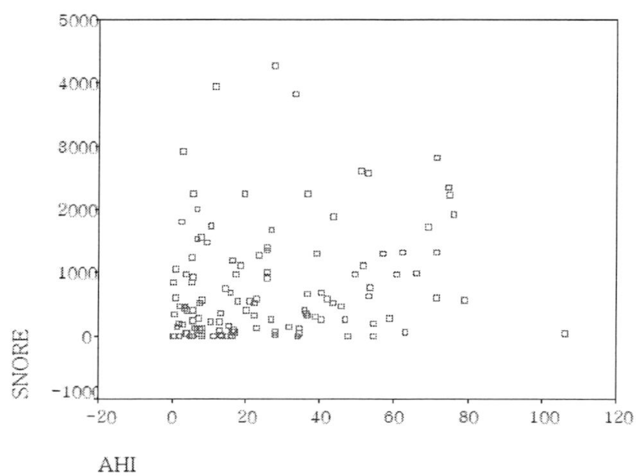

Figure 4. Spearman's Rank Correlation between snore and AHI (With AHI along the x-axis, and snore along the y-axis).

Results

Sleep Apnea in This Study

Sleeping apnea was found in 81.3% (n=103) with age of 65 years and over suffering from sleep disturbance found in this Taiwanese study according to the result of PSG. This author reports that 15.3% (n=19) of his subjects had an AHI > 5; 14 out of 19 were obese and each had an AHI >30. There are also 45 overweighted subjects (39.52%), 7 of them have AQHI > 5; a half (16 out of the rest of overweight subject have AHI > or = 30).

Correlation between age and AHI within this study's data revealed that a Pearson correlation coefficicient (r) of -.099, even with the total number of the subjects studied, n=124.

AHI: Decile and the Quartile Distributions

1) Decile Distributions

In this study, a decile, instead of quartile, distribution of height was found with a significant negative correlation with AHI. There was no such a relationship found between the latter (AHI) and weight. In the tenth decile (D10), the range of height was from 169 to 174 cm, and that of AHI was from 0.3 to 106 per hour. In D10, the Spearman rank correlation test revealed that, between height and AHI, there was a significant negative correlation coefficient rs-0.74, with two tailed p-value = 0.0058, which was significant.

2) Quartile Distribution

It has, as well, shown that a quartile distribution by the height with the Spearman correlation coefficient being-0.34, albeit the two tailed p-value was marginal insignificant (p = 0.06). Another novel finding reported here is that AHI relates with weight only but not with height.

A Novel Finding *of AHI*: A Quantile-Quantile Plots *(Q-Q Plots)*

The quantile-quantile plots (q-q plots) of AHI suggest deviations from normality, supported with kurtosis and skewness that are greater/ less than a -2 to +2 range when their standard deviations are considered, implying that the assumption of normality is not met; notwithstanding, none of the diagnostics of kurtosis and skewness on multiple linear regressions of AHI being significant or worrisome. (Figures 1-4)

AHI and BMI

For the correlation between AHI and BMI, after checking with Spearman's rank correlation coefficient (rs), it reveals 0.295. This rs is close to 0.330, that is, the rs between AHI and neck circumstance. As another novel application, the relationship between AHI and Neck Circumference can be used to approximate with that between AHI and BMI. It is noted that in Taiwan, distributions of body composition are usually generated for children, adolescent, and middle-aged groups, but not for the older people. (Table 2).

In this study the AHI mode was 9.4 per hour; its count for 40 among the total 124 subjects. The highest number of AHI was 106 per hour in this study. There was an insignificant effect on frequency of snoring by grouping factor of AHI. Spearman's rank correlation coefficient (rs) test reveals a mildly positive correlation between AHI and snoring (rs was 0.26).

Table 1. Comparison between study and control populations

Characteristics	Study population M2 SD2	Control population M1	Effect size (By using Study population SD) ES: (M2-M1)/SD2
Minimal age (years)	65 (sample size=124)	65 (sample size=1,014)	
Mean of ages (years)	71.69	71.97 ± 5.86 (Sample size=1014)	-0.048
Mean of height (cms)	159.76	157.03 ± 8,87 (Sample size=665)	0.31
Mean of weight (Kgs)	65.5	58.7 ± 10.59b (Sample size=665)	0.64
Ratio of males to females	1(m): 2(f)	528(m): 525(f) 50.1(m): 49.9(f)	
Snorers/ total number of subjects	50.8%(63/124)	----	
Diastolic blood pressure greater than 90 mmHg, and AHI greater than 5	32.7%(54/124)	----	
Diastolic blood pressure great than 90 mmHg, and AHI less than 5	8.87%		
Mean of BMI	25.68	23.77 ± 3.66 (Sample size=665)	0.52
Mean for effect size	M2	M1	
Mean of cervical circumference (cm)	36.45	----	
Mean of systolic blood pressures (mmHg)	140.08	140.64 ± 22.39 (Sample size=1014)	-0.025
Mean of diastolic blood pressures (mmHg)	79.87	79.17 ± 13.56 (Sample size=1014)	0.05

Table 2. Total 124 patients: means and standard deviations of age, body height and weight, BMI, and cervical circumference

	Age	Height	Weight	BMI	NC
Mean	71.9	159.8	65.5	25.7	36.7
SD	4.85982	7.753368	11.96101	4.54209	3.385379

Table 3. Comparison between female and male patients: means and standard deviations of age, body height and weight, BMI, and cervical circumference

Female	Age	Height	Weight	BMI	NC
Mean	72.28611	152.4881	61.76977	26.55061	34.79311
SD	5.476353	5.605407	12.56028	5.309538	2.848923
Male	Age	Height	Weight	BMI	NC
Mean	71.60027	163.8267	67.648	25.19525	37.88961
SD	4.474771	5.481846	11.1309	3.999075	3.15668

Table 4. Tabulation of AHI scores < 15: means, variances, number (Count), T-test result, and P values

Data	Mean	Variance	N	T	P
Without Snoring, and AHI < 5	2.1	2.32154	14	-0.10234	0.9195
With Snoring, and AHI < 5	2.0285	2.16905	7		
Without Snoring, And AHI between 5 and 10	6.71	1.21433	10	-0.05301	0.958
With Snoring, And AHI between 5 and 10	6.68	1.98844	10		
Without Snoring, And AHI between 10 and 15	12.51429	2.0181	7	-0.22884	0.8241
With Snoring, And AHI between 10 and 15	12.3	2.66	4		

*Statistical significance was assigned to probability (P) values less than 0.05.

Table 5. T-test on without Snoring, and AHI < 15 vs with snoring and AHI < 15

Data	Mean	Variance	N	T	P
Without Snoring, and AHI < 15	33.876	361.989	30	2.309	0.023
With Snoring, and AHI < 15	44.516	378.238	42		

*Statistical significance was assigned to probability (P) values less than 0.05.

Table 6. Apnen-Hypopnea-Index (In this article, this author uses the following criteria for clinical diagnosis)

AHI < 5	normal
AHI ≥ 5	likely with respiratory distress
AHI 30-50 events per hour or greater	severe cases of sleep disordered breathing

Table 7. Apnen-Hypopnea-Index (In this article, the author uses the following criteria as the indication for clinical treatment or not)

AHI < 5	SDB can probably ruled out
10< AHI < 15	gray zone
AHI = 15	Warrented treatment
30<AHI<50	definitely needed to be treated clinically

Table 8. Correlations between BMI, Snore, and AHI

			BMI	SNORE	AHI
Spearman's mo	BMI	Correlation coefficient	1.000	.551**	.295
		Sig. (2-tailed)	.	.000	.001
		N	116	116	116
	SNORE	Correlation coefficient	.511**	1.000	.260**
		Sig. (2-tailed)	.000	.	.004
		N	116	124	124
	AHI	Correlation coefficient	.295**	.260	1.000
		Sig. (2-tailed)	.001	.004	.
		N	116	124	124

**Correlation issignificant at the .01 level (2-taoled).

Characteristics of BMI in subjects with sleep disturbance are to be presented as following; Females had mean height, weight and cervical circumference less than the counterparts of the males. The female subjects had mean BMI of 26.550 kg/m2, and male 25.19. The mean BMI of total 124 subjects was 25.574 $\pm$ 4.521 kgs/m2, while that of control was 23.768 $\pm$ 3.662 kg/m2, the difference was significant (the p-value < 0.0009). The Pearson's correlation coefficient between height and BMI was significantly different between males (-0.227) and females (-0.0854). In these data, the female subjects were about one year older than the male in the average. However, the female subjects had average (mean) height, weight and cervical circumference, which were respectively less than the counter parts of the males. The 95% confidential interval for mean BMI was 24.746-26.402.

The minimum BMI was 15.20, and maximum was 25.33. (Tables 2 and 3, and Figure 5) Note that the difference of mean BMI between non-snores and snorers is 2.02.

AHI was correlated with BMI, snoring, body height and weight, and cervical circumference respectively. There was a highly significant correlation between BMI and snoring. The subjects whose BMIs were more than 25 had more frequent snoring than those whose BMIs were less than 25 in the studied population. BMI of patients was higher than control subjects. There were significant positive correlation between AHI and BMI;

Spearman's rank correlation test revealed that the relationship between snoring and BMI was highly significant.

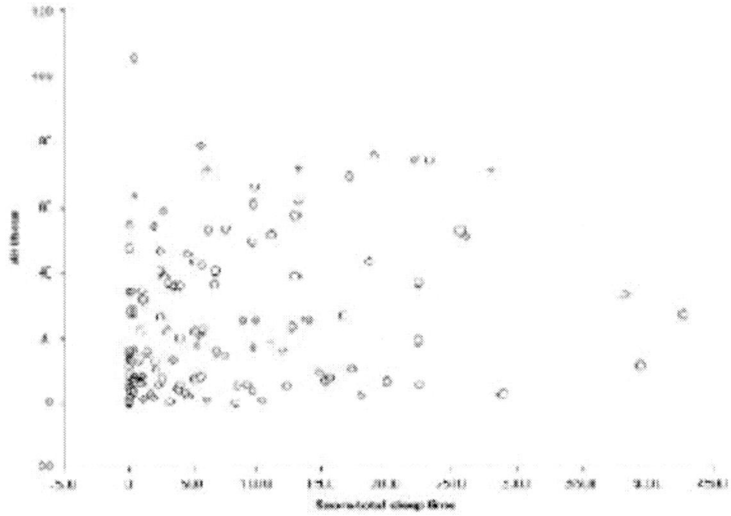

n-106 |cases excluded: 18 due to missing values|.
rs statistic- 0.26.
95% CI- 0.09 to 0.42.
2-tailed p- 0.0035 |t approximation, corrected for tiles|.

Figure 5. Spearman rank correlation between AHI and snore.

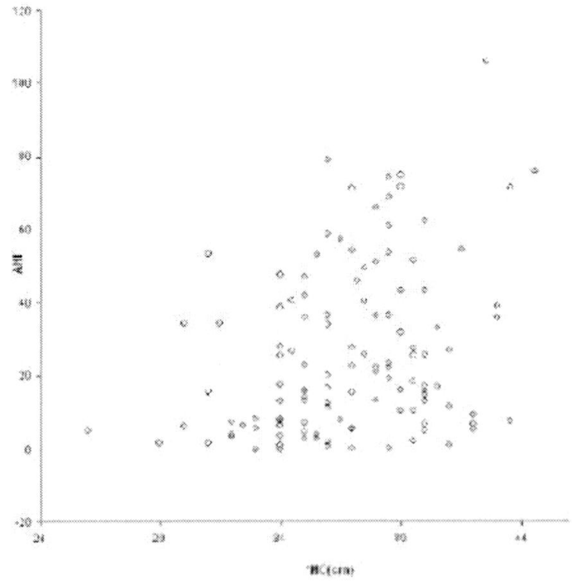

n-122 |cases excluded: 20 due to missing values|.
rs statistic-0.33.
95% CI- 0.16 to 0.48.
2-tailed p-0.0002 |t approximation, corrected for tiles|.

Figure 6. Spearman Rank correlation between AHI and NECK Circumference.

The Effect of BMI Grouping on the Frequency of Snoring

If BMIs were further classified as following 3 groups; group 1 (BMI > 30), group 2 (BMI between 25 and 30), and group 3 (BMI < 25). Mann-Whitney test reveals as follows. The two groups of BMI were selected for comparison. Group difference between group 1 and group 2 was insignificant (Grouping factor BMI: Mann-Whitney test, degree of freedom (d.f.) = 1, p = 0.432).

Group differences either between group 1 and group 3, or group 2 and group 3 were significant (Grouping factor BMI: Mann-Whitney test, d.f. = 1, p < 0.001, p = 0.001, respectively) . Noticeably, the groups of subjects with their BMI > 30, and a BMI between 25 and 30 snored more frequent than those in the group with a BMI < 25. (Table 9)

Table 9. Three group difference

Group	BMI	Mann-Whitney Test d.f.	p	Group difference
Gr 3	< 25	1	= 0.001	Between Gr 2 and Gr 3 was significant
Gr 2 more snoring > gr 3	25~ 30			Between Gr 1 and Gr 2 insignificant
Gr 1 more snoring > gr 3	BMI > 30	1	< 0.001	Between Gr 1 and Gr 3 was significant

Correlation between age and AHI within this study's data revealed that a Pearson correlation coefficicient (r) of -.099, even with n=124; a suggestion of a correlation when the data are further broken down by gender.

The 78 males show essentially no correlation between age and AHI, whereas the 46 females show a correlation of r=-.236. With n=46, this r=-.236 is non-significant (although it does come close to the p level of significance (0.0500) for a one-tail test: p=.0566); nonetheless, it might contain something interesting there.

Discussion

It is worthwhile to evaluate the difference between 1999 and 2007, two different, American Academy of Sleep Medicine (AASM) criteria. The difference between AASM's 1999 and 2007 criteria in AHI follows. For example, the recommended hypopnea definition of 2007 has to meet the following criteria:

1 The nasal pressure signal excursions drop by not less then 30% of pre-event baseline.
2 The duration of this drop occurs for a period lasting 10 seconds.
3 There is a not less than 4% desaturation for pre-event baseline.
4 At least 90% of the event's duration must meet the amplitude reduction of criteria for hypopnea.

The aforementioned definition of AASM's 2007 criteria is somehow different from that of 1999's criteria. It would cause some hypopnea cases to be deleted in the classification, if one is using the criteria based on 2007's instead on 1999's criteria. Such a consideration in the aforementioned difference is applicable to all the clinical classification [9].

Obstructive Sleep Apnea (OSA)

Notwithstanding the fact that obstructive sleep apnea (OSA) may be common in Asian and Southern Pacific patients as compared with what has been reported for Western patients, there is hardly any information of large study. No prior publication relating height measurement with AHI exists. Apnea (elevated AHI) and snoring are associated with obesity and increased neck size has been widely reported in the literature, as well as has the relationship of AHI to hypertension [11-20].

The size of suprachiasmatic nucleus (SCN) decreases in normal aging, which in turn, changes circadian rhythms and has been thought to affect and disturb sleep by increasing sleep fragmentation. To further investigate this issue, a scatter plot of data in the current study failed to present any inverted U-shaped curve (not shown here, since it failed to present the curve). Hence, there nothing appeared looking like a low level of AHI in the 60-year-olds, then a higher level in the 70-year-olds, and then a lower AHI again in the 80-year-olds. The means of AHI were then segmented by age groups and this does show a fall off, but shows that for 29 participants who were 75 years of age, the AHI was 19.66 (95% CI 14.20-29.43), there was a obvious fall off as compared with the 95 participants who were younger than 75 year-old.

Hence, in this study, it conclude that aging per se does not contribute to the intrinsic variability of respiratory control mechanisms, although there may be a lower probability of finding elderly people without respiratory instability. This might be one of the important findings of this study.

Nevertheless, there is little information of the prevalence and severity of OSA in Asian snorers although OSA may not be uncommon in Asian patients. There is no information on snoring and OSA in Asian population either. Nevertheless, there is a single report on Singaporean population, which is predominantly Chinese [21]. The authors contend that the sleep apnea syndrome is more common than the 2-4% prevalence frequently quoted [22]. The authors studied the snorers in an adult population in Singapore and then went ahead to evaluate how many snorers suffer from pathological apnea as well as sleep apnea syndrome. Their examinees age ranged from 30-60 years. There were 106 consecutive habitual loud snorers of a similar age group in the same population studied with PSG in their sleep laboratory. In that laboratory, an AHI greater than 5 was regarded as pathological. The authors have found that 24.1% were loud habitual snorers. Among the latter, 87.5% of loud habitual snorers had significant OSA on PSG and 72% of these apneics complained of excessive daytime sleepiness (EDS). Assuming the clinical observation that all apneics snored, and by extrapolating these figures, the authors conjecture that sleep apnea syndrome affects about 15% of the population. EDS in their cases were validated with clinical hypersomnia.

Hypersomnolence was significantly related to the poor delta wave sleep. OSA occurred mainly in stage 1 and 2 non-rapid eye movement (NREM) sleep instead of in REM sleep.

Frequently, the arousals prevented sleep from going beyond stage 1 and 2. With such a higher-than-expected prevalence of sleep apnea syndrome in Singaporean population aged 30-60 years is likely because of the evidence that the people in that population, aged 30 to 60 years, suffer more hypersomnolence, which is associated with the repression of delta wave sleep by apnea occurring taking place mostly in stage 1 and 2 NREM sleep [21]. The authors report that the prevalence of OSA reveals as follows. The 19% of 1,775 subjects with a mean age of 71 years, SD 10.5, range 40-100 had OSA [23].

In Taiwan, there appears to be no such type of study found yet. While in the US, it was 9 percent for women and 24 percent for men-male sex and obesity were strongly associated with the presence of sleep-disordered breathing, according to a US study [22].

In the US, a rather high percentage of over-65-years subjects have AHI > 5; this is the population norm, knowing that Sleep-disordered breathing (SDB) is defined as having AHI score of 5 or higher as aforementioned.

In fact, all elders do not need to be treated if their AHIs are greater than 5. An AHI > 5 have conventionally been a cutoff for the presence of SDB. A higher cut-off of AHI > 15 have been used, especially for the elderly, in most of sleep studies. An AHI > 20 might be better to distinguish those requiring treatment. There are not enough prospective studies to indicate any treatment responses at different levels of AHI. It is unclear that AHI is a risk factor for those aged over 65 years.

It is important to acknowledge that a report in English by Ancoli-Israel et al. [24] presented that it was largely central apnea, not obstructive apnea. Hence, central apnea predicts mortality above age 65 years; others also published similar data [7,25]. Nevertheless, to the interest of researchers, Asian's small upper airway merits attention. With respect to the risk factors for SDB, for example, a report from Singapore indicated that the risk factors associated with habitual snoring and SDB were largely similar to those reported in other populations [21]. However, differential risks may underscore the importance of ethnicity in determining the burden of SDB.

Hence, it is mandatory for health care development and research on SDB in Asia and the answer will come with hard work towards endorsement of alertness of this circumstance in both lay and professional communities [26].

Snoring is an independent and significant risk factor for a vascular disease especially for stroke and myocardial infarction. It is not uncommon that patients in Sleep Studies are not aware of their own snoring, which was consistent with the original indication of sleep-problem referrals to one of the Taiwanese studied population that patients were not referred solely for sleep apnea, or for snoring alone. For all practical purpose, they were referred rather for mixed sleep problems in most of Sleep Medicine Laboratories. Many patients are not aware of their own heavy snoring and nocturnal arousals. Any excessive and inappropriate assumption that subjects who did not give an account of snoring and sleepiness were lacking OSA is almost definitely erroneous. We should always be ready to lend a hand by questioning the bedroom partner of a potential patient with chronic sleepiness and fatigue, or other type of sleep disturbances. For examples, one of the US studies revealed that Tractenberg et al.'s findings [27,28] were as follows. Their 2003 and 2006 articles were based on a large sample recruited from the community, and not from a sample with sleep disturbances. Their data may be useful as an external (independent) reference for the samples cited in this study. Hence, comparing studies referred in this article, Tractenberg et al. had the analysis based on the intake interviews of 399 healthy non-demented elderly whose data were achieved as of September 2002 [27,28]. In their studies, those participants who were non-demented elderly people had the frequency rating of heavily snoring from 0.7 ± 1.1 to 0.9 ± 1.2. Noticeably, on the frequency measure of heavy snoring, its rating was defined as following: 0, for rating for no snoring in previous month, that is, never; 1, less than once per month; 2, at least once per month; 3, at least once per week; and 4, nearly everyday per day/ night.

Snoring

As a comparison, it is interesting to notice that in the US;

1 From Tractenberg et al.'s reports [27,28], the criteria of the frequency of snoring were defined as: 0, for never; 1, for less than once per month; 2, for at least once per month; 3, at least once per week; and 4, nearly everyday per day/ night.
2 Those reports encounter the issue of the level of awareness of the sleep disturbance problems in the reporters at nights.
3 Unlike this autor's using PSG in his study, Tractenberg et al. used reporter for recording snoring or not in their study.

Habitual Snoring

Conversely, the literature estimates the prevalence of habitual snoring in general population ranges from 3.6% to 35.7% [22,29-41]. In this study, which was investigating the subjects of a Sleep-Medicine-Laboratory based cohorts, who lived in the dwelling communities of Taichung area, Mid-Taiwan, where is adjacent to Taichung area. The snoring prevalence in Taichung area, Taiwan, is 47.8 % for males, whereas 37.2% for females with age ranging from 10 to older than 60 years, all together 1,252 people who were successfully interviewed, whereas 606 were males and 646 were females, according to a report by Liu and Liu [31] on the prevalence of snoring in a regional area of Mid-Taiwan.

Snoring as a risk factor for stroke has been reported. Sawit has extensively reported with its relevance to sleep apnea. [6] Snoring is a not only an independent risk factor for acute vascular disease, but also a factor for stroke and myocardial infarction.

In addition, habitual snorers are more at risk for acute vascular disease compared to occasional snorers. Taiwan is short of such studies except one by Liu and Liu [31]. They merely reported the prevalence of snoring in a regional area of Mid-Taiwan. Further studies to inquire on the relationship of sleep apnea with snoring in Taiwan should be done.

Asian BMI

On the other hand, BMI has relevance to OSA (upper limit of normal BMI in Far-East Asian is 23.5 kg/m2 from current WHO data.)

The updated census data for Taiwanese BMI [37] was 18.5-24 (applied to all ages), issued by the Health Department of Taiwan. An earlier (1993-1996) survey conducted by the same department concluded that BMI for subjects with 19-44 year-old > 26.4 would be classified as obese [38]. The census data of BMI in the period from 2001 to 2003 was 20-24.9.

Noted that in one of the Taiwanese study [7], the difference of BMI between non-snores and snorers is 2.02; (Table 3), which is clinically significant. When this author used the all data of AHI values in this study to further analyzed by dichotomizing them at a level of 5, then this author found that BMI was a factor that has become marginally not significant upon independent sample t-test (p=0.069) in this study; this merits evaluation as follows.

The p-value suggests a "very small amount of evidence in support of the null hypothesis". It does not suggest that the difference (the alternative hypothesis) is true, because p-values are strictly related to the null, and not the alternative, hypotheses. Thus, a p-value=0.069 is "marginally not significant" is the most appropriate; one should not attribute any characteristic apart from marginality to the p-value.

The Importance of Gender Factor

The prevalence of OSA in patients up to the age of 60 is two times higher in men than in women [41]. The gender data on BMI in this study is consistent with the literature.

The Significance of Asian BMI

Researchers modified WHO expert committee's suggestion and used the criterion of > 27 instead of > 30 for Asian BMI. The upper limit of normal BMI in Far-East Asian is 23.5 kg/m2 from new WHO data. Obesity in Asia, as in Singapore nationwide, is defined as BMI > 23 kg/m2. Asian nutritionists are inclined to apply a limit of BMI > 23 kg/m2 for obesity due to lower stature and body weight in Asians. Asian BMI allows for a smaller skeleton (physical) frame, of most Asians. For every height, weight is set lower to compensate for the smaller frame.

The current WHO data shows that the upper limit of the normal Asian BMI is 23.5 kg/m2; this has been as well applied in Singapore nationwide. Singapore nationwide protocol reveals that the scales that have been compiled using a nationwide survey data in Singapore. If one is not an Asian or an Asian descendent, Asian BMI will not be applicable. For example, census data of BMI from 2001 to 2003 in one of Taiwanese studies [42] is 20-24.9 kg/m2.

Obstructive sleep apnea (OSA) may not be uncommon in Asian and Southern Pacific patients [43] compared to what has been reported for Western patients [44,45]. Several studies comparing OSA between Caucasians and Asians have shown that Asian subjects have a greater severity of illness, as indicated by high AHI, when contrasted with Caucasian patients matched with age, gender, and BMI [11,12].

Asians generally have a higher percentage of body fat than Caucasians. Hence, making cross-ethnic comparison of body structures by using absolute BMI values may be misleading. Obesity is associated with an increased risk of OSA [11,12,14,46,47]. Therefore, OSA's association with BMI in Asia should be evaluated with Asian BMI criteria.

In this study, author used Asian BMI criteria, for those subjects who were underweight: with the BMI of 18.5 included 0.85% of the population. For those with the BMI < 18.5, it was three times greater, 2.56%. This author can never be over-stressing the importance of Asian BMI, hence the upper limit of normal BMI in Far-East Asian is 23.5 kg/m2 is not only useful but as well helpful indeed.

The Limitation of This Study

The standard PSG used in most of sleep studies does not include the measurement of esophageal pressure (Pes), which represents the intrathoracic pressure. Hence, the upper airway resistance syndrome is often overlooked because the severity of OSA has already been evaluated by the AHI.

Most of the sleep studies discussed in this article are cross-sectioned. [48-51] Most of such variables studied are restricted by the application of a dichotomous measure of urbanicity. Modifications do happen in sleep disturbance and apnea, along with metabolism and dieting habit during the life span; such modifications may decrease the odds of being overweight for those who survive to the age of 70 years and over.

Given the cross-sectional nature of most of the studies referred here, elderly participants who live long enough to join the studies naturally were the survivors. On the tapis of such a cross-sectional nature of studies cited in this article, it is difficult to justify that poverty, culture, changing metabolism, dieting and nutrition habits across the life span may mitigate the effects of being obese or overweight for those who survive to the older age. [52-55] Hence, there is a need for the proper longitudinal survey for the future work in this area.

CONCLUSION

The study of sleep disturbance in the elderly is a new area of scientific direction for fruitful areas of future research. Hopefully, this article provides some highlight in focusing on an individual patient-oriented approach in a very near future.

The study of sleep disturbance in the elderly is a new area of scientific research. This is a research field that has largely been neglected for this age group. Probable reasons for this are numerous. One of the most likely factors may be due to a combination of the lack of documented data, together with the complexity of human sleep itself. This field is an undiscovered entity, especially in sleep research on the age groups mentioned in this article. All patients are vulnerable, especially those who are elderly. Thus, the difficulty is also due to the problem faced in obtaining volunteers as well as the philosophical and theological undertones people associate with "the sleep when they are at the end stage of their life". Moreover, there is as well the misconception of sleep study on "those who haven't many years left anyhow" in general. Additionally, there is difficulty in recognizing cross-cultural differences among the elderly people's sleep disturbance. Hence, as aforementioned, this article has served to provide a broader understanding of the complicated relationship of various aspects of sleep apnea in the age group studied. Results from this article may provide direction for fruitful areas of future research. Hopefully, this article provides some highlight in focusing on an individual patient-oriented approach [56-60] to sleep disturbance. This is in contrast to a population-based approach in which the entire distribution of risk factors in total population in Taiwan and Asia hopefully can soon be expected [61-65] to deal with sleep disturbance.

REFERENCES

[1] Kamble 1979, trans. by Cauri Deshpaude. In. *No place in the inn: Voice of minority people of Asia*. Hong Kong: CCA-Urban Rural Mission; 1979: 70.

[2] Bliwise, D. L., King, A. C., Harris, R. B., et al. Prevalence of self-reported poor sleep in a healthy population aged 50-65. *Soc. Sci. Med.* 1994; 34: 49-55.

[3] Prinz, P. N., Peskind, E. R., Vitaliano, P. P., et al. Changes in the sleep and waking EEGs of nondemented and demented elderly subjects. *J. Am. Geriatr. Soc.* 1982; 30: 86-93.

[4] Prinz, P. N., Vitaliano, P. P., Vitiello, M. V., et al. Gender, insurance status and culture in attitudes and health behavior in a US Chinese student population. *Ethn. Health* 2001; 6: 197-209.

[5] Vitiell, M., Borson, V. Sleep disturbances in patients with Alzheimer's disease: Epidemiology, pathophysiology and treatment. *CNS Drugs* 2001; 15: 777-96.

[6] Sawit, S., Cabildo, J. E. S., Roxas, A. A., Jr. Snoring as a risk factor for stroke and myocardial infarction. *Philipp. J. Intern. Med.* 2006; 44: 117-24.

[7] Tang, B. Living by a biological clock: Age-related functional changes of sleep homeostasis in people aged 65-88.5 years. *Sleep Biol. Rhythms* 2007; 5: 180-95.

[8] American Academy of Sleep Medicine Task Force. Sleep-related breathing disorders in adults: Recommendations for syndrome definition and measurement techniques in clinical research: The report of an American Academy of Sleep Medicine Task Force. *Sleep* 1999; 22: 667-89.

[9] Tang, B. The Correlation between snoring and apnea/ hypopnea index in the elderly, along with the Octogenarians' comparison study. Accepted for publication in 2009.

[10] Rechtschaffen, A., Wolpert, E. A., Dement, W.C. Nocturnal sleep of narcoleptics. *Electroencephalogr. Clin. Neurophysiol.* 1963; 15: 599-609.

[11] Pawar, R., Smith, M., Wright, J. R., et al. Blood pressure regulation and sleep apnea. *Cardiovasc. Rev. Rep.* 1996; 16: 8-24.

[12] Fletcher, E. C. The relationship between systemic hypertension and obstructive sleep apnea: Facts and theory. *Am. J. Med.* 1995; 98: 118-28.

[13] Hader, C. A., Schroeder, A., Hinzi, M., et al. Sleep disordered breathing in the elderly: Comparison of women and men. *J. Appl. Physiol.* 1998; 84: 1055-62.

[14] Hoffstein, M. D., Chan, C. K., Slutsky, A. S. Sleep apnea and systemic hypertension: A causal association review. *Am. J. Med.* 1991; 91: 190-6.

[15] Wilcox, I., Grunstein, R. R., Hedner, J. A. Effects of nasal continuous positive airway pressure during sleep on 24-hour blood pressure in obstructive sleep apnea. *Sleep* 1993; 16: 539-44.

[16] Shahar, E., Redline, S., Young, T., et al. Hormone replacement therapy and sleep-disordered breathing. *Am. J. Respir. Crit. Care Med.* 2003; 167: 1186-92.

[17] Johnson, M., Anch, A., Remmers, J. Induction of the obstructive sleep apnea syndrome in a woman by exogenous androgen administration. *Am. Rev. Respir Dis.* 1984; 129: 1023-5.

[18] Tang, B. A comparison study on senior sleep apnea with biostatistical and biomedical evaluations *Med. J. Southern Taiwan* 2008; 4: 73-90.

[19] Tang, B. Roles of apolipoprotein-e4, and suprachiasmatic nucleus: Age related cognitive decline and sleep disordered breathing/ sleep fragmentation. *In Vivo* 2008; 30: 10-8.

[20] Tang, B. A comment on predicting continuous positive airway pressure from a modified split-night protocol in moderate to severe obstructive sleep apnea-hypopnea syndrome. *Intern. Med.* 2009; 48: 1489.

[21] Puvanendran, K., Goh, K. L. From snoring to sleep apnea in a Singapore population. *Sleep Res. Online* 1999; 2: 11-4.

[22] Young, T., Palta, M., Dempsey, J., et al. The occurrence of sleep-disordered breathing among middle-aged adults. *N. Engl. J. Med.* 1993; 328: 1230-5.

[23] Gottlieb, D. J., DeStefano, A. L., Foley, D. J., et al. APOE epsilon4 is associated with obstructive sleep apnea/ hypopnea: The Sleep Heart Health Study. *Neurology* 2004; 63: 664-8.

[24] Ancoli-Israel, S., Engler, R. L., Friedman, P. J., et al. Comparison of patients with central sleep apnea: With and without Cheyne-Stokes respiration. *Chest* 1994; 106: 780-6.

[25] Lavie, L., Lavie, P. Elevated plasma homocysteine in older shift-workers: A potential risk factor for cardiovascular morbidity. *Chronobiol. Int.* 2007; 24: 115-28.

[26] Lam, B., Lam, D. C., Ip, M. S. Obstructive sleep apnoea in Asia. *Int. J. Tuberc. Lung Dis.* 2007; 11: 2-11.

[27] Tractenberg, R. E., Singer, C. M., Cummings, J. L., et al. The sleep disorders inventory: An instrument for studies of sleep disturbance in persons with Alzheimer's disease. *J. Sleep Res.* 2003; 12: 331-7.

[28] Tractenberg, R. E., Singer, C. M., Kaye, J. A. Characterizing sleep problems in persons with Alzheimer's disease and normal elderly. *J. Sleep Res.* 2006; 15: 97-103.

[29] Rice, D. H., Persky, M. Snoring: Clinical implications and treatment. *Otolaryngol. Head Neck Surg.* 1986; 95: 28-30.

[30] Koskenvuo, M., Kaprio, J., Partinen, M., et al. Snoring as a risk factor for hypertension and angina pectoris. *Lancet* 1985; 1: 893-6.

[31] Liu, S. A., Liu, H. Y. Prevalence of snoring in Taichung area: An epidemiological study. *J. Chin. Med. Assoc.* 2004; 67: 32-6.

[32] Ishikawa, K., Takanohashi, T., Wako, H. Compulsory milk feeding and the cystine content of the hair in children as a subclinical standard of nutritional status. *Paediatr. Indones* 1965; 5: 535-9.

[33] Johnston, B. F., Greaves, J. P. Manual on food and nutrition policy. *FAO Nutr. Stud.* 1969; 22: 1-95.

[34] Kelley, A. C., Williamson, J. G. Sources of growth methodology in low-income countries: A critique. *Q. J. Econ.* 1973; 87: 138-47.

[35] Hayami, Y., Ruttan, V. W. *Agricultural development: An international perspective.* Baltimore, Maryland: Johns Hopkins University Press; 1971.

[36] Conde-Corpuz, F. Q. Predictive value of oxygen desaturation in the diagnosis of obstructive sleep apnea. *Philipp. J. Intern. Med.* 2003; 41: 197-202.

[37] Health Department of Taiwan. Taiwanese National Health and Nutritional Survey from 1999 to 2000. Available at http://www.doh. gov.tw/ Access date 30 August 2009.

[38] Health Department of Taiwan. Taiwanese National Health and Nutritional Survey from 1993 to 1996. Available at http://www.doh. gov.tw/ Access date 30 August 2009.

[39] Stunkard, A. J., Albaum, J. M. The accuracy of self-reported weights. *Am. J. Clin. Nutr.* 1981; 34: 1593-9.

[40] Troy, L., Hunter, D., Manson, J., et al. The validity of recalled weight among younger women. *Int. J. Obes. (Lond.)* 1995; 19: 570-2.

[41] Black, D. R., Taylor, A. M., Coster, D. C. Stepped Approach Model (SAM) component assessment: Accuracy of self-reported body weight. *Health Educ. Res.* 1998; 13: 301-7.

[42] Tang, F. C., Cheng, H. L. Correlations of exercise, diet, and bone metabolism in elderly. 18th International Congress of Nutrition, South Africa; 2005.

[43] Tan, W. C., Koh, T. H. The prevalence and severity of OSA in Asian snorers. *Respir. Med.* 2004; 98: 557-66.

[44] Duran, J., Esnaola, S., Rubio, R., et al. Obstructive sleep apnea-hypopnea and related clinical features in a population-based sample of subjects aged 30 to 70 yr. *Am. J. Respir Crit. Care Med.* 2001; 163: 685-9.

[45] Redline, S., Kirchner, H. L., Quan, S. F., et al. The effects of age, sex, ethnicity, sleep-disordered breathing on sleep architecture. *Arch. Intern. Med.* 2004; 164: 406-18.

[46] Young, T., Peppard, P. E., Gottlieb, D. J. Epidemiology of obstructive sleep apnea: A population health perspective. *Am. J. Respir. Crit Care Med.* 2002; 165: 1217-39.

[47] Franco, C., Brandberg, J., LoBin, L., Andersson, B., Bengtsson, B., Johannsson, G. Growth Hormone Treatment Reduces Abdominal Visceral Fat in Postmenopausal Women with Abdominal Obesity: A 12-Month Placebo-Controlled Trial. *The Journal of Clinical Endocrinology and Metabolism* 2004; 3204; 466-1474.

[48] Tseng, C. H., Chong, C. K., Tseng, C. P., et al. Mortality, causes of death and associated risk factors in a cohort of diabetic patients after lower-extremity amputation: A 6.5-year follow-up study in Taiwan. *Atherosclerosis* 2008; 197: 111-7.

[49] Jenkins, K., Johnson, N., Ofstedal, M. Pattern and associations of body weight among older adults in two Asian societies. *J. Cross Cult. Gerontol.* 2007; 22: 83-99.

[50] FAO Corporate Document Repository of the Philippines. Available at http://www.fao.org/ Access date 30 August 2009.

[51] Sa, Z., Larsen, U. Social disparities in overweight and obesity among nearly elderly and elderly people in Taiwan. Available at http://paa2007.princeton.edu/default.aspx {Population Association of America (PAA) 2007 Annual Meeting Program Website} Access date 30 August 2009.

[52] Tsai, A. C., Chang, J. M., Lin, H., et al. Assessment of the nutritional risk of > 53-year-old men and women in Taiwan. *Public Health Nutr.* 2004; 7: 69-76.

[53] Huang, K. C., Lee, M. S., Lee, S. D., et al. Obesity in the elderly and its relationship with cardiovascular risk factors in Taiwan. *Obes. Res.* 2005; 13: 170-8.

[54] Monteiro, C. A. Socioeconomic status and obesity in adult populations of developing countries: A review. *Bull. World Health Organ* 2004; 82: 940-6.

[55] Drewnowski, A., Specter, S. E. Poverty and obesity: The role of energy density and energy costs. *Am. J. Clin. Nutr.* 2004; 79: 6-16.

[56] McCamy, T. NIH Funding and The Sleep Apnea Crisis. Available at http://www.dailykos.com/story/2007/4/18/05313/1000 {Wrote on April 17, 2007, 09:53:13 PM PST} Access date 30 August 2009.

[57] Von Braun, J., Pachauri, R. K. The promises and challenges of biofuels for the poor in developing countries. Available at http://www.ifpri.org/publication/promises-and-challenges-biofuels-poor-developing-countries {IFPRI 2005-2006 Annual Report Essay, November 2006} Access date 30 August 2009.

[58] Tang, B. Sleep disordered breathing in Pakistani population. *J. Ayub. Med. Col. Abbottabad* 2005; 17: 91-3.

[59] Tang, B. Special communication: Support with clarity: A proper trend in medical statistics. *J. Ayub Med. Coll. Abbottabad.* 2006; 2: 178-9.

[60] Tang, B. Nonapneic and apneic snorers during sleep. *Chest E-letter*; 5 April 2006.

[61] Champion, A. G., Hugo, G. New forms of urbanization: Beyond the urban-rural dichotomy. Aldershot, Hants, England; Burlington, VT: Ashgate xxiv; 2004: 420.

[62] McDade, T. W., Adair, L. S. Defining the "urban" in urbanization and health: A factor analysis approach. *Soc. Sci. Med.* 2001; 53: 55-70.

[63] Vlahov, D., Galea, S. Urbanization, urbanicity, and health. *J. Urban Health* 2002; 79: S1-S12.

[64] Yach, D., Mathews, C., Buch, E. Urbanisation and health: Methodological difficulties in undertaking epidemiological research in developing countries. *Soc. Sci. Med.* 1990; 31: 507-14.

[65] Chien, K. L., Hsu, H. C., Su, T. C., et al. Consistency of genetic inheritance mode and heritability patterns of triglyceride vs high density lipoprotein cholesterol ratio in two Taiwanese family samples. *BMC Genet.* 2003; 4: 7.

In: Asia Pacific Life Sciences Compendium
Editor: Bandit Chumworathayi
ISBN: 978-1-62257-338-7

Chapter 22

HEAVY METALS CONCENTRATION IN WATER, SEDIMENT AND SOME SELECTED SPECIES OF FISHES FROM PANCHGANGA RIVER, KOLHAPUR (MAHARASHTRA, INDIA)

Sagar Anandrao Vhanalakar * ***and Deepak Vishwanatrao Muley*** †
Department of Zoology, Shivaji University,
Kolhapur, Maharashtra, India

ABSTRACT

The heavy metal content from the non-polluted zone of the Panchganga River, Kolhapur (MS, India) water was found below detectable level. Samples from polluted zone showed variations in their heavy metal content in river water, sediment, and fish tissues. The bioaccumulation of heavy metals is found to be highest in liver, followed by gill, and muscle, of the selected fish species, from three selected stations of river Panchganga. From present study, it is concluded that the higher concentrations of heavy metals at selected stations is due to the heavy load of industrial waste in addition to domestic and agricultural waste. The industrial waste contains high amount of textile mill wastes, including dyes and untreated chemicals, which caused severe damage to the fisheries of Panchganga River.

Keywords: Panchganga River, heavy metal, water, sediment, fish

BACKGROUND

The increasing amount of waste, generated by over population, urbanization, and industrialization, finds its way in aquatic ecosystems. Many of these persistent pollutants get

---

* Corresponding author: Sagar Anandrao Vhanalakar. E-mail: sagarayan36@rediffmail.com.
† Department of Zoology, Shivaji University, Kolhapur, Maharashtra, 416004, India.

locked in food chain and reach at higher levels with increasing concentrations, some times far away from source of pollution. Aquatic animals are capable of taking up and accumulating heavy metals to a degree many times than that present in surrounding medium. [1]

The effect of heavy metals on aquatic organism is currently attracting wide spread attention, particularly in the study of pollution. Heavy metals are being introduced into an aqueous environment through industrial and urban effluents, soil leaching and rainfall. The widespread contamination of aquatic ecosystem with heavy metals is the increasing concern of environmental scientists. [2] It was demonstrated that, toxic heavy metals from industrial wastewater are resistant to biodegradation and get accumulated in the water and sediment and considerably affects the water quality and aquatic life. [3] The heavy metals from industrial wastewater on sludge from Dandesara, GIDC area of Surat, were detected. The concentration of heavy metals like copper, cadmium, nickel, lead, zinc, mercury, and iron, was determined and few metal ion levels was found beyond the ISI permissible limits. [4] The waste water of electroplating industry contains nickel and chromium, especially, the latter was more toxic.

Considering the load of heavy metals in the aquatic environment through many aspects, the present study is aimed to quantify the level of some heavy metals in the river water, sediment, and tissues of economically important fishes from Panchganga River at three selected sites.

METHOD

Fortnightly sampling of water, sediment, and fish, was done from three different stations (A, B, and C) along the river Panchganga, near Kolhapur [5] Station A (Shiroli Bridge) is situated at boundary line of Kolhapur town and receives domestic and industrial wastes. Station B (Near village Chandur, 17 km away from Kolhapur), receives textile and agricultural waste. Station C is 1½ km away from the town Ichalkaranji, which is famous for textile industry in Maharashtra. Huge amount of textile mill waste is drained directly in river Panchganga from this town. For present study, three economically important fish species were selected, i.e., *Cyprinus carpio, Cirrhina mrigala,* and *Oreochromis mossambicus*. Gill, liver, and muscle tissues were selected as target tissue for heavy metal bioaccumulation study.

From three stations, water, sediment, and fish samples, were collected and polled samples of gills, liver and muscle tissues were acid digested by using perchloric acid and Nitric acid. [6] All samples were analyzed for heavy metals by using AAS (Chemito model). Results are expressed in mg/l for water and mg/g for tissue samples and sediment.

RESULT

The results of heavy metal detection in water samples at three stations showed highest concentration of zinc (14.1 mg/l, 11.2 mg/l, and 11.3 mg/l) followed by copper, nickel, iron, manganese, and lowest concentration of chromium (0.015 mg/l, 0.02 mg/l, and 0.015 mg/l). (Table 1) The similar pattern was observed in sediment at three stations having highest concentration of zinc (18 mg/g, 16.1 mg/g, and 14.6 mg/g) followed by copper, nickel, iron, manganese, and lowest concentration of chromium (0.4 mg/g and 0.7 mg/g). (Table 1)

Mercury was found to be absent in water and sediment at all three stations. The heavy metal content in fish tissue showed lowest concentration of zinc (0.025 mg/g to 1.8 mg/g) at three stations in liver followed by gill and lowest in muscles of three selected fish species.

While the highest concentration of nickel (8.5 mg/g to 4.7 mg/g) at station A, and of iron (12.6 mg/g to 9.2 mg/g) at station B and C were found in liver, gill, and muscles of selected fish species, mercury was at below detectable level (BDL). (Table 2)

Discussion

From these results, the maximum bioaccumulation was found in liver at three stations followed by gill, while minimum bioaccumulation was found in muscles of three selected fish species. Maximum bioaccumulation was found in *Cyprinus carpio,* followed by *Cirrhina mrigala,* and *Oreochromis mosambicus.*

Table 1. Heavy metal contents in water and sediments samples collected from polluted zones of Panchganga River

Station	Heavy metals	Mean values	
		Water (mg/l)	Sediment (mg/g)
A	Mn	0.3	0.8
	Ni	1.3	2.7
	Fe	0.5	1.7
	Cu	1.4	4.4
		Water (mg/l)	Sediment (mg/g)
	Zn	14.1	18.0
	Cr	0.015	0.4
	Hg	BDL	BDL
B	Mn	0.35	1.45
	Ni	0.3	3.15
	Fe	0.25	2.1
	Cu	0.35	4.5
	Zn	11.22	16.1
	Cr	0.002	0.4
	Hg	BDL	BDL
C	Mn	0.4	1.7
	Ni	0.3	1.1
	Fe	0.3	1.7
	Cu	0.85	2.3
	Zn	11.34	14.6
	Cr	0.015	0.7
	Hg	BDL	BDL

BDL = Below Detectable Level.

Table 2. Heavy metal contents in gill, liver, and muscle of freshwater fishes from Panchganga River

Station	Heavy metals	Mean values (mg/g)								
		Oreochromis mossambicus			*Cirrhina mrigala*			*Cyprinus carpio*		
		Gill	Liver	Muscle	Gill	Liver	Muscle	Gill	Liver	Muscle
A	Mn	1.6	1.75	1.5	1.9	2.05	1.55	2.75	2.9	1.55
	Ni	5.4	6.3	4.7	7.4	8.5	6.2	6.2	6.7	5.8
	Fe	1.95	2.55	1.5	2.3	3.1	1.9	2.1	2.35	1.7
	Cu	0.5	0.6	0.2	0.8	1.0	0.3	0.5	0.7	0.3
	Zn	0.02	0.03	0.15	0.02	0.02	0.01	0.01	0.03	0.01
	Cr	2.25	2.4	1.5	3.05	3.35	2.65	3.25	3.25	2.85
	Hg	BDL	BDL	BDL	BDL	BDL	BDL	BDL	BDL	BDL
B	Mn	2.85	2.95	1.8	2.75	3.1	1.75	2.85	3.05	1.9
	Ni	10.0	11.9	5.8	10.5	10.6	9.5	8.4	8.5	7.3
	Fe	10.4	11.5	9.2	10.85	12.28	9.3	10.5	10.6	9.2
	Cu	1.75	2.05	1.2	1.45	1.5	1.2	1.85	2.05	1.0
		Gill	Liver	Muscle	Gill	Liver	Muscle	Gill	Liver	Muscle
	Zn	1.45	1.85	1.15	1.2	1.6	6.9	1.85	1.7	1.05
	Cr	4.0	4.8	3.0	3.95	4.2	3.3	1.82	0.2	1.1
	Hg	BDL	BDL	BDL	BDL	BDL	BDL	BDL	BDL	BDL
C	Mn	2.85	3.5	2.0	2.6	2.8	1.95	2.75	3.2	2.05
	Ni	10.8	11.3	9.25	10.5	10.8	9.55	8.8	8.5	8.0
	Fe	11.8	12.6	10.0	11.8	11.85	10.0	10.7	11.2	9.9
	Cu	2.25	2.25	1.7	1.85	2.2	1.05	1.75	2.15	1.1
	Zn	1.55	1.8	1.0	1.45	1.75	1.4	1.25	1.65	1.0
	Cr	4.2	4.7	8.2	4.9	5.0	3.0	1.9	2.2	1.8
	Hg	BDL	BDL	BDL	BDL	BDL	BDL	BDL	BDL	BDL

BDL = Below Detectable Level.

A series of changes can be observed in the lotic water ecosystem, which might be due to variety of factors. Though, these ecosystems are influenced by the seasonal changes and some man made activities, such as addition of industrial wastes, agricultural run off, sewage etc. [7]. Heavy metals are considered to be serious contaminants of aquatic system due to their extended biological half-lives, inherent toxicities to nature at low concentration, and high rates of bio-accumulation. [8,9]

Direct dumping of wastewater into water bodies causes the transfer of pollutants into natural environmental, hereby disturbing the ecosystem. [10,11] This is how heavy metals are transferred to aquatic system in dissolved from and by their interaction with organisms and their consumers ultimately leading to metal toxicity by bio-magnification. [12,13]

Present study showed highest concentration of zinc in water and sediment samples. In present study, pollution is mainly due to receiving heavy load of untreated sewage and industrial waste at station A. At station B, pollution is due to industrial and agricultural waste. At station C, discharge of various dyes and untreated chemicals from textile mills is found in major.

In all observations, maximum bio-accumulation of heavy metals was found at station C followed by B, and A, in three economically important fish species from Panchganga River.

From the above studies, it is clear that heavy metal content from the water, sediment, and fish tissues, from three economically important fish species from Panchganga River is high in polluted zone, which causes hazardous effects on the fauna of river and may also affect the users of river water and fish consumers in the long run.

REFERENCES

[1] Riley, J. P., Roth, I. The distribution of trace element in some species of phytoplankton grown in culture. *J. Mar. Biol. Assn.* 1971; 51: 63-72.

[2] Mahapatra, S. P. Heavy metal concentration in the industrial effluents discharge to the Thane Creek. *Indian Environ. Prot.* 1987; 41: 284-6.

[3] Shrivastava, V. S. Heavy metal in industrial wastewater and sludge by Inductively coupled plasma atomic emission spectrophotometer. *Indian J. Environ. Prot.* 1996; 15(5): 328-9.

[4] Singh, R. K., Singh, S. P. Effect of industrial effluent on the food fishes of river Sone. *Proceeding of 83rd session of Indian Science Congress*, Patiala; 1996.

[5] Trivedy, R. K., Goel, P. K., Trisal, C. L. Practical methods in ecology and environmental science. Karad (M.S.) *Enviro. Media Publication*; 1998; 1-125.

[6] Lithnor, G. *Methods for detection measurement and monitoring of water pollution.* Rome: FAO; 1975.

[7] Whitton, B. A. Zinc and plants in river and streams. In: Narriagu, J. O., editor. *Zinc in the environment part II: Health affects.* New York: John Wiley; 1980: 363-400.

[8] Wong, M. H. Effects of zinc and copper salts on *Cyprinus carpio* and *Ctenophyrengodon idellus*. *Acta Anta* 1970; 99(4), 450-454.

[9] Amundson, P. A., Staldvik, F. J., Lukin, A. A. Heavy metal contamination in freshwater fish from the border region between Norwey and Russia. *Sci. Total Environ.* 1997; 18(3); 211-24.

[10] Vernberg, W. B., Vernberg, F. J. *Pollution and Physiology of Marine Organisms*. New York: Academic press, Inc; 1974: 381-425.

[11] Saikia, P. K. Heavy metals in river sediments of Upper Ganga. *Indian J. Environ. Hlth.* 1988; 31(1): 11-7.

[12] Enk, M. D., Mathis, B. J. Distribution of Cadmium and Lead in a stream eco-system. *Hydrobiologia* 1977; 52: 153-8.

[13] Baccini, P. *Regulation of trace metal concentrations in freshwater systems*; Sigel, H., Ed.; Dekker: New York, 1984; 18:239-286.

INDEX

A

B

C

D

E

F

G

H

I

J

K

N

Q

R

S

V

W

Y

Z